Rolf H. J. Schlegel, PhD

Encyclopedic Dictionary of Plant Breeding and Related Subjects

Pre-publication
REVIEWS,
COMMENTARIES,
EVALUATIONS . . .

"I have read through Dr. Schlegel's book, *Encyclopedic Dictionary of Plant Breeding and Related Subjects.* I was impressed with the wide range of terms and with the other information related to crop breeding, cytogenetics, and taxonomy. Almost any conceivable term related to these disciplines is defined within Schlegel's dictionary. I was especially impressed with his treatment of nomenclature for wheat cytogenetic stocks. I expect this work will become a valuable desk reference for plant breeders, geneticists, and students of these disciplines."

Eric N. Jellen, PhD
Associate Professor of Cytogenetics,
Department of Plant and Animal Sciences,
Brigham Young University

"Dr. Schlegel has produced a landmark dictionary of prodigious scope. It is targeted to plant breeding and is generous in its coverage of botanical terms and other terms that plant scientists or workers in other fields can use to get a quick definition of a term or concept. A highlight of the book is a large list of crop plants by common and scientific names. Another feature is a large collection of schematics showing reproductive biology, plant breeding methods, field designs, and other topics. Finally, the author demonstrates the principles of genetics, genomes, and taxonomy, to name a few, in a series of tables. I believe the book will be especially useful to workers whose first language is other than English because technical terms are easily found and simply defined. Students will want to know about this book and have ready access to it."

Calvin O. Qualset, PhD
Professor Emeritus,
Division of Agriculture
and Natural Resources,
University of California,
Davis

More pre-publication
REVIEWS, COMMENTARIES, EVALUATIONS . . .

"During the past decade plant breeding has transformed from pure agronomic genetics to an interdisciplinary research field. Sometimes it is not easy to understand your colleagues' agronomic jargon especially when your background is in biochemistry and your knowledge about the possible life forms on the nearby star system is more extensive than how to breed sunflowers. I was lucky enough to have preliminary access to the *Encyclopedic Dictionary of Plant Breeding and Related Subjects* and it proved to be extremely helpful to me. Professor Schlegel has successfully completed the difficult task of gathering in a single volume all practical terms in the vast field of plant breeding. I am sure this book will become an important reference source for students and researchers engaged in plant breeding work."

Dr. Ivelin Y. Pantchev
Assistant Professor,
Department of Biochemistry,
Faculty of Biology,
Sofia University,
Bulgaria

"This is a massive effort to collate terms and definitions of direct and indirect relevance to plant breeding and plant biotechnology. It brings together under one roof a diversity of terminology, ranging from the agricultural through breeding, botany, biochemistry, genetics, and cytology to molecular biology. The volume also includes a substantial compendium of tables and illustrative figures, which certainly adds value and will distinguish it from its competitors."

Robert Koebner, PhD
Senior Scientist,
John Innes Centre,
Norwich, United Kingdom

Encyclopedic Dictionary of Plant Breeding and Related Subjects

Encyclopedic Dictionary of Plant Breeding and Related Subjects

Rolf H. J. Schlegel, PhD

Viva Books Private Limited
New Delhi • Mumbai • Chennai • Kolkata • Bangalore

First Indian Edition 2004

VIVA BOOKS PRIVATE LIMITED

- 4262/3 Ansari Road, Daryaganj, New Delhi 110 002
  E-mail: viva@vivagroupindia.net
- 215-216 New Apollo Estate, 2nd Floor, Mogra Lane, Andheri (E), Mumbai 400 069
  E-mail: vivamumbai@vivagroupindia.net
- Jamal Fazal Chambers, 26 Greams Road, Chennai 600 006
  E-mail: vivachennai@vivagroupindia.net
- B-103, Jindal Towers, 21/1A/3, Darga Road, Kolkata-700 017
  E-mail: vivakolkata@vivagroupindia.net
- 7 GF, Sovereign Park Apts., 56-58 K. R. Road, Basavanagudi, Bangalore 560 004
  Email: vivabangalore@vivagroupindia.net

Published by arrangement with

The Haworth Press, Inc.
10 Alice Street, Binghamton
NY 13904-1580 USA

ISBN 81-7649-786-X

Published by Vinod Vasishtha for Viva Books Private Limited, 4262/3 Ansari Road, New Delhi 110 002 and printed at Rajkamal Electric Press, G. T. Karnal Road, New Delhi.

CONTENTS

ABOUT THE AUTHOR

Rolf H. J. Schlegel, PhD, is Director of Research & Development at HYBRO-TEC in Aschersleben, Germany. He is also a professor of cytogenetics and applied genetics, with over thirty years of experience in research and teaching in advanced genetics and plant breeding. Dr. Schlegel is the author of more than 150 research papers and other scientific contributions, a co-coordinator of international research projects, and a member of several international scientific societies. He has also been a scientific consultant at the Bulgarian Academy of Agricultural Sciences for several years. He received his master's degree in agriculture and plant breeding, and his PhD and DSc in genetics and cytogenetics from the Martin-Luther University Halle/S. in Germany. Later, he became Head of Laboratory of Chromosome Manipulation and the Department of Applied Genetics and Genetic Resources at the Institute of Plant Genetics and Crop Plant Research in Gatersleben, Germany, and Head of the Genebank at the Institute of Wheat and Sunflower Research in General Toshevo/Varna, Bulgaria, as well as at the Institute of Plant Biotechnology and Genetic Engineering in Kostinbrod/Sofia, Bulgaria.

Preface

As a complex subject, plant breeding makes use of many scientific and technological disciplines, such as agronomy, horticulture, forestry, seed production, genetics, molecular genetics, biotechnology, biochemistry, mutagenesis, population genetics, biometry, botany, cytology, cytogenetics, ecology, plant evolution or genetic conservation, and laboratory technologies. Often it is difficult to know the precise meaning of many terms and to accurately interpret specific concepts of plant breeding. Most dictionaries and glossaries available are highly specific or fragmentary. Until now, no attempt has been made to provide a comprehensive compilation of plant breeding terms. This dictionary includes the specific terms of plant breeding and terms that are adjusted from other disciplines.

Moreover, the intent was to create a book that includes not only present terms, but also some terms used during the long history of plant breeding. In addition, this book offers most of the words used by a plant breeder and seed producer together in one source. In order to also serve students, teachers, and research workers, the book is supplemented with breeding schemes, tables, examples, and a list of crop plants, including a few details.

Completeness could not be achieved. The modern subjects, such as biotechnology, molecular genetics, cytogenetics, and genetic engineering, were included as far as possible. Selection was made on the basis of the author's experience and his knowledge of the breeder's requirements. This is a modest effort to serve the scientific community.

Plant breeding is a rapidly developing subject, particularly through the achievements of modern genetics and genetic engineering. Therefore, the definitions or descriptions given here may subsequently be modified. The author, therefore, accepts no responsibility for the legal validity, accuracy, adequacy, or interpretation of the terms given in this book. Suggestions or errors brought to his attention will be considered whenever possible.

Acknowledgments

I express my thanks to Mariana Atanasova, MSc; Kliment Piperkov of Doubroudja Agricultural Institute; General Toshevo/Varna (Bulgaria); and Viktor Korzun, PhD, of Lochow-Petkus GmbH, Einbeck (Germany) for substantial contributions to the manuscript and proofreading, as well as providing several photographs for the cover.

User's Guide

This dictionary provides a representative selection of technical terms from the huge vocabulary of plant breeders, seed producers, and all those who work in related fields. Different terms included in this book have been arranged alphabetically on a word-by-word basis. When Greek letters were necessary in association with some words, they were translated into English and also arranged according to alphabetical order.

Names of scientists and/or family names are in all capital letters. However, when used adjectively (e.g., Mendelian) or as a unit of measure (e.g., Ångström), only the first letter is capitalized.

Explanations to a given term may be more or less extensive. Several definitions have simply been separated by a semicolon. Terms that are either very general in nature or self-explanatory have been avoided. Alternate names of terms have been given within the definition.

Cross-references have been provided wherever necessary for economizing space, demonstrating interrelationships, and organizing the material in a clear manner. Cross-referenced terms are indicated by the symbol >>>.

Several terms are glossed and supplemented by tables, figures, and illustrations. Cross-reference figures and tables are also indicated by the symbol >>>, together with the number of the figure or table.

Since plant breeders and specialized scientists make use of different meanings of identical terms or vice versa, the terms were categorized according to the scheme given under abbreviations. Whenever it was necessary or possible the term was associated with one or more abbreviations of a scientific field and/or category.

Figures, tables, and a list of crops, weeds, and some other important plants are included in separate sections. The latter shows the common names and the scientific descriptions, together with the chromosome number, genome constitution, DNA content, and other details as far as possible.

Abbreviations

agr	agriculture
anat	anatomy
bio	biology
biot	biotechnology
bot	botany
bp	base pair(s)
cf	confer, compare
chem	chemistry
cyto	cytology
eco	ecology
env	environment
eth	ethology
evol	evolution
fore	forestry
gene	genetics
hort	horticulture
Lat	Latin
meth	methods, methodology
micr	microscopy
phy	physics
phys	physiology
phyt	phytopathology
pl	plural
prep	preparation, techniques
seed	seed science, seed production
sero	serology
stat	statistics, experimental planning
syn	synonymous
tax	taxonomy
tech	technology
US	term used in the United States
zoo	zoology
>>>	see

DICTIONARY OF TERMS

A

A chromosome: any of the standard chromosomes of a given genome *cyto*

A line: the seed-bearing parent line used to produce hybrid seed that is male sterile; in wheat hybrid seed production, a male-sterile parent line used to produce hybrid seeds and, hence, the seed-producing parental line *seed* >>> Figure 2

ABA >>> abscisic acid

abaxial: the surface of a leaf facing away from the axis or stem of a plant, as opposed to adaxial *bot*

aberrant: having uncommon characteristics or not strictly true to the phenotype *gene*

aberration: variation of chromosome structure caused by induced or spontaneous mutations; in general, a nontypical form or function *cyto;* in microscopy, failure of an optical or electron-optical lens to produce exact geometric (and chromatic) correspondence between an object and its image *micr*

aberration rate: the portion of chromosomal changes as compared to normal chromosomes *cyto*

abiotic: factors or processes of the nonliving environment (climate, geology, atmosphere)

abjection: the separating of a spore from a sporophore or sterigma by a fungus *bot*

abjunction: the cutting off of a spore from a hypha by a septum *bot*

ablastous: without germ or bud *bot*

abnormal: unusual variance from the natural habit

abort: to fail in the early stages of formation; the collapse or disappearance of seeds or cells *bot*

abortive: defective or barren *bot*

abortive infection: when pathogenic microorganisms fail to become established in the tissue of the host *phyt*

aboveground biomass >>> biomass

abscise: separate by abscission, as a leaf from a stem *bot*

abscisic acid (ABA): a growth regulator or plant hormone such as auxins, gibberellins, or cytokinins; it occurs in various tissues and seeds; the substance is thought to play an important role in the regulation of certain aspects of seed growth and development, as well as being involved in fruit growth, rejection of plant organs such as leaves and fruits, and certain other physiological phenomena *phys*

abscisin(e) >>> abscisic acid

abscissa: in monovariable distributions the abscissa (x-axis) is used for plotting the trait, while the ordinate (y-axis) gives the frequencies *stat*

abscission: rejection of plant organs (e.g., of leaves in autumn) *bot phys* >>> abscisic acid

absorbency: a measure of the loss of intensity of radiation passing through an absorbing medium *meth*

absorption: uptake of substances, usually nutrients, water, or light, by plant cells or tissue; in soil science, the physical uptake of water or ions by a (soil) substance *phys;* in microscopy, the interaction of light with matter, resulting in decreased intensity across entire spectrum or loss of intensity from a portion of the spectrum *micr*

absorption spectrum: a graph that shows the percentage of each wavelength of light absorbed by a pigment (e.g., chlorophyll) *phys*

absorptive: the state or process of being absorbed

abundance: the estimated number of individuals of a species in an area or population *phyt*

acantha: a spine, thorn, or prickle *bot*

acanthocarpous: a fruit showing prickles *bot*

acaricide: a pesticide used to kill or control mites or ticks *phyt*

acarides: (*pl*) related to the spider animals; more than 10,000 species are known; many species parasitically live on plants *zoo phyt*

acarpous: describes a plant that is sterile *bot* >>> sterile

acaudate: not having a tail *phyt*

acauline: not having a culm *bot*

accelerated aging test >>> aging test

acceleration: the time rate of change of velocity with respect to magnitude or direction *phy*

acceptor: an atom that receives a pair of electrons to form a chemical bond *chem phys*

accession: a distinct sample of germplasm (cultivar, breeding line, population) that is maintained in a gene bank for conservation and evaluation; in order to represent the genetic variation of a sample, ideally 4,000 seeds are needed for genetically homogeneous lines and about 12,000 seeds for heterogeneous

accession number: a unique identifier assigned to each accession when it is registered within a gene bank

accessory bud: buds that are at or near the nodes but not in the axils of leaves *bot*

accessory chromosome: a chromosome that is present in addition to the normal chromosome complement *cyto* >>> B chromosome

accessory DNA: surplus DNA present in certain cells or cell stages due to gene amplification *gene*

accidental host: that type of host on which the pathogen or parasite lives only for a limited time; it has no particular importance for the reproduction of the pathogen or parasite *phyt* >>> host

accidental sample: a sampling technique that makes no attempt to achieve representativeness, but chooses subjects based on convenience and accessibility *stat*

acclimation >>> acclimatization

acclimatization: changes involving the synthesis of proteins, membranes, and metabolites that occur in a plant in response to chilling or freezing temperatures that protect tissues or confer tolerance to the cold *phys*

acclimatized: a state of physiological adjustment by plants to changed environmental or stress conditions *phys*

accommodation: the act of adjusting the eye to bring objects that are closer to the eye into focus *micr*

accumbent: used to describe the first sprouts of an embryo when they lie against the body of the seed *bot*

accumulation center: an area where a great deal of variation of a given species or crop plant may be found, but which is not considered a center of origin

acellular: describing tissue or organisms that are not made up of separate cells but often have more than one nucleus *bot*

acentric: chromosome, chromosome segments, or chromatids that show no centromere; applied to a chromosome fragment formed during cell division that lacks a centromere; this fragment is unable to follow the rest of the chromosomes in migration toward one or the other pole, as it has lost its point of attachment to the cell spindle *cyto* >>> Figure 11

acephalous: not having a head *phyt*

acephate: a systemic insecticide that is used to control pests (e.g., aphids, scale, and thrips) *phyt*

acerose: needlelike and stiff, like pine needles *bot*

acervate: growth in heaps or groups *bot*

acetabuliform: saucerlike in form *bot*

acetaldehyde: a simple aldehyde that is a bridge product of alcoholic fermentation *chem*

acetate: a salt or ester of acetic acid *chem*

acetic acid (ethanoic acid): a carboxylic acid, CH_3COOH, and simple fatty acid; a final product of several fermentation, oxidation, or rot processes; plays a crucial role in energy metabolism *chem*

aceto-carmine staining: used as a dye for staining of chromosomes; usually cells or tissues are prepreated (fixation) for 12-24 hours with a mixture of alcohol and acetic acid (3:1) or alcohol, acetic acid, and chloroform after CARNOY; prior to squashing the material it is stained with aceto-carmine *cyto micr* >>> opuntia >>> CARNO's fixative

acetone: a simple but most important ketone that is often used as a lipid solvent *chem*

aceto-orcein: a fluid consisting of the dye orcein dissolved in acetic acid that is used in chromosome staining *cyto micr*

acetylation: introduction of an acetyl group into a compound *chem phys*

achene: a small, usually single-seeded, dry, indehiscent fruit formed from a single carpel (e.g., the feathery achene of Clematis); variants of the achene include caryopsis, cypsela, nut, and samara *bot*

achiasmate: meiosis and/or chromosome pairing without crossing over and chiasma formation *cyto* >>> Figure 15

achlorophyllous: a plant or leaf without chlorophyll *bot*

achromat: a microscope objective corrected for axial chromatic aberration *micr*

achromatic: parts of the nucleus not stainable by common chromosome dyes *cyto*

achromatic aplanatic condenser: a well-adapted microscope condenser lens that is corrected for chromatic and spherical aberrations *micr*

achromatic lens (achromat): a lens cluster whose foci and power are made the same for two wavelengths; the simplest achromat is a doublet that combines two single lenses with different dispersions and curvatures to achromatize the combination *micr*

achromatin: that part of the nucleus that does not stain with basic dye *cyto*

achromycin: the trade name for tetracycline; an antibacterial antibiotic from *Streptomyces* spp. *biot*

acicular: pointed or needle shaped *bot*

acicular leaf: a pointed or needle-shaped leaf (e.g., in conifers) *bot*

acid phosphatase (Acph): an enzyme that is a member of hydrolases; because of its variability it is sometimes used as a biochemical marker in genetic studies *chem gene*

acid soil: specifically a soil with pH value < 7.0, which is caused by the presence of active hydrogen and/or aluminum ions; the pH value decreases as the activity of these ions increases *agr*

acid(ic) dye: an organic anion that stains positively charged macromolecules and acts on protoplasm *micr*

acidity: in soil science, a measure of the activity of the hydrogen and aluminum ions in wet soil, usually expressed as pH value; crop plants show specific requirements; some of them grow reasonably well on acid soils (e.g., cowpea—pH 5.0-7.0; oats, rye—pH 5.5-7.5; maize, sorghum, wheat—pH 6.0-7.5; barley—pH 6.5.-7.5; alfalfa—pH 7.0-8.0) *agr*

acidophilic: in cytology, having an affinity for acid stains (eosinophilic) *cyto;* in ecology, thriving in or requiring an acid environment *eco*

acidophilous >>> acidophilic

acinaceous: consisting of or full of kernels *bot*

acinaciform: shaped like a scimitar (e.g., the shape of the pods of some beans) *bot*

aconitase (Aco): an enzyme (dehydratase) that catalyzes the production of iso-citric acid from citric acid; because of its variability it is sometimes used as a biochemical marker in genetic studies *chem gene*

acorn: the nonsplitting, one-seeded fruit of, for example, an oak tree *bot*

acquired character: a nonheritable modification of structures or functions impressed on an individual by environmental influences during its development *gene* >>> modification

acquired immunity: an immunity that may be induced by means of preimmunization *phyt*

acquired resistance: plant resistance to a disease activated after inoculation of the plant with certain microorganisms or treatment with certain chemical compounds *phyt*

acre: 1 acre = 0.4047 ha = 4,840 square yards = 10 square chains; 640 acres = 1 square mile (also called a section)

acreable: in terms of an acre or per acre *agr*

acreage: extent or area in acres *agr*

acridine: a chemical that is capable of causing frameshift mutations in the DNA sequence; several derivatives of acridine, such as acridine orange, are used as dyes or biological stains *chem gene cyto* >>> fluorescence staining >>> mutagen

acridine orange: an acridine dye that functions as both a fluorochrome and a mutagen *micr cyto gene* >>> fluorescence staining >>> mutagen

acrocarpic: fruits and/or seeds are formed on the top of a stem of a plant *bot*

acrocarpous >>> acrocarpic

acrocentric: the centromere is present on the end or close to the end of a chromosome *cyto* >>> Figure 11

acropetal: toward the apex; the opposite of basipetal *bot*

acrosyndesis: incomplete end-to-end chromosome pairing *cyto*

actinomycin: an antibiotic produced by *Streptomyces chrysomallus* that prevents the transcription of mRNA *micr cyto* >>> tetracycline

activator: in enzymology, a proteinlike substance that is able to stimulate developmental processes *phys;* in molecular biology, a protein upstream from a gene on which the DNA binds; it activates the transcription of the gene *gene*

active collection: a collection of germplasm used for regeneration, multiplication, distribution, characterization, and evaluation; ideally germplasm should be maintained in sufficient quantity to be available on request; it is commonly duplicated in a base collection and is often stored under medium- to long-term storage conditions *meth*

active immunity: all means and reactions that enable a plant to prevent an interaction with a pathogen *phyt*

active ingredient: in any pesticide product, the component that kills or controls target pests *phyt* >>> active substance

active site: that portion of an enzyme where the substrate molecules combine and are transformed into their reaction products *phys*

active substance: pesticides, herbicides, etc., are usually mixtures of different substances; among them the active substance is the most important one as it attacks the pathogen *phyt* >>> active ingredient

active transport: the passage of substances across a cell membrane against a concentration gradient that requires energy *phys*

activity rhythm: an individual's daily pattern of physiological activity *phys*

acyclic: not cyclic; in botany, an acyclic flower *bot;* in chemistry, of or pertaining to a chemical compound not containing a closed chain or ring of atoms *chem*

adaptability: the potentiality for adaptation; the ability of an individual or taxon to cope with environmental stress; the range and extent of reaction is genetically determined *phys*

adaptable: capable of being adapted or able to adjust oneself readily to different environmental conditions *phys* >>> adaptability

adaptation: the process of changes of an individual's structure, morphology, and function that makes it better suited to survive in a given environment *phys*

adapted race >>> physiological race

adaptedness: the state of being adapted *phys*

adapter: synthetic double-stranded oligonucleotide; specific type of linker, usually wiground th one blunt and one sticky end; applied to attach sticky ends to a blunt-ended DNA molecule *biot*

adaptive: changes of a plant that act to preserve its full development *phys*

adaptive capacity: the genetically set range or flexibility of reactions of a plant and/or population enabling it to respond in different ways to differing conditions *eco genet*

adaptive character: a functional or structural characteristic of an organism that enables or enhances the probability of survival and reproduction *gene*

adaptive fitness >>> adaptedness

adaptive reaction >>> adaptive capacity

adaptive selection: the evolution of comparable forms in separate but ecologically similar areas *eco gene evol*

adaptive trait >>> adaptive character

adaptive trials: multilocational coordinated field experiments across the crop-growing regions in one or more countries that test adaptation of varieties or breeding strains under specific ecological conditions *meth*

adaptive value: a measure of the reproductive efficiency of an organism or genotype compared with other organisms or genotypes *gene* >>> selective value

adaptiveness >>> adaptedness

adaptivity >>> adaptability

adaxial: toward the axis *syn* ventral *bot* >>> abaxial

addition line: a cell line or line of individuals carrying chromosomes or chromosome arms in addition to tground he normal standard chromosome set *cyto*

additive effects (of genes): gene action in which the effects on a genetic trait are enhanced by each additional gene, either an allele at the same locus or genes at different loci *gene* >>> Tables 6, 20, 21

additive genes: gene interaction without dominance (if allele), or without epistasis (if nonallele); the expression of any genetic trait is enhanced to the simple sum of the individual genetic or allelic effects contributing to that character *gene* >>> Tables 6, 20, 21

additive variance (VA): the proportion of the genetic variance due to additive effects *stat*

adelphogamy: sib pollination or pollination involving a stigma and pollen belonging to two different individuals that are vegetatively derived from the same mother *gene*

adenine (A): a purine base that occurs in both DNA and RNA *chem gene*

adenine sulfate: a growth factor used in some tissue culture media *biot*

adenose: having glands or glandlike organs *bot*

adenosine: the nucleoside formed when adenine is linked to ribose sugar *chem gene*

adenosine diphosphate (ADP): high-energy phosphoric ester (nucleotide) of the nucleoside adenosine that functions as the principal energy-carrying compound in the living cell *phys*

adenosine triphosphate (ATP): high-energy phosphoric ester (nucleotide) of the nucleoside adenosine that functions as the principal energy-carrying compound in the living cell *phys* >>> mitochondrion

adhesion: the molecular attraction between substances causing their surfaces to remain in contact *meth*

adjacent distribution: the orientation and distribution of adjacent chromosomes in the ring or chain configuration of translocation heterozygotes *cyto* >>> Figure 15

adjacent segregation: a reciprocal translocation heterozygote in which during meiosis the segregation of a translocated and a normal chromosome happens together, giving unbalanced gametes with duplications and deficiencies leading to nonviable zygotes; adjacent segregation is of two kinds depending on whether nonhomologous (adjacent-1) or homologous (adjacent-2) centromeres segregate together; adjacent-1 segregation is the usual type of adjacent segregation and adjacent-2 segregation is rare *cyto*

adjacent-1 segregation: segregation of nonhomologous centromeres during meiosis in a reciprocal translocation heterozygote such that unbalanced gametes with duplications and deficiencies are produced, as opposed to alternate segregation and adjacent-2 segregation *cyto*

adjacent-2 segregation: segregation of homologous centromeres during meiosis in a reciprocal translocation heterozygote such that unbalanced gametes with duplications and deficiencies are produced, as opposed to alternate segregation and adjacent-1 segregation *cyto*

adosculation: the fertilization of plants by pollen falling on the pistils *bot*

adpressed: lying flat against (e.g., the rachilla against the palea in the grain of, e.g., barley and oats) *bot*

adsorption: the physical binding of a particle of a particular substance to the surface of another by adhesion or penetration *phy;* attachment of phage to host bacterium (e.g., phage lamda adsorbs to a maltose binding protein) *biot* >>> lamda phage

adsorption complex: the various substances in the soil that are capable of adsorption (e.g., clay or humus) *agr*

adspersed: to have a wide distribution; scattered *bot*

adult: having attained full size and strength; mature *phys*

adult resistance: resistance not expressed at the seedling stage; it increases with plant maturity *phyt*

adult stage >>> adult

adulthood >>> adult stage

aduncate: hooked, crooked, and bent *bot*

advance crop >>> forecrop

advanced character: a feature that shows a real deviation from the ancestral trait or type (e.g., an agronomic trait such as brittle rachis)

adventitious: growing from an unusual position (e. g., roots from a leaf or stem) *bot* >>> Figure 28

adventitious bud: a bud appearing in an unusual place (e.g., a bud on leaves) *bot*

adventitious embryony: a condition in a seed in which the embryo arises from somatic rather than reproductive tissue; the development of a diploid embryo from nucellary or integumentary tissue (sporophyte tissue); common in certain grasses and often results in multiple embryos *bot* >>> twin seedling >>> Figure 28

adventitious plant: an individual that arises from somatic rather than reproductive tissue *bot*

adventitious root: arising from any structure other than a root (e.g., from a node of a stem or from a leaf) *bot*

adventive: a plant that has been introduced but is not yet naturalized *agr*

aeolian soil: a type of soil that is transported from one place to another by the wind *agr*

aerating tissue >>> aerenchyma

aeration: bringing air into a substance, tissue, or soil (e.g., by earthworms or digging and turning the soil to loosen) *phys agr*

aerator: any implement that is used for breaking up compacted soil to facilitate air and gas exchange *agr*

aerenchyma: plant tissue containing large, intercellular air spaces *bot*

aerial pathogens: antagonistic microorganisms that inhibit numerous fungal pathogens of aerial plant parts (e.g., *Tilletiopsis* spp. parasitize the cucumber powdery mildew fungus *Spaerotheca fuligena*); present in crop soils and exert a certain degree of biological control over one or many plant pathogens *phyt* >>> biological control

aerial pest control: pest control by utilization of aircraft and helicopters in order to be more efficient, to prevent damages of the crop, or to cope with difficult landscape and soil conditions *phyt*

aerial root: in some epiphytic orchids, the leaves and the shoot axis are reduced or missing; then flatted and green roots take over the fixation of the plant and the function of the leaves (photosynthesis) *bot*

aerial shoots: shoots growing high above the ground (e.g., trees, bushes, etc.) *bot*

aerobe: an organism needing free oxygen for growth *bot*

aerobium >>> aerobe

aerosol: a colloidal substance that is suspended in the air *phyt*

aesculin >>> glucoside

afflux: the act of flowing to or toward some point or organ in a plant *phys*

afforest: converting bare or cultivated land into forest *eco*

afforestation: the establishment of forest by natural succession or by the planting of trees on land where they formerly did not grow *eco*

aflatoxin: one of a group of mycotoxins produced by fungi of the genus *Aspergillus phys*

AFLP >>> amplified fragment length polymorphisms >>> random amplified polymorphic DNA (RAPD) technique

African cassava mosaic virus (ACMV): this disease continues to be the major constraint to both the commercial and subsistence production of cassava across central and southern Africa, causing up to 100 percent crop losses; the causative agent of the disease has been shown to be a geminivirus of the *Begomovirus phyt*

after-ripening: a term for the collective changes that occur in a dormant seed that make it capable of germination; it is usually considered to denote physiological changes *seed*

AG complex: the complete set of factors assumed to be responsible for the formation of sexual organs *bot*

agameon: a plant species reproducing exclusively by apomixis *bot*

agamete: any nonreproducing germ cell *bot*

agamic: reproducing asexually *bot*

agamic complex: it refers to hybrids or their derivatives that are partially or entirely reproduced by asexual seed formation *bot*

agamogenesis: any reproduction without the male gametes *bot*

agamogony: a type of apomixis in which cells undergo abnormal meiosis during megasporogenesis, resulting in a diploid embryo sac rather than the normal haploid embryo sac *bot* >>> Figure 28

agamont: an asexual individual in whose agametangia the agametes are formed *bot*

agamospecies: populations morphologically differentiated from one another and reproducing apomictically *bot* >>> Figure 28

agamospermy: a type of apomixis in which sporophytic tissue is formed, ultimately leading to seed development; it may occur through adventitious embryony or gametophytic apomixis *bot* >>> twin seedling >>> Figure 28

agar: a complex polysaccharide obtained from certain types of seaweed (red algae); when it is heated with water and subsequently cooled to about + 45°C, it forms a gel *prep*

agar culture: cells, organs, tissue, or embryos artificially grown on a solid medium composed of agar together with certain nutrients, hormones, etc. *biot*

agar gel: gels for electrophoresis that were produced from agar *prep*

agar medium >>> agar culture

agar-agar >>> agar

agarose (starch) gel: an inert matrix used in electrophoresis for the separation of nucleic acids based on their size or conformation; the molecules are visualized in the gel by ultraviolet fluorescence of ethidium bromide, which is either included in the gel or in the running buffer or used to stain the gel after electrophoresis *prep*

agent: a natural force, object, or substance producing or used for obtaining specific results *prep*

agglomerate: in biotechnology, a mass of cells clustered together *biot;* in soil science, rock composed of rounded or angular volcanic fragments *agr*

agglutinate: fixed together as if with glue *prep*

agglutination: the clumping of cellular components *prep*

agglutinin: any antibody capable of causing clumping of types of cells *prep*

agglutinogen: an antigen that causes the production of agglutinins *chem biot*

aggregate: in soil science, a cluster of soil particles forming a pad *agr*

aggregate fruit: a fruit development from several pistils in one flower, as in strawberry or blackberry *bot* >>> composite fruit

aggressiveness: the ability of a pathogen to infect a plant, to break its resistance, to become a parasite on a plant, or to use a host plant for reproduction; the degree of aggressiveness can be estimated only when a pathogen meets a resistant host; when the aggressiveness copes with the resistance of the host it is termed virulence *phyt*

aging test: a method to originally evaluate seed storability; it subjects unimbibed seeds to conditions of high temperature (+41 °C) and relative humidity (~100 percent) for short periods (three to four days); the seeds are then removed from the stress conditions and placed under optimum germination conditions; the two environmental variables cause rapid seed deterioration; high vigor seed lots will withstand these extreme stress conditions and deteriorate at a slower rate than low vigor seeds *seed*

agnesis: the absence of development *bio phys*

agriculture: the science of transforming sunlight energy into plant and animal products that can be utilized in humans; the selective breeding of crop and farm animals has had an enormous impact on productivity in agriculture; modern varieties of crop plants have increased nutritional value and resistance to disease; recent developments in genetic engineering have enabled the potential use of transgenic organisms in agriculture to be explored *agr*

Agrobacterium rhizogenes: a species of Gram-negative, rod-shaped soil bacteria, often harboring large plasmids, called Ri plasmids. It can cause a tumorous growth known as hairy root disease in certain plants *bot biot* >>> Figure 27

Agrobacterium tumefaciens: a species of soil bacteria that can infect the stem of many plants and form crown gall tumors when it contains a Ti plasmid *bot biot* >>> Figure 27

***Agrobacterium*-mediated transformation:** *Agrobacterium* is the generic name of a soil bacterium that frequently causes crown gall in many plant species; besides *A. rhizogenes*, *A. tumefaciens* is one species that is most used in DNA transfer by manipulating the Ti (tumor inducing) plasmid, which is harbored by these bacteria *biot* >>> Figure 27 >>> *Agrobacterium rhizogenes* >>> *Agrobacterium tumefaciens*

agrobiology: the scientific study of plant life in relation to agriculture, especially with regard to plant genetics, cultivation, and crop yield *agr bio*

agroinfection: infection of plants via soilborne pathogens *phyt*

agronomist: the person who is doing farm management and organizes and/or realizes the production of field crops *agr*

agronomy: the part of agriculture devoted to the production of crops and soil management; the scientific utilization of agricultural land *agr*

air layering: a method of plant propagation in which roots are induced to form around a stem; a very narrow strip of bark is removed from around the branch or stem; a sliver of wood can be inserted into the cut to keep it open; a bundle of moist sphagnum moss is tied securely around the cut area; the moss must remain moist and the roots of the plant somewhat dry; new roots will sprout from the incision; the new plant is then cut off below the moss, potted and kept in a humid atmosphere until it is established *bot*

air screen cleaner: the basic piece of equipment for cleaning seed, utilizing airflow and perforated screens *seed*

airlock: an airtight chamber permitting passage to or from a space *seed*

akaryotic: without a nucleus; a stage in the nuclear cycle before meiosis in which no or little chromatin is seen in the nucleus *cyto*

akinete >>> akinetic

akinetic: a nonmotile reproductive structure (e.g., a resting cell) *cyto*

alanine (Ala): an amino acid present in almost all proteins *chem*

alate: winged *bot*

albido: the white tissue beneath the peel of citrus *bot*

albinism: in plants, a deficiency of chromoplasts *bot*

albino: a plant lacking chromoplasts *bot*

albumen: starchy and other nutritive material in a seed, stored as endosperm inside the embryo sac, or as perisperm in the surrounding nucellar cells; in general, any deposit of nutritive material accompanying the embryo *phys* >>> endosperm >>> albumin(e)

albumin(e): any of certain proteins soluble in distilled water at neutral or slightly acid pH and in dilute aqueous salt solution; they coagulate by heat (e.g., leucosins in cereal grains, ricin in rice, or legumelins in pulsė seeds, which are mainly enzymes) *chem* >>> Table 15

albuminoid: containing or resembling albumen or albumin *bot*

albuminous seed: a seed having a well-developed endosperm or perisperm *seed* >>> Table 15

alepidote: having no scales of scurf, smooth *bot*

aleurodid >>> whitefly

aleuron(e): a granulated protein that forms the outermost layer of a cereal grain *bot* >>> aleuron(e) grain >>> Table 15

aleuron(e) grain: small protein grains present in cells of storage tissue *bot*

aleuron(e) layer: a layer of cells below the testa of some seeds (e.g., cereals), which contains hydrolytic enzymes (e.g., amylases and proteases) for the digestion of the food stored in the endosperm; the production of enzymes is activated by gibberellins when the seed is soaked in water prior to germination *bot*

aleuroplast: a leucoplast in which protein granules are present as a main storage product *bot*

alien addition line: a line of plants with one or more extra chromosomes of an alien species *cyto*

alien chromosome: a chromosome from a more or less related species transferred to a crop plant *cyto*

alien chromosome transfer: cytogenetic methods that facilitate the transfer of individual chromosomes from one species to another *meth cyto*

alien gene transfer: the transfer of genes between species or genera by different means *gene biot*

alien germplasm: genes introduced from a wild relative or nonadapted species *cyto gene*

alien substitution line: a line of plants in which one or more alien chromosomes from a certain donor species replace one or more chromosomes of a recipient species *cyto*

aliquot: a part, such as a representative sample, that divides the whole without a remainder; two is an aliquot of six because it is contained exactly three times; loosely it is used for any fraction or portion *prep*

alkali: a substance capable of furnishing hydroxyl-OH ions to its solution; the most important alkali metals are potassium and sodium *chem*

alkaline phosphatase: an enzyme that is a member of hydrolases; it cleaves from linear DNA or RNA molecules the 5'-terminal phosphate group; dephosphorylated 5' DNA or RNA ends cannot be joined by ligase to 3' ends; polynucleotide kinase reverses the reaction; sometimes used as a biochemical marker in genetic studies *phys biot gene*

alkaline soil: specifically a soil with pH value > 7.0 caused by the presence of carbonates of calcium, magnesium, potassium, and sodium; commonly used for soils showing a pH value 8.5 *agr*

alkaloid: one of a group of basic, nitrogenous, normally heterocyclic compounds of a complex nature; alkaloids occur in several plants (e.g., coniine in hemlock, morphine in poppy fruits, strychnine in seeds of *Strychnos nux vomica*, atropine in nightshades, colchicine in meadow saffron, caffeine in coffee and tea, nicotine in tobacco leaves, theobromine in cacao) *chem phys*

alkaloidity: the alkaloid content of cell, tissue, organs or individuals of plants *phys*

alkylating agent: a chemical agent that can add alkyl groups (e.g., ethyl or methyl groups to another molecule; many mutagens act through alkylation) *chem*

allele: one of two or more alternate forms of a gene occupying the same locus on a particular chromosome; currently, different alleles of a given gene are usually recognized by phenotype rather than by comparison of their nucleotide sequences *gene* >>> allelism

allele frequency: a measure of the commonness of an allele in a population of alleles *gene* >>> allele

allele shift: a modification of allele frequency in a population due to either natural or artificial selection *gene*

allele trend: a directed change in allele frequency of populations per time unit *gene*

allele-specific associated primers (ASAP): a PCR variant in which the sequence of the deca-mer oligo is derived from normal RAPD, which generated an absence and/or presence polymorphism; these polymorphisms do not require electrophoretic separation of the sample; the presence of an amplification product is detected by measuring fluorescence of ethidium-bromide stained DNA (Gu et al., 1995) *biot*

allele-specific oligo (ASO): a special kind of oligo for an AC-PCR reaction; the sequence of the oligo is designed in such a way to allow and/or inhibit hybridization at the spot where the mutant (resistant) allele differs from the wild-type (susceptible) allele *biot*

allelic complementation: the production of a nonmutant phenotype when two independent mutations at the same gene locus, but on different homologous chromosomes, are introduced *gene*

allelism: the common shortening of the term "allelomorphism"; one of the two or more forms of a gene arising by mutation and occupying the same relative position (locus) on homologous chromosomes *gene*

allelobrachial: changes of chromosome structure in which the arms of homologous chromosomes are included *cyto*

allelogenous: females that produce only males or only females in different progenies *bot*

allelomorph: a term that is commonly shortened to "allele" *gene* >>> allele

allelopathy: chemical substance released into the environment by an organism that acts as a germination or growth inhibitor to another organism *eco*

allelotype: the genetic composition (i.e., allele frequency) of a breeding population *gene*

alliaceous: onionlike in smell or form *bot*

allocompetition (intergenotype competition): cultivation at high plant density implies the presence of strong interplant competition *eco;* the individual plants, clones, lines, or families are evaluated when being subjected to intergenotypic competition; also called intergenotype competition *stat*

allocycly: differences in chromosome coiling caused by environmental or genotypic effects *cyto*

allodiploid: cells or individuals in which one or more chromosome pairs are exchanged for one or more pairs from another species *cyto*

allogamous: cross-fertilizing in plants, as opposed to autogamous *bot* >>> Table 35

allogamy: cross-fertilization as opposed to autogamy *bot* >>> Table 35

allogene >>> recessive allele

allogenetic: cells or tissues related but sufficiently dissimilar in genotype to interact antigenically *phys*

allogenic: applied to successional change due to a change in abiotic environments *eco*

allograft: a graft of tissue from a donor of one genotype to a host of a different genotype but of the same species *hort*

allohaploid: a haploid cell or individual derived from an allopolyploid and composed by two or more different chromosome sets *cyto* >>> Figure 3

alloheteroploid: heteroploid individuals or cells whose chromosomes derive from various genomes *cyto* >>> Figure 3

alloiogenesis: growth of a part of an organism in relation to the growth of the whole organism or some other part of it *phys*

allometric: growth in which the growth rate of one part of the plant differs from that of another part or of the rest of the plant *bot*

allometry >>> alloiogenesis

allopatric: applied to species that occupy separate habitats and that do not occur together in nature (*cf* parapatric and sympatric) *eco*

allophene: a phenotype not due to the mutant genetic constitution of the cell or the tissue in question; such a cell or tissue will develop a normal phenotype if it is transplanted to a wild-type host *gene*

allophenic: characteristics that arise by intercellular gene action *gene*

alloplasmic: an individual having the common nucleus, but an alien cytoplasm (e.g., alloplasmic rye containing a wheat cytoplasm); usually leads to meiotic disturbances and sterility *gene*

alloploid (allopolyploid): a plant that arises after natural or experimental crossing of two or more species or genera; they may contain genomes of the parents in one or more copies *cyto* >>> amphiploid >>> Figures 3, 8

alloploidy >>> alloploid >>> allopolyploid

allopolyploid: plants with more than two sets of chromosomes that originate from two or more parents; the sets contain at least some nonhomologous chromosomes *cyto* >>> Figure 3 >>> Table 17

allopolyploidy >>> allopolyploid

allosome: a chromosome deviating in size, form, or behavior from the other chromosomes (autosomes), such as the sex chromosome or B chromosome *cyto* >>> heterochromosome

allosteric: an enzyme whose activity is altered when its structure is distorted by an organic compound at a nonsubstrate site *phys chem*

allosteric effect: the binding of a ligand to one site on a protein molecule in such a way that the properties of another site on the same protein are affected *chem*

allosteric transition: a change from one conformation of a protein to another conformation *chem*

allosubstitution: the replacement of a chromosome or chromosome arm by an alien chromosome or chromosome arm *cyto*

allosynapsis >>> allosyndesis

allosyndesis: chromosome pairing of completely or partially homologous (homoeologous) chromosomes *cyto*

allotetraploid *syn* **amphidiploid:** a plant that is diploid for two genomes, each from a different species *cyto* >>> Figure 8

allotetraploidy >>> allotetraploid

allotopic: a type specimen of the sex opposite that of the holotype *bot;* in immunology, an antibody that acts as an antigen to other antibodies of the same species that have variant molecular sites *meth*

allozygosity: homozygosity in which the two alleles are alike but unrelated *gene*

allozygote: a zygote heterozygous for different mutant alleles *gene*

allozyme: isoenzymes of protein nature whose synthesis is usually controlled by codominant alleles and inherited by monogenic rations; they show a specific banding pattern if separated by electrophoresis *phys* >>> Table 29

alluvial soil: soils developed on fairly recent alluvium (a sediment deposited by streams and varying widely in particle size); usually they show no horizon development *agr*

alpha amylase >>> falling number

alpha complementation (of beta-galactosidase): pUC18 and similar vectors contain only a small part of the whole gene for beta-galactosidase; this small part gives rise to a truncated protein that forms an enzymatically active hetero dimer with a specific mutant beta-galactosidase *biot*

alpha helix: the right-handed, or less commonly left-handed, coillike configuration of a polypeptide chain that represents the secondary structure of some protein molecules *gene*

alpha level >>> significance level

alpha-bromonaphthalene: a chemical agent that is used for artificial chromosome condensation; for several hours root tips are treated with a saturated water solution prior to staining *cyto*

Alternaria: a genus of fungi; it forms yellowish-brown conidia that are divided by transverse and longitudinal septa; there are many species, including important plant pathogens (early blight of potato, *Alternaria solani;* black rot of carrot and rape, *A. radicina* and *A. brassica; Alternaria* of wheat) *phyt*

alternate: not opposite to each other on the axis, but borne at regular intervals at different levels (e.g., of leaves) *bot*

alternate host >>> alternative host

alternate segregation: at meiosis in a reciprocal translocation heterozygote, the segregation of both normal chromosomes to one pole and both translocated chromosomes to the other pole, giving genetically balanced gametes, or segregation of centromeres during meiosis in a reciprocal translocation heterozygote such that genetically balanced gametes are produced *cyto*

alternating dominance: a change of dominance from one allele to the other ($A1a2 \rightarrow a1A2$) of a pair of alleles during ontogenetic development of a heterozygous hybrid; the phenotypic expression of the alleles acts one after another *gene*

alternation of generation: the alternation of two or more generations, reproducing themselves in different ways (i.e., alternation of gametophyte [sexual reproductive] and sporophyte [asexual reproductive] stages in the life cycle of a plant) *bot*

alternative disjunction: the distribution of interchange chromosomes at anaphase I of meiosis is determined by their centromere orientation; in the case of alternative disjunction, chromosomes locàted alternatively in the pairing configuration are distrubuted to the same spindle pole, as opposed to adjacent disjunction *cyto* >>> adjacent disjunction >>> translocation

alternative host: a host that harbors a pest or disease while the primary host is absent or out of season *phyt* >>> host

***Alu* element:** a repetitive DNA element approximately 300 base pairs long that is abundantly dispersed throughout the genome of primates; the name derived from the *AluI* restriction enzyme cleavage site that is within most *Alu* elements *biot*

aluminum (Al): has no specific importance in the metabolism of higher plants; small amounts of uptake favor the imbibition of the cytoplasm; higher concentrations in the soil may cause severe inhibition of plant growth; aluminum tolerance is a main task of plant breeding in several regions of the world *chem phys agr* >>> Table 33

AMBA >>> American Malting Barley Association

amber codon: amber suppresser mutation that changes anticodon of amino acid-carrying tRNA to UAG *biot*

ambisexual: a plant that has the reproductive organs of both sexes *bot* >>> bisexual

ambivalent gene: genes with both advantageous and disadvantageous effects *gene*

ambosexual >>> ambisexual

ameiosis: the failure of meiosis and its replacement by nuclear division without reduction of the chromosome number *cyto*

ameiotic parthenogenesis: parthenogenesis in which meiosis has been entirely suppressed *gene cyto* >>> parthenogenesis

ament(um) >>> catkin

amidase: an enzyme that catalyzes the hydrolysis of an acid amide *chem phys*

amide: a compound derived from ammonia by replacement of one or more of the hydrogens with organic acid groups *chem*

amine: an organic base derived from ammonia by replacement of one or more of the hydrogens with organic radical groups *chem*

amino acid: an organic compound containing an acidic carboxyl group (–COOH) and a basic amino group (NH_2); amino acid molecules combine to form proteins; they are the fundamental constituent of living matter; they are synthesized by autotrophic organisms, such as green plants *chem*

amino acid sequence: the sequence of amino acid residues in a polypeptide chain that represents the primary structure of a protein; the sequence is unique to each protein and influences the protein structure (secondary, tertiary, quaternary) *gene*

aminoacetic acid >>> glycine

aminoacyl-tRNA: a tRNA molecule covalently bound to an amino acid via an acyl bond between the carboxyl group of the amino acid and the 3'-OH of the tRNA *chem*

aminoacyl-tRNA ligase: an enzyme that synthesizes a specific aminoacyl-tRNA molecule, employing a specific amino acid (e.g., alanine, its cognate tRNA and ATP to form (e.g., alanyl-tRNA alanine) *chem phys biot*

aminopeptidase (Amp): an enzyme that catalyzes the hydrolysis of amino acids in a polypeptide chain by acting on the peptide bond adjacent to the essential free amino group *chem phys*

amitosis: nuclear division by a process other than mitosis *cyto*

amixis: reproduction in which the essential events of sexual reproduction are absent *bot*

ammonia: compound that plays an important role in the natural nitrogen pathways *chem agr*

ammonium fixation: in soil science, adsorption of ammonium ions by clay minerals rendering them insoluble and nonexchangeable *agr*

ammonium phosphate: used as a mineral fertilizer *chem agr*

ammonium sulfate: used as a mineral fertilizer with nitrogen content of about 21 percent *chem agr*

amorph: a gene that is inactive—an amorphic gene; sometimes used to refer to something that lacks a discernable shape and can thus be described as "amorphous" *gene bot*

amorphous >>> amorph

amphibiotic: applied to an organism that can be either parasitic on or symbiotic with a particular host organism *phyt*

amphibivalent: a ringlike interchange configuration of four chromosomes *cyto*

amphicarpous: plants producing two classes of fruit that differ either in form or in time of ripening *bot*

amphidiploid >>> allotetraploid >>> didiploid

amphihaploid >>> allohaploid

amphikaryon: the nucleus of the zygote produced after fertilization *cyto*

amphimict: a species or individual that is reproduced by fusion of nuclei during sexual reproduction *bot*

amphimictic >>> amphimict

amphimixis >>> amphimict

amphiplasty (nucleolar dominance): morphological changes of chromosomes after interspecific hybridization; occurs when the genomes of the parental species are spatially separated in the hybrid nucleus in a concentric fashion; the genome occupying the central position has an active NOR, while the NOR of the peripherical genome is suppressed (e.g., in hexaploid wheat, NORs on chromosome 1B and 6B function preferentially, although 1A and 5D also carry NORs; however, when the NOR on 1B is deleted, then the NORs on 5D and/or 1A are used to a greater extent) *cyto*

amphiploid >>> alloploid

amphiploidy >>> alloploidy; allopolyploidy

amphitene >>> zygotene

amphitoky: in some insects (e.g., butterfly), progeny of both sexes may develop from unfertilized eggs *zoo*

amphitropous ovule: a type of ovule arrangement in which the ovule is slightly curved so the micropyle is near the funicular attachment *bot*

ampicillin (beta-lactamase): an antibiotic substance; the resistance to ampicillin is sometimes used as a screening marker in genetic experiments with bacteria (e.g., the cloning vector pBR322 carries a gene for ampicillin resistance; it interferes with bacterial cell wall synthesis; the *bla* gene is used as a selective marker in many vectors, including pBR322 and pUC18) *biot*

amplexicaul: a leaf whose base wholly or partly surrounds the stem *bot*

amplification: the intrachromosomal or extrachromosomal production of many DNA copies from a certain region of DNA; it can happen spontaneously or it can be done by molecular techniques (e.g., PCR) *cyto*

amplified fragment length polymorphisms (AFLPs): polymorphic DNA fragments are amplified through PCR procedure; their differences are used for genotype identification and linkage studies *biot*

amplify >>> amplification

ampoule: a bottle with a bulbous body and narrow neck *prep*

ampulliform: flasklike in form *bot*

amygdaliform: almond shaped *bot*

amygdalin >>> glucoside

amylaceous >>> amyliferous

amylase: a member of a group of enzymes that hydrolyze starch or glycogen by splitting of glucosidic bonds, giving rise to sugars, glucose, dextrin, or maltose; they occur particularly in germinating seeds in which the amylase mobilizes food reserves for the growth of the seedling *chem phys*

amyliferous: containing starch *bot*

amylopectin: larger, highly branched chains of glucose molecules *phys*

amyloplast: a plastid that synthesizes and stores starch to the exclusion of other activities *bot*

amylose: relatively short, unbranched chains of glucose molecules; a polysaccharide consisting of linear chains of between 100 and 1,000 linked glucose molecules; a constituent of starch; in water, amylose reacts with iodine to give a characteristic blue color *chem phys*

anabiose: the situation in the life cycle of some plants in which there is no visible metabolic activity (resting period) *bot*

anabolic: pertaining to an enzymatic reaction leading to the synthesis of a more complex biological molecule from a less complex one *chem phys*

anaerobe >>> anaerobic

anaerobic: an organism able to grow without free oxygen *phys*

anagenesis: a mode of evolution characterized by cumulative changes in an evolutionary lineage *eco*

analysis of variance (ANOVA): a statistical method that allows the partitioning of the total variation observed in an experiment among several statistically independent possible causes of the variation; among such causes are treatment effects, grouping effects, and experimental errors; the statistical test of the hypothesis that the treatment had no effect is the F-test or variance-ratio test; if the ratio of the mean square for treatments to the mean square for error exceeds a certain constant that depends on the respective degrees of freedom of the two mean squares at a chosen significance level, then the treatments are inferred to have been effective *stat*

anandrous: free of anthers (male sexual organs) *bot*

anaphase (A): a stage that occurs once in mitosis and twice in meiosis and that involves the separation of chromosomal material to a given two sets of chromosomes, which will eventually form part of new cell nuclei; the separation is controlled by the spindle; in anaphase of mitosis and anaphase II of meiosis, the centromere becomes functionally double and daughter chromosomes separate from the equator, moving toward the opposite poles of the cell; the spindle then elongates and pushes the two groups of chromosomes further apart; in anaphase II of meiosis the centromere does not divide *cyto*

anaphase movement: the movement of chromosomes and/or chromatids toward the cell poles during mitotic or meiotic anaphase *cyto* >>> Figure 15

anaphase separation: the disjunction of the chromatids of each chromosome during mitosis and anaphase II of meiosis or the separation of chromosomes in anaphase I of meiosis *cyto*

anaphragmic: mutations that lead to increased enzyme activity by removal of inhibitors *phys gene*

anatropous ovule: a type of ovule arrangement in which the ovule is completely inverted, having a long funiculus with the micropyle adjacent to the base of the funiculus *bot*

ancestor: the form or stock from which an organism has descended or the actual or assumed earlier type from which a species or other taxon evolved *gene*

ancestry: a series of ancestors; ancestral descent; lineage *gene*

anchorage dependence: describes the normal eukaryotic cell's need for a surface to attach to in order to grow in culture *biot*

ancient: dating from a remote period; very old; aged

androdioecious: describes a species having male and hermaphroditic flowers on separate individuals *bot*

androecious: applied to a plant that possesses only male flowers *bot*

androecium: a collection of stamens that form the male reproductive organs of flowering plants *bot*

androecy: applied to a plant that possesses only male flowers *bot*

androgenesis: development of a haploid embryo from a male nucleus *bot*

androgenetic embryo: an embryo that contains two paternally derived sets of chromosomes and no maternally derived chromosomes *biot*

androgenous: producing only male offspring *bot*

andromonoecious: a species having male and hermaphroditic flowers on the same individual *bot*

andromonoecism: plants with staminate and perfect hermaphroditic flowers *bot*

androphore: the stalk or column supporting the stamens of certain flowers; usually formed by a union of the filaments (e.g., in Leguminosae) *bot*

androsome: any chromosome exclusively present in the nucleus of the male *cyto*

anemochory: distribution of seeds by the wind *bot*

anemophilous: windborne pollen *bot* >>> Table 35

anemophily: pollination of a flower in which the pollen is carried by the wind (e.g., in grasses) *bot* >>> anemophilous >>> Table 35

aneucentric: applied to an aberrant chromosome possessing more than one centromere *cyto*

aneuhaploid: when the chromosome number deviates from the haploid standard chromosome number of the species or individual *cyto*

aneuploid: a cell or organism whose nuclei possess a chromosome number that is greater or smaller by a certain number than the normal chromosome number of that species; an aneuploid results from nondisjunction of one or more pairs of homologous chromosomes *cyto*

aneuploid reduction: reduction of the genetic variability by decreasing the chromosome number *cyto*

aneuploidy >>> aneuploid

aneusomatic: individuals whose cells exhibit variable numbers of chromosomes *cyto*

aneusomic: individuals or cells that contain unequal sets of the individual chromosomes *cyto*

angiocarpous: having a fruit enclosed within a distinct covering (e.g., a filbert within its husk) *bot*

angiosperm: seeds formed within an ovary *bot*

angiospermous >>> angiosperm

angiosperms: any vascular plant of the phylum, having the seeds enclosed in a fruit, grain, pod, or capsule and comprising all flowering plants *bot* >>> Table 32

Ångstrom (Å): a unit of length equal to 10^{-10} meter; formerly used to measure wavelengths and intermolecular distances but has now been replaced by the nanometer (1 Å = 0.1 nanometer); the unit is named after Swedish pioneer of spectroscopy A. J. ÅNGSTRÖM (1814-1874)

angustifoliate: with small leaves *bot*

anhydride: a compound formed by removing water from a more complex compound *chem*

anion: an ion that carries a negative electrical charge *chem*

anisogamete: a gamete that differs in size, appearance, structure, or sex chromosome content from the gamete of the opposite sex *bot*

anisogamy: unequal gametes fusing during fertilization *bot*

anisomeric: nonequivalent genes that interact to produce particular phenotypes *gene*

anisoploid: an individual with an odd number of chromosome sets in somatic cells *cyto*

anisoploid seeds: a mixture of seeds of different ploidy levels (e.g., in sugarbeet [2x, 3x, 4x]) *seed cyto*

anisotrisomic: a mixture of seeds or individuals that are not only trisomic *cyto*

anneal: in molecular genetics, heating that results in the separation of the individual strands of any double-stranded nucleic acid helix, and cooling that leads to the pairing of any molecules that have segments with complementary base pairs; synonymously used for hybridization *biot*

annouline: the fluorescent protein pigment exudated by the roots of ryegrass seedlings; the fluorescent nature of this material is useful in distinguishing annual and perennial ryegrass *seed*

annual: a plant that completes its life cycle within a single growing season *bot* >>> biennial >>> perennial

annual ring: in woody plants, the layer of wood produced each year that can be seen when the wood is cut into a cross section; the number of rings equals the age of the tree *fore*

annulus: applied to any of a number of ring-shaped nuclear pores *bot*

anorthogenesis: adaptive changes of evolutionary significance based on preadaptation *evol*

ANOVA >>> analysis of variance

***Antennaria* type:** mitotic diplospory where the megaspore mother cell does not enter meiosis but proceeds directly into the first mitosis; the megaspore mother cell thus functions as an unreduced gamete *bot*

antephase >>> prophase

anther: the terminal portion of a stamen of a flowering plant;. the pollen sacs containing pollen are borne on the anther; the number of anthers in a flower varies from 3 to 10 among most species *bot*

anther culture: culturing of anthers containing pollen or of single pollen grains; the method is used for the production of haploid plants, for the production of doubled haploids (which are homozygous) after spontaneous or induced rediploidization or for breeding on a lower ploidy level *biot* >>> microspore culture >>> Figures 17, 26 >>> Table 7

antheridium: the male sex organ or gametangium within which male gametes are formed *bot*

anthesis: the time of flowering in a plant, the opening of a flower bud, or the time when the stigma is ready to receive the pollen *bot*

anthocyan >>> anthocyanin

anthocyanin: water-soluble, nitrogenous pigments that contribute to the autumnal colors of the leaves of temperate-climate plants; red- and purple-colored matter found in various parts of the plants (e.g., in the auricles, awns, nodes, and coleoptiles of many cereals) *chem phys*

anthoxanthin: the yellow- or orange-colored matter of yellow flowers and fruits *bot*

anthracnose: a general term for any of several plant diseases in which symptoms include the formation of dark and often sunken spots on leaves, fruits, etc. (e.g., caused by *Colletotrichum lindemuthianmum* in dwarf bean, *Elsinoe ampelina* in grape, *Kabtiella caulivora* in clover, *Gloeosporium ribis glossulariae* in gooseberry, *Colletotrichum linicolum* in linseed, *C. oligochaetum* in cucumber, *C. graminicolum* in rye and maize, or *C. orbiculare* in melons) *phyt*

antibiosis: the phenomenon whereby a natural organic substance secreted by one organism has an injurious effect on normal growth and development of another organism when the two organisms are brought together *eco*

antibiotic: a substance that is produced by some organisms; it may inhibit the growth of another organism or even kill it; several antibiotics are produced by fungi; they act against bacteria *phys*

antibiotic resistance gene: a gene that encodes an enzyme that degrades or excretes an antibiotic, so conferring resistance *gene phyt*

antibody: a protein, usually found in serum, whose presence can be shown by its specific reactivity with an antigen; it binds with high affinity to antigens and thereby destroys them; in molecular biology, it is used in Western blots; a normal immuneserum contains different antibodies, which recognize all the many different antigens an animal normally produces; monospecific antiserum is purified to exclusively contain antibodies that recognize epitopes of a single macromolecule; monocolonal antibodies contain only one particular type of antibody that is specific for a single epitope; they are produced by hybridoma cell cultures *phys biot*

anticlinal: referring to a layer of cells running orthogonal to the surface of a plant part *bot*

anticline >>> anticlinal

anticodon: a triplet sequence of nucleotides in tRNA that during protein synthesis binds by base pairing to a complementary sequence, the codon, in mRNA attached to a ribosome *gene*

antigen: a molecule, normally of a protein although sometimes of a polysaccharide, usually found on the surface of a cell, whose shape causes the production in the invaded organism of antibodies that will bind specifically to the antigen *phys biot*

antigenicity >>> antigen

antimetabolite: a substance that resembles in chemical structure some naturally occurring compounds and which specifically antagonizes the biological action of such compounds *phys*

antimitotic: substances that may lead to the cessation of mitosis *cyto*

antimorph: a mutant allele that acts in a direction opposite to the normal allele *gene* >>> amorph >>> hypermorph >>> neomorph

antimutagenic: substances that can reduce the rate of mutations *gene*

antimutator gene: mutant genes that decrease the mutation rates *gene*

antinutritive character: a crop product containing substances causing either diseases or negative influences in humans and animals (e.g., the pentosans of rye and triticale) *agr*

antioxidant: a substance that inhibits oxidation *chem*

antipodal cells: three haploid nuclei that are formed during megasporogenesis in plants; all are located opposite the micropylar end of an ovule *bot* >>> Figures 25, 35

antipodal nuclei: three of the eight nuclei that result from the megaspore by mitotic cell divisions within the developing megagametophyte (embryo sac); they are usually located at the base of the embryo sac and have no apparent function in most species *bot* >>> Figure 25

antipodes >>> antipodal cells

antisense DNA: noncoding DNA of one of the double-stranded DNA, as opposed to sense DNA, which is the coding DNA (i.e., which is transcribed as mRNA) *gene biot*

antisense gene: a gene construct placed in inverted orientation relative to a promoter; when it is transcribed it produces a transcript complementary to the mRNA transcribed from the normal orientation of the gene *biot*

antisense RNA: a complementary RNA sequence that binds to a naturally occurring (sense) mRNA molecule; in this way it thus blocks its translation *biot*

antiserum: a serum that contains antibodies *meth*

antixenosis: a nonhost preference by the pest; estimated by the degree of colonization on the plants *phyt*

anucleolate: without a nucleus *cyto*

AOSA: Association of Official Seed Analysts

apatite: an important inorganic phosphate of soil, but only of limited use for plant nutrition *chem agr*

aperture: in an optical instrument, the opening of a lens *micr*

aperture diaphragm: an adjustable diaphragm in illumination path that regulates amount of excitation intensity (numerical aperture of excitation light) *micr*

aperture plane: in a microscope adjusted for KOEHLER illumination, the conjugate planes that include the light source, the condenser iris diaphragm, the objective lens back aperture, and the eye point *micr*

apetalous: a plant showing no flowers *bot*

apex: extreme point or distal end *bot*

apex culture >>> shoot-tip culture

apex of the leaf >>> leaf tip

aphid vector: any of numerous tiny soft-bodied insects of the family *Aphididae* transferring viruses *phyt*

aphids: any of numerous tiny soft-bodied insects of the family *Aphididae* that suck the sap from the stems and leaves of various plants; they are also called plant louses *zoo phyt*

aphyllous: leafless; applied to flowering plants that are naturally leafless (e.g., many species of cactus) *bot*

aphylly: leafnessness *bot*

apical: at the end; related to the apex or tip *bot*

apical cell: a meristematic initial in the apical meristem of shoots or roots of plants; as this cell divides new tissue is formed *bot*

apical dominance: a condition in plants where the stem apex prevents the development of lateral branches near the apex *phys*

apical meristem: an area of actively dividing cells at the tip of a shoot, branch or root; it is also the precursor of the primary tissue of root or shoot *bot*

apical placentation: a type of precentral placentation in fruit where the seeds are attached near the top of the central ovary axis *bot*

apical segment (of spike): the uppermost segment of the rachis *bot*

apical spikelet: the spikelet occurring at the apex (tip) of the ear or panicle *bot*

apiculture: beekeeping for the sale of honey and for pollination of crop plants *seed*

aplanate: lying in a plane; leaves may be displayed on the twigs of a plant to give aplanate foliage; flattened *bot*

apoamphimict: a plant that reproduces predominantly by apomixis but also sexually *bot*

apocarp >>> apocarpy

apocarpy: the condition in which the female reproductive organs (carpels) of a flower are not joined to each other (e.g., in buttercup) *bot* >>> syncarpy

apochromat: a microscope objective corrected for spherical and chromatic aberration *micr*

apoenzyme: the portion of a conjugated enzyme that is a protein *phys*

apogamety: autonomous development of a vegetative cell, not the egg cell, into an embryo in an apomict *bot*

apogamy: a type of apomixis involving the suppression of gametophyte formation so that seeds are formed directly from somatic cells of the parent tissue *bot* >>> apogamety >>> Figure 28

apomeiosis: sporogenesis without reduction of chromosome number during meiosis and giving rise to apomixis *bot*

apomictic: plants that form asexual progenies *bot* >>> Figure 28

apomixis: asexual reproduction in plants without fertilization or meiosis *bot* >>> Figure 28

apomorphy: a derived state of a character that has changed in state in relation to its predecessor *bot*

apophase: the period of postmeiotic reconstruction of the cell *cyto*

apoplast: areas of a plant that lie outside the plasmalemma, such as cell walls and dead tissue of the xylem; the apoplast may represent one of the main pathways of water through the plant *bot*

apoptosis: the process by which a cell dies in a programmed way, or in other words, kills itself; it is the most common form of physiological (as opposed to pathological) cell death; it is an active process requiring metabolic activity by the dying cell; often characterized by shrinkage of the cell and cleavage of the DNA into fragments *bot phys*

apospory: the development of a diploid embryo sac in some plants by the somatic division of a nucellus or integument cell without meiosis; a sort of agamospermy in which a seed is produced without fertilization *bot* >>> Figure 28

apostatic selection: a selection of very rare genotypes (e.g., a macromutation) *gene*

apothecium: a roughly cup-shaped or dishlike ascocarp, in which the asci line the inner surface and are thus exposed to the atmosphere *bot*

appendage: a process of outgrowth of any sort *bot*

apple scab: a common disease of apple trees in which the most obvious symptom is the appearance of superficial, dark, corky scabs on the fruit *phyt*

applied genetics: studying genetic factors and processes for utilization in agriculture and other branches of science *gene*

appressed: lying close and flat against *bot*

appressorium: in certain parasitic fungi, an attachment organ consisting of a flattened hypha that presses closely to the tissue of the host as a preliminary stage in the infection process *bot*

aquatic: living in water *bot*

arabinose (Ara): an aldose that contains five carbon atoms; a member of pentoses (e.g., it occurs in sugarbeet) *chem phys*

arable: capable of producing crops by ploughing or tillage *agr*

arable land: land capable of producing crops by ploughing or tillage *agr*

arachin >>> globulin

arachnoid: covered with hairs or fibers or formed of hairs or fibers *bot*

arborescent: treelike shape *bot*

archegonium: female sex organ of liverworts, mosses, ferns, and most gymnosperms; usually a flasklike organ, comprising a swollen base or center containing a single egg cell and a slender elongated neck containing one or more layers of cells *bot*

archesporial: the differentiated cell situated in the nucellar tissue of the ovule that is destined to undergo meiosis and give rise to the haploid generation *bot*

archesporium: cells formed by mitosis of the micro- and macrospore mother cells *bot*

arcuate: curved or arched *bot*

arginine (Arg): an aliphatic, basic, polar amino acid that contains the guanido group *chem phys*

arid: dry climates or dry regions (<250 mm rainfall in temperate climates and <350 mm rainfall in tropical climates); commonly applied to a region or a climate in which precipitation is too low to support crop production *eco*

aril(lus): a usually fleshy and often brightly colored outgrowth from the funicle or hilum of a seed, a third integument; arils probably often aid seed dispersal by drawing attention to the seed after the fruit has dehisced and by providing food as an attractant and reward to the disperser *bot*

arista >>> awn

aristate: showing awns; bearded; bristle tipped *bot* >>> Figure 34

arithmetic mean: an average; the number found by dividing the sum of series by the number of items in the series *stat*

arnautka: strains of durum wheat *(Triticum durum) agr* >>> Table 1

arrect: stiffly upright *bot*

arrhenotoky: in insects (e.g., bees), the development of male progenies from nonfertilized eggs *zoo*

arrow: the inflorescence of sugarcane *agr*

arsenic: it may be found in plants in very low dosage; higher concentrations in soil may be toxic for plants *chem agr*

articulate: jointed; having a node or joint *bot*

artifact: a human-made object; something observed that usually is not present but that has arisen as a result of the process of observation or investigation *meth*

artificial chromosome: a chromosome experimentally created and constituted, in addition to genetically coding DNA sequences, by ligating origin of replication, autonomous replicating sequences, and telomeric and centromeric sequences *biot*

artificial light: light other than sunlight; often from fluorescent tubes; used to grow plants in greenhouses or growing chambers usually out of season *hort agr*

artificial seeds: based on embryogenic suspension cultures, embryos may be coated with water-soluble hygrogels and other substances in order to guarantee a proper germination even under field conditions *seed biot*

artificial selection: plant selection by human or agronomic means

artioploid: it refers to polyploids with even sets of genomes *cyto*

ASAP >>> allele-specific associated primers

ascendent >>> ancestor

ascending: rising somewhat obliquely, or curved upward *bot*

ascogonium: the female gametangium of some fungi *bot*

ascorbic acid: the water-soluble vitamin C that occurs in large quantities in fruits and vegetables *chem phys*

ascospore: a sexually produced, haploid spore formed within an ascus by some fungi *bot*

ascus: a minute, baglike structure within which ascospores develop in some fungi *bot*

aseptic: free from living microorganisms or pathogens; sterile *pre*

asexual: any reproductive process that does not involve the fusion of gametes *bot* >>> Figure 28

asexual reproduction: a propagation without formation of zygotes by sexual organs and genetic recombination; in plants, there are two types of asexual reproduction: vegetative propagation (by stolons, rhizomes, tubers, tillers, bulbs, bulbils, or corms) and apomixis (by vegetative proliferation or agamospermy) *bot* >>> Figure 28

ASO >>> allele-specific oligo

asparagine (Asn): an amino acid found in storage proteins of plants (e.g., in peas and beans); the designation derived from the presence in asparagus *chem phys*

asparagus knife: a tool for prying and pulling out long-rooted plants *syn* dandelion weeder *syn* fishtail weeder *hort*

aspartic acid (Asp): an aliphatic, acidic, polar alpha-amino acid *chem phys*

aspirator: an air-blast separator; a seed conditioning (cleaning) machine that uses air to separate according to specific gravity (weight) and resistance to air flow *seed* >>> Table 11

assimilation: in plant physiology, the production of an organic substance from anorganic elements and compounds via photosynthesis *phys* >>> carbon dioxide

assortative mating: occurs if the plants mating resemble each other, with regard to some traits *gene* >>> disassortive mating

ASTA: Americ̦an Seed Trade Association

asymmetric fusion: a cell formed by the fusion of dissimilar cells; referred to as a heterokaryon *biot* >>> cell fusion

asynapsis: chromosomes of meiosis I in which pairing either fails or is incomplete *cyto*

asynaptic >>> asynapsis

asyndesis >>> asynapsis

atavism: the reappearance of a character after several generations; the reversion to an ancestral or earlier type of character; the character being the expression of a recessive gene or of complementary genes *bio*

atrophy: reduced or diminished organ size, shape, or function, usually a deteriorate change *bot phys*

atropine: a poisonous crystalline alkaloid, $C_{17}H_{23}NO_3$, it can be extracted from deadly nightshade and other solanaceous plants; used in medicine to treat colic, to reduce secretion, and to dilate the pupil of the eye *chem phys*

***att* site:** loci on a phage and the bacterial chromosome at which site-specific recombination takes place *biot*

attar: a term used for a perfume from flowers (an essential oil, e.g., the attar of roses) *hort*

attenuator: a nucleotide sequence, located in the leader region between the promoter and the structural genes of some operons, that causes RNA polymerase to cease transcription in the leader region before transcribing the structural genes of the operon *gene biot*

atypical: having no distinct typical character; not typical; not conformable to the type *meth*

auger: a tool for boring a hole in the soil; used in planting or transplanting seeds, seedlings, or bulbs or in fertilizing shrubs, trees and ground covers *prep hort;* in seed production, a long pipe with a twist inside to carry the grain up into a grainery *seed*

auricle: clawlike outgrowth arising at the junction of the leaf blade and sheath (e.g., present in wheat and barley, absent in oats) *bot* >>> Table 30

auriculate: furnished with auricles *bot* >>> auricle

aurofusarin: an orange-yellow pigment of the fungus *Fusarium culmorum* *phyt*

authorship claim to a variety >>> seed breeder's rights

autoallopolyploid: cells or individuals whose genomes show characteristics of both auto- and alloploidy *cyto*

autoallopolyploidy >>> autoallopolyploid

autobivalent: a bivalent of meiosis I that is formed from two structurally and genetically completely identical sister chromosomes *cyto* >>> Figure 15

autochthon(e): one of the indigenous plants of a region *eco*

autochthonous: applied to material that originated in its present position *eco* >>> indigenous

autoclave: an apparatus in which media, glassware, etc., are sterilized by steam and/or pressure *meth*

autoecious: describing rust fungi completing the life cycle on one host plant (e.g., rusts) *phyt*

autofertility >>> autogamy >>> self-fertility

autofluorescence: fluorescence from objects in a microscope sample other than from fluorophores *micr*

autogamous >>> autogamy

autogamy: obligatory self-fertilization *bot* >>> Table 35

autogenomatic: genomes that are completely homologous and pair normally in meiosis *cyto*

autograft: a graft of tissue from a donor of one genotype to a host of the same genotype; the graft usually takes place from one part to another part of the same individual *hort*

autolysis: the destruction of a cell or some of its components through the action of its own hydrolic enzymes *phys*

automatic selection >>> unconscious selection

automatic weighing machine >>> automatic scales

automixis: obligatory self-fertilization *bot*

automutagen: any mutagen formed by the organism itself that may induce mutations *gene*

automutation: a mutation that arises without exogenic application of mutagens *gene* >>> spontaneous mutation

autonomous apomixis: agamic seed formation that does not depend on pollination *bot* >>> diploid parthenogenesis >>> parthenogenesis

autophene: a genetically controlled character that is manifested by the cell's own genotype and which shows special behavior in transplants and explants *gene*

autophyte: any organism that synthesizes its own food, as a photosynthetic plant, as opposed to heterophyte *bot*

autoploid: a cell or individual with genomes characteristic of the species itself *cyto*

autopolyploid: a polyploid organism that originates by the multiplication of a single genome of the same species *cyto*

autopolyploidization: the occurrence of doublings of chromosome number by failure of chromosomes to divide equationally in a mitosis following chromosome replication; plants seem to have commonly used autopolyploidization as an evolutionary tool *cyto* >>> C mitosis

autopolyploidy >>> autopolyploid

autoradiography: a method of determination of amounts and distributions of radioactive substances using photographic material, which is blackened when it is exposed to radiation; usually tritium, the radioactive isotope of hydrogen, or radioactive phosphorus are incorporated into molecules instead of hydrogen or common phosphorus, in this way certain compounds can be traced *meth biot*

autoreduplication: biological systems that generate the template for their own reproduction and duplicate themselves *bio*

autoregulation: a regulatory system of gene expression in which the product of a structural gene modulates its own expression *gene*

autosegregation: the occurrence of changes in the chromosome complement during the formation of the egg cell *cyto*

autosomal gene: a gene located on an autosome (i.e., a chromosome that is not a sex-determining chromosome) *gene*

autosome: any chromosome in the cell nucleus other than a sex chromosome *cyto*

autosyndesis: the pairing of complete or partial homologues of chromosomes *cyto*

autotetraploid: an autopolyploid with four similar genomes; if a given gene exists in two allelic forms *A* and *a,* then five genotypic classes can be formed: *AAAA* (quadruplex), *AAAa* (triplex), *AAaa* (duplex), *Aaaa* (simplex), and *aaaa* (nulliplex) *cyto* >>> Table 3

autotetraploidy >>> autotetraploid

autotroph >>> autotrophic

autotrophic: cells or organisms that synthesize cell components from simple chemical substances *phys*

autotropism: the ability of plants to self-regulate a stimulated crooking of an organ in a way that the previous shape is reestablished *bot*

autumn wood >>> late wood

auxin: a hormone that promotes longitudinal growth in the cell of higher plants; in combination with cytokinin, auxin is required for the sustained proliferation of many cultured plant tissues *chem phys*

auxotroph(ic): fail to grow on a medium containing the minimum nutrients essential for the growth of the wild type *phys*

availability: it describes the amount of a nutrient and/or water in fertilizer or the soil, respectively, that a plant can immediately absorb; it can be different from the actual amount of the nutrient present *agr phys* >>> available water capacity >>> available water

available water: that part of the water in the soil that can be taken up by plant roots *agr*

available water capacity: the weight percentage of water that a soil can store in a form available to plants; it is about equal to the moisture content at field capacity minus that at the wilting point *agr*

avenacin >>> saponin

average: a quantity or rating that represents or approximates an arithmetic mean *stat*

avidin: a glycoprotein component of egg white that binds strongly to the vitamin biotin; proteins and nucleic acids can be linked to biotin (biotinylated) and the avidin-biotin reaction can then be used in a number of assay methods, such as antigen-antibody reactions or DNA hybridization (e.g., enzymes conjugated with avidin can be used to bind to biotinylated antibodies) *chem cyto micr*

avirulence: the inability of a pathogen to infect *phyt*

avirulent: a strain of a parasite unable to infect and cause disease in a host plant *phyt*

AVRDC: Asian Vegetable Research and Development Center

awn: the bristlelike projection arising from the top of the glume and lemma (e.g., in barley, the top of the lemma in wheat and from the back of the lemma in oats) *bot* >>> Figure 34

awned >>> aristate

axenic: a pure culture of one species; it implies that cultures are free of microorganisms *biot* >>> aseptic

axil: the angle between the upper surface of a leaf and the stem that bears it *bot*

axil placentation: the type of ovule attachment within a fruit in which the seeds are attached along the central axis at the junction of the septa *bot*

axillary: in or related to the axis *bot*

axillary bud: develops in the axil of a leaf; the presence of axillary buds distinguishes a leaf from a leaflet *bot*

axillary tiller: a tiller may form a bud located at the coleoptilar node (coleoptilar tiller) and at each crown node (axillary tiller); the coleoptilar tiller can emerge at any time, independent of the number of leaves on the main stem; axillary tillers usually begin to emerge when the plant has three leaves; rarely are more than five axillary tillers formed on a cereal plant *bot*

axis (of cereal plants): the stem or central column upon which other parts are borne; in general, the central part of a longitudinal support on which organs or parts are arranged *bot*

5-azacytidine: a drug that may activate the expression of rRNA genes by reduction of their methylation level *biot*

azotobacter: bacteria living in soil and water that are able to bind and incorporate atmospheric nitrogen into their cells *bio agr*

B

B chromosome: any chromosome of a heterogeneous group of chromosomes present in several plant species, which differ in their morphology, numerical variation, meiotic pairing, and mitotic behavior from normal A chromosomes; they are also called supernumerary chromosomes, accessory chromosomes or extra chromosomes; a B chromosome derives from the A chromosome complement by aberrant division processes and subsequent modifications; up to 12 and more B chromosomes were observed in addition to the diploid A chromosome complement (e.g., in rye) *cyto*

B line: the fertile counterpart or maintainer line of an A line; does not have fertility restorer genes; used as the pollen parent to maintain the A line; used in hybrid seed production *seed* >>> Figure 2

BAC >>> bacterial artificial chromosomes

BAC vector: an *Escherichia coli* vector for DNA fragments; larger than cosmids; alternative to YAC vectors *biot*

bacciferous: berry load-bearing, producing berries *bot*

bacciform: berry-shaped *bot*

Bacillus thuringiensis: a bacterium that kills insects; a major component of the microbial pesticide industry and a subject in biotechnology >>> *Bt*

back mutation: a reverse mutation in which a mutant gene reverts to the original standard form and/or wild type; it is rare to forward mutations, but often strongly selected for; the AMES test relies on back mutation for the detection of mutagens *gene*

backcross: a cross of an F1 hybrid or heterozygote with an individual of genotype identical to that on one or the other of the two parental individuals; matings involving a hybrid genotype are used in genetic analyses to determine linkage and crossing-over values *meth* >>> Figures 2, 31 >>> Tables 27, 35

backcross breeding: a system of breeding whereby recurrent backcrosses are made to one of the parents of a hybrid, accompanied by selection for a specific character(s) *meth* >>> Figure 31 >>> Tables 27, 35

backcross (donor) **parent:** that parent of a hybrid with which it is again crossed or with which it is repeatedly crossed; backcrosses may involve individuals of genotype identical to the parent rather than the parent itself *meth* >>> Figures 2, 31 >>> Tables 27, 35

backcross method >>> backcross breeding

backcross-assisted selection (BCAS): a method that allows the selection of plants carrying a favorable recessive allele at each generation, limiting the need for a progeny test, which is common in traditional backcrossing; in cases where the traditional means of selection are limited by environmental conditions (e.g., the presence of an abiotic or a biotic stress such as drought) this selection strategy is superior to conventional ones; particularly in genetic transformation approaches, where the transgenes can be used as markers, BCAS may show a considerable advantage *meth biot* >>> Table 35

backcrossing >>> backcross

backfill: filling in a planting hole around roots with a soil mix for better establishing the plant *meth*

backhoe: a shovel mounted on the rear of a tractor, hydraulically operated to dig trenches or pits in soil *agr*

backward selection: selection of parent plants based on results from a progeny test *meth*

bacterial artificial chromosomes (BAC): pieces of plant DNA that have been cloned inside living bacteria; they can be used as probes to detect complementary DNA sequences within large pieces of DNA via hybridization techniques, or for marker-assisted selection by faster selection of segregant-bearing genes for a particular trait and to develop future crop varieties faster *biot*

bacterial diseases: diseases caused by specialized bacteria *phyt* >>> disease

bactericidal: killing or hampering bacteria *phyt*

bacteriocide: a chemical compound that kills bacteria *phyt*

bacteriocin(s): bactericidal substances produced by certain strains of bacteria and active against some other strains of the same or closely related species *phyt*

bacterioid: bacteria cells that are not normal shaped, usually found in root nodules of legumes *agr*

bacteriology: the branch of science for bacteria *phyt*

bacteriolysis: the lysis of bacterial cells, usually induced by antibodies formed by the host organism *phyt*

bacteriophage: a virus that infects bacteria; consists of a polyhedral head containing DNA or RNA enclosed in a protein coat (e.g., the bacteriophages T4, M13, P1 and PS8 are used in genetic engineering) *biot*

bacteriosis >>> bacterial diseases

bacteriostatic: a chemical or physical agent that prevents multiplication of bacteria without killing them *phyt*

bait: a material used to lure insects; it is often added to pesticides (e.g., against snails) *meth*

Bakanae disease: seedling disease of rice caused by fungus producing gibberellins *phyt* >>> gibberellin

balanced design: an experimental design in which all treatment combinations have the same number of observations *stat*

balanced diallelic: the genotype involving a multiple allelic locus in an autotetraploid where two different alleles are represented an equal number of times *gene*

balanced incomplete block design (BIB): a design in which one constant value for the residual variance of the difference between candidates for all pairs of candidates is indicated *stat meth*

balanced lattice: a special group of balanced incomplete block; allows incomplete blocks to be combined into one or more separate complete replicates *stat meth*

balanced lethal(s): recessive lethals at different loci, so that each homologous chromosome carries at least one lethal, and associated with inversions, so that no recombination occurs between the homologous chromosomes *gene*

balanced polymorphism: a genetic polymorphism that is stable, and is maintained in a population by natural selection, because the heterozygotes for particular alleles have a higher adaptive fitness than either homozygote; it is referred to as overdominance, as opposed to underdominance, where the heterozygotes have a lower fitness, giving rise to unstable equilibrium *gene*

balanced tertiary trisomic (BTT): a specific interchange trisomic spontaneously selected or experimentally designed in a way that it is heterozygous (*Aaa*), its trisomic progeny after selfing is genetically similar to the parent; the dominant allele is present on the translocated chromosome linked to the break point; BTTs were thought to be used for hybrid seed production in barley *cyto meth* >>> Figure 14

baler: a machine that picks up dry hay or straw after harvest and bundles it into big rectangular or round bales; the bales are tied together with baler twine *agr*

ball metaphase: a form of mitosis with characteristically clumped chromosomes *cyto*

balm >>> balsam

balsam: a mixture of resins and ethereal oils of sticky consistency, secreted by some plants *micr*

band: specific heterochromatic regions along a chromosome that can be stained by different banding methods *cyto*

band application: the spreading of fertilizer or other chemicals over, or next to, each row of plants in a field, as opposed to broadcast application *meth agr*

banding: a special staining technique for chromosomes, which results in a longitudinal differentiation (e.g., Giemsa staining, which is a complex of stains specific for the phosphate groups of DNA) *cyto;* in agriculture, placing fertilizer in continuous narrow bands and then covering it with soil *agr;* in horticulture, encircling part of a plant (e.g., a trunk) or a portion of a garden with some type of material that traps, kills, or keeps out pests (e.g., poisonous baits or copper stripping) *hort meth* >>> C banding >>> G banding

banding pattern: the linear pattern of deeply stained bands and weakly staining interbands that results from more or less defined local differences in the degree of DNA compactation along the chromosome *cyto*

band-pass: a microscopic filter that passes light of a certain restricted range of wavelength *micr*

band-seeding: placing forage crop seed in rows directly above but not in contact with a band of fertilizer *agr*

bar chart >>> bar diagram >>> histogram

bar code: a pattern of light and dark lines on labels that can be read by a light pen for direct entry into a computer; used for tagging and labeling of plants, seed accessions, etc. *meth*

bar diagram >>> histogram

***bar* gene:** a gene from *Streptomyces hygroscopices* that encodes the enzyme phosphinothricin acetyltransferase; it confers resistance to "Bialaphos" herbicide; used in genetic transformation studies as a marker gene for selection of successful transformants *biot*

barb: a stiff bristle or hair terminating an awn or prickle *bot*

barbate: having one or more groups of hairs; bearded *bot*

bark: the outer skin of a tree trunk, outside the secondary, vascular cambium; it is composed of phloem tissue, which occurs as living inner and dead outer zones; the outer zone is penetrated by the cork layers formed from the cork cambia *bot*

bark ringing: a method used for forcing fruit trees to flower; a complete ring is cut around the trunk below the lowest branch and another ring is cut right below the first; the bark between the rings is removed; the scar should be covered with grafting wax *meth hort*

barley yellow dwarf virus (BYDV): it infects all cereal species but barley and oats are usually more severely affected than wheat; plants are most vulnerable to infection early in growth; infection results in stunting, discoloration and substantial yield loss; the virus is transmitted by several species of cereal aphids (mainly bird-cherry aphid, *Rhopaosiphum padi,* or grain aphid, *Sitobion avenae*) *phyt* >>> aphid vector

barn: a building for storing straw, hay, grain, etc. *agr*

barren glume >>> spikelet glume

basal node: the node or joint at the base of the stem *bot*

basal placentation: a type of free-central placentation in which the seeds are attached at the bottom of the central ovary axis *bot*

basal rosette: in some plants, a cluster of leaves around the stem on or near the ground *bot*

base: a compound that reacts with an acid to give water (and a salt); a base that dissolves in water to produce hydroxide ions is called an alkali *chem*

base analogues: a purine or pyrimidine base that differs slightly in structure from normal base, but that because of its similarity to that base may act as a mutagen when incorporated into DNA (e.g., uracil, 5-bromouracil, 5-fluorouracil, 5-methylcytosin, 5-bromocytosin, hypoxanthin) *chem gene*

base collection: a collection of germplasm that is kept for long-term, secure conservation and is not to be used as a routine distribution source; seeds are usually stored at subzero temperatures and low moisture content *meth*

base pair (bp): the nitrogenous bases (adenine-thymine/uracil; guanine-cytosine) that pair in double-stranded DNA or RNA molecules; 1,000 bp = 1 kb *gene*

base pairing: a complementary binding by means of hydrogen bonds of a purine to a pyrimidine base in nucleic acids *gene*

base seed: particularly valuable seeds, usually derived from highly productive single plants (elite plants), which are used for seed production of commercially grown material; seed stock produced from breeder's seed by, or under the control of, an appropriate agricultural authority; the source of certified seed, either directly or as registered seed *seed* >>> Table 28

basic chromosome set: the standard chromosome number of a given species *gene cyto*

basic form >>> primitive form

basic number (of chromosomes): the haploid number of chromosomes in diploid ancestors of polyploids, represented by "x" *gene cyto* >>> basic chromosome set >>> genome

basic seed >>> base seed >>> elite >>> super-elite

basidiospore: from a basidium of *Basidiomycetes*-produced haploid spore that is formed after meiosis and exogenously laced up from a steringma *bot phyt*

basidium: a stand-like cell, mostly club-shaped, from which exogenously laced up haploid spores after karyogamy *bot*

bast: any of several strong, woody fibers, such as flax, hemp, ramie, or jute, obtained from phloem tissue *bot* >>> phloem

bast plant: crop plants used for fiber production, such as flax or hemp *agr*

bastard: the product of crossing two sperm cells of genetically different constitution *gene* >>> hybrid

batch culture: a cell suspension grown in liquid medium of a set volume; inocula of successive subcultures are of similar size and cultures contain about the same cell mass at the end of each passage; cultures commonly exhibit five distinct phases per passage (a lag phase follows inoculation, then an exponential growth phase, a linear growth phase, a deceleration phase, and finally a stationary phase) *biot*

batch drying: drying seeds in relatively small quantities held in a stationary position (as opposed to drying in a continuous moving line) *meth seed*

BC1, BC2, BC3, etc.: symbols indicating the first, second, third, etc. backcross generation *meth*

BCAS >>> backcross-assisted selection

beak: the extension of the keel at the tip of the glume or lemma in wheat *bot*

bean yellow dwarf virus (BYDV): a disease occurring in French bean, that can cause up to 90 percent losses in yield *phyt*

beat up: to replace dead trees with new ones, especially during the early years of the establishment or reestablishment of a plantation *fore*

Becquerel (Bq): the SI unit of radioactivity; the unit is named after the discoverer of radioactivity, A. H. BEQUEREL *phy*

bed: an area within a garden or lawn in which plants are grown *meth*

bedding plant: a plant grown for its flowers or foliage that is suited by habit for growing in beds or masses *hort*

beet: any of various biennial plants of the genus *Beta,* of the goosefoot family, especially *B. vulgaris,* having a fleshy red or white root and dark-green red-veined leaves; sugarbeet derived from *B. vulgaris* by selection for high sugar content *bot hort*

behavior flexibility: all means of plant behavior permitting temporary adaptation to environmental conditions *eco*

behavior genetics: a branch of genetics dealing with the inheritance of different types and/or forms of behavior *gene*

belowground biomass >>> biomass

berry: a simple, fleshy or pulpy and usually many-seeded fruit that has two or more compartments and does not burst open to release its seeds when ripe (e.g., banana, tomato, potato, grape) *bot*

berry-bearing >>> bacciferous

berry-shaped >>> bacciform

best linear prediction (BLP): a statistical method that utilizes matrix algebra to predict the breeding values for any trait or selection index; in BLP fixed effects are assumed to be known; BLP is especially suited for analyses of messy or unbalanced data *meth stat*

best linear unbiased prediction (BLUP): a statistical method that predicts breeding values for any trait or selection index *meth stat*

beta-DNA: the normal form of DNA found in biological systems, which exists as a right-handed helix *gene*

bevel (of lemma): a depression variable in depth in the base of the lemma, rounded in barley, transverse in oats *bot*

bias: a consistent departure of the expected value of a statistic from its parameter *stat*

biased >>> bias

bi-cropping: a method of growing cereals in a leguminous living mulch; it could potentially reduce the need for synthetic inputs to cereal production while preventing losses of nutrients and increasing soil biological activity; also a method of low input production system for cereals *agr meth*

biennial: a plant that lives for two years; during the first season food may be stored for the use during the flower and seed production in the second year *bot* >>> annual >>> perennial

biennial crop >>> biennial

bifloral: showing two flowers *bot*

bifoliate: showing two leaves *bot*

bifurcate: forked *bot*

bigerm: having two seeds *bot*

bimitosis: the simultaneous occurrence of two mitoses in binucleate cells *cyto*

bimodal distribution: a statistical distribution having two modes *stat*

binary scale: a scale for scoring data where there are only two possible responses *meth*

binemic: chromosomes that contain two DNA helices per metaphase chromatid *cyto*

binomial nomenclature: the system of naming organisms using a two-part Latinized (or scientific) name that was devised by the Swedish botanist Carolus LINNAEUS (1707-1778); the first part is the generic name (genus), the second is the specific epithet or name (species); the Latin name is usually printed in italics, starting with a capital *tax bot*

binucleate: cells with two nuclei *cyto*

bioassay: the use of living cells or organisms to make quantitative and qualitative measurements *meth*

biocatalysis >>> biotechnology

biocatalyst: a biological substance used to cause a particular chemical or biochemical reaction *phys chem*

biochemical genetics: a branch of genetics dealing with the chemical nature of hereditary determinants *gene*

biochemistry: the chemistry of life; the branch of chemistry that is concerned with biological processes *chem*

biocide: a natural or synthetic substance toxic to living organisms *phyt*

biocoenosis: a community of organisms and its interaction with abiotic factors of habitat *eco*

biocontrol >>> biological control

biodiversity: the existence of a wide variety of species (species diversity), other taxa of plants or other organisms in a natural environment or habitat, or communities within a particular environment (ecological diversity), or of genetic variation within a species (genetic diversity); genetic diversity provides resources for genetic resistance to pests and diseases; in agriculture, biodiversity is a production system characterized by the presence of multiple plant and/or animal species, as contrasted with the genetic specialization of monoculture *evol eco agr*

bioethics: a field of study and counsel concerned with the implications of certain genetic and medical procedures, such as organ transplants, genetic engineering, and care of the terminally ill *bio*

biogenesis: the production of living organisms from other living organisms *bio*

biolistic gene gun: "biolistic" derived from a contraction of the words "biological" and "ballistic"; it refers to a projectile fired from a gun; it is used to shoot pellets that are loaded with genes into plant seeds or tissues, in order to get them integrated and/or expressed in the foreign background; the gun uses an actual explosive to propel the material; compressed air or steam may also be used as the propellant *biot*

biological assay >>> bioassay

biological containment: precaution taken to prevent the spread of recombinant DNA molecules in the natural environment; disabled host organisms (e.g., with stable auxotrophic requirements or defective cell walls) together with nontransmissible cloning vectors are used; biological containment is especially important when toxin genes from pathogens are expressed in *Escherichia coli* or other vectors *biot*

biological control: the practice of using beneficial natural organisms to attack and control harmful plants, animal pests, and weeds is called biological control, or biocontrol; this can include introducing predators, parasites, and disease organisms, or releasing sterilized individuals; biocontrol methods may be an alternative or complement to chemical and gene-engineered pest control methods *phyt* >>> *Bt* >>> *Bt* gene

biological pesticide: a chemical which is derived from plants, fungi, bacteria, or other natural synthesis and which can be used for pest control *phyt*

biological species concept: a system in which organisms are classified in the same species if they are potentially capable of interbreeding and producing fertile offspring *evol*

biological yield: the total yield of plant material (i.e., the total biomass including the economic yield, e.g., the grain yield); the larger the biological yield, the greater the photosynthetic efficiency *phys*

biom: interactive groups of individuals of one or more species *eco*

biomass: the total weight of organic material in a given area or volume; it can be divided into aboveground and belowground biomass *phys*

biometrical genetics >>> quantitative genetics

biometry: mathematical statistics applied to biological investigations *stat*

biopesticide >>> biological pesticide >>> *Bt*

biopiracy: the collecting and patenting of plants and other biological material formerly held in common and their exploitation for profit *biot*

bioreactor: a culture vessel used for experimental or large-scale bioprocessing *biot*

bioseeds: seeds produced via genetic engineering of existing plants *biot seed*

biosome: any autonomous cell constituent multiplying by autoreduplication *gene*

biostatistics: the application of statistics to biological data *stat* >>> biometry

biosynthesis: the synthesis of the chemical components of the cell from simple precursors *phys*

biotechnology: any technique (e.g., recombinant DNA methods, protein engineering, cell fusion, nucleotide synthesis, biocatalysis, fermentation, cell cultures, cell manipulations etc.) that uses living organisms or parts of them to make or modify products, to improve organisms or to make them available for specific uses *biot*

biotin(e): it functions as coenzyme; is a part of the vitamin B complex; it is also called vitamin H; it is present in all living cells, bound to polypeptides or proteins; it is important in fat, protein, and carbohydrate metabolism; it is a common addition to plant tissue culture media *biot phys*

biotinylated probe: a DNA sequence in which biotinylated dUTP is incorporated and labeled with biotin; it is used in DNA-DNA hybridization experiments, such as Southern transfer or in situ hybridization, with chromosomes; the detection of hybrid molecules is realized by a complex of streptavidin, biotin, and horseradish-peroxidase; if there is a hybridization then the complex shows a green fluorescence color *micr cyto biot*

biotope: a portion of a habitat characterized by uniformity in climate and distribution of biotic and abiotic components *eco*

biotrophic pathogen: a parasitic organism that obtains its nutrient supply only from living host tissue regardless of whether or not it can be artificially cultured *phyt*

biotype: a group of genetic identical individuals; sometimes, a physiologic race *gene*

bird netting: different types of mesh used as a drape to keep birds out of fruit trees, berry patches, vegetable gardens, or field experiments *meth hort agr*

bird pollination >>> ornithophily

birdscare: scarecrow

birimose: opening by two slits (e.g., anthers of plants) *bot*

bisexual: species comprises individuals of both sexes or a hermaphrotide organism in which an individual plant possesses both stamens and pistils in the flower *bot* >>> Table 18

bisulfite genomic sequencing: a procedure in which bisulfite is used to deaminate cytosine to uracil in genomic DNA; conditions are chosen so that 5-methylcytosine is not changed; PCR amplification and subsequent DNA sequencing reveals that cytosines are methylated in genomic DNA *biot*

bivalent: two homologous chromosomes when they are paired during prophase-metaphase of the first meiotic division *cyto* >>> Figure 15

bivalent formation: the association of two homologous chromosomes as a ring or rod configuration depending on chiasma formation *cyto* >>> Figure 15

bivalent interlocking >>> interlocking

***bla* gene:** beta-lactamase gene conferring resistance to ampicillin; commonly used as selective marker for plasmid vectors *biot*

black leg (of beets): a number of diseases (e.g., caused by *Pythium debaryanum*) in which symptoms include blackening of the base of the stem, often followed by the collapse of the stem *phyt*

black leg (of potato): a bacterial disease (*Erwinia carotovora* ssp. *atroseptica*) causing severe yield loss, particularly in wet conditions *phyt*

blade: the expanded portion of a leaf, petal, or sepal *bot* >>> Table 30

blade joint: the flexible union between the leaf blade and the leaf sheath *bot*

blanch: a method to whiten or prevent from becoming green by excluding light; blanching is applied to the stems or leaves of plants (e.g., celery, lettuce, and endive); it is done either by banking up the soil around the stems, tying the leaves together to keep the inner ones from light, or covering with pots, boxes, etc. *meth hort*

blasting: a plant symptom characterized by shedding of unopened buds; leads to a failure of producing fruits or seeds *phyt agr*

bleeding: exudation of the contents of the xylem stream at a cut surface due to root pressure *bot*

blend: a term applied to mechanical seed mixtures of different crop varieties or species that have been mixed together to fulfill a specific agronomic purpose *seed*

blended variety >>> multiline variety >>> blend

blending >>> blend

blending inheritance: inheritance in which the characters of the parents appear to blend into an intermediate level in the offspring with no apparent segregation in later generations *gene*

blight: a disease characterized by rapid and extensive death of plant foliage, and applied to a wide range of unrelated plant diseases caused by fungi, when leaf damage is sudden and serious (e.g., fire blight of fruit trees, halo blight of beans, potato blight, etc.) *phyt*

blind: without flowers; sterile *bot*

blind floret >>> blind

blindfold (trial): a trial to study soil heterogeneity (i.e., variation in the soil fertility); all plots contain the same genetically uniform plant material; the study may show that the growing conditions provided by a particular field may appear homogeneous when observed in some season and for some trait of a crop, but they may appear heterogeneous when observed in a different season or for some trait of a different crop; for a given crop, different traits may differ with regard to their capacity to show soil heterogeneity *stat meth*

block: a number of plots that offer the chance of equal growing conditions; comparisons among the entries, which are tested in the same block, offer unbiased estimates of genetic differences *stat meth* >>> Tables 25, 26 >>> biased

blocking: the procedure by which experimental units are grouped into homogeneous clusters in an attempt to improve the comparison of treatments by randomly allocating the treatments within each cluster or block *stat* >>> block >>> thinning >>> Table 25

bloom: the white powdery deposit often present on the surface of the stem, leaves, and ears of cereals or sorghum; often of a waxy nature; in general, the flower of a plant or the state of blossoming *bot* >>> waxiness

blooming: in grasses, the period during which florets are open and anthers are extended *bot agr*

blossom: the flower of a plant, especially of one producing an edible fruit; the state of flowering *bot*

blot: the transfer of DNA, RNA, or proteins to an immobilizing binding matrix, such as nitrocellulose, or the autoradiograph produced during certain blotting procedures (Southern blot, Northern blot, Western blot, etc.) *meth gene biot*

blotch: a disease characterized by large and irregularly shaped spots or blots on leaves, shoots, and stem *phyt agr*

blunt end ligation: ligation of DNA with blunt ends requires higher concentration of DNA ligase than sticky end ligation; it is inhibited by ATP concentrations > 1 mM *biot*

blunt ends: DNA fragments that are double-stranded paired over the whole length, usually produced by certain types of restriction enzymes *gene*

BOERNER divider >>> conical divider

boleless: without a trunk *bot*

boll: the fruit of cotton *bot*

boll size: weight in grams of seed cotton from one boll *agr meth*

bolt: formation of an elongated stem or seedstalk; in the case of biennial plants, this generally occurs during the second season of growth *bot* >>> bolter

bolter: they develop in long cold springs with morning frosts and low temperatures, not exceeding +5°C, causing vernalization of the plants (e.g., in sugarbeet) *phys*

bolting: production of seed stalks the first season in a biennial crop (e.g., in beets) *phys* >>> bolter

bolting resistance >>> bolter >>> bolting

bonsai: a tree or shrub grown in a container or special pot and dwarfed by pruning, pinching, and wiring to produce a desired shape *meth hort*

boot: the lower part of a cereal plant *bot*

boot(ing) stage: it refers to the growth stage of grasses at the time the head is enclosed by the sheath of the uppermost leaf *agr*

border effect: the environmental effect on plots that are on the edge of an experimental area *stat meth*

border strip: a demarcation surrounding a plot, usually given the same treatment as the plot; it is arranged in order to minimize border effects *meth agr hort*

boron (B): a nonmetallic element occurring naturally only in combination, as in borax or boric acid; boron can cause toxicity in several crop plants; as a micronutrient, deficiency of boron can be as severe *chem agr*

botanical pesticides: pesticides whose active ingredients are plant-produced chemicals such as nicotine, rotenone, or strychnine *phyt* >>> biological control

botany: the science of plants; the branch of biology that deals with plant life; the plant life of a region; the biological characteristics of a plant or group of plants *bot*

bottleneck: a period when a population becomes reduced to only a few individuals *gene eco*

botuliform: cylindrical with rounded ends, sausagelike in form *bot*

bough: the main arm or branch of a (fruit) tree *bot hort*

boundary mark >>> landmark

bouquet stage: a meiotic prophase stage of some organisms during which the chromosome oriented by one or both ends toward one point in the nuclear envelope *cyto*

Boyage system >>> chopping

bp >>> base pair

brachyomeiosis: an abnormal meiosis characterized by omission of the second meiotic division *cyto*

bract: a modified leaflike structure occurring in the inflorescence *bot*

bracteole: a little bract borne on the flowerstalk above the bract and below the calyx *bot*

bran: compromises aleuron and pericarp cell layers; the bran and germ are separated during milling *agr meth*

branch: an axillary (lateral) shoot or root *bot*

branch >>> ramify

branching >>> ramification

branching agent: a substance inducing and/or increasing branching *hort*

brand: a legal trademark registered by a particular company or distributor for its exclusive use in marketing; a product such as seeds or plants *seed agr;* in plant pathology, a leaf disease caused by a microscopic fungus (e.g., a rust or smut); sometimes names the fungi *phyt*

brassinosteroid: brassinosteroids are endogenous, plant-growth-promoting natural products with structural similarities to animal steroid hormones; they affect cell elongation and proliferation, distinct from that of auxins, cytokinins, and gibberellic acids, although they interact with them *phys* >>> biological control

breakage-reunion hypothesis: the classical and generally accepted model of crossing-over by physical breakage and crossways reunion of broken chromatids during meiosis *gene*

breakpoint: when chromosome mutations occur, the site at which the single or double strand of DNA breaks along a chromosome and/or chromatid *cyto gene*

breathing root >>> pneumatophore

breed: an artificial mating group derived from a common ancestor or for genetic analysis; in breeding, a line having the character type and qualities of its origin; in general, a group of plants, developed by humans that will not keep their characteristics in the wild

breeder: a person who raises plants primarily for breeding purposes

breeder('s) seed: seed or vegetative propagating material increased by the originating or sponsoring plant breeder or institution; it represents the true pedigree of the variety; it is used for the production of genetically pure, foundation, registered, and certified seeds *seed* >>> Table 28

breeder's collection >>> working collection >>> stock

breeder's preference: a general impression by the breeder concerning a material that is under selection *meth* >>> Table 33

breeder's rights: varietal protection; the legal rights of a breeder, owner, or developer in controlling seed production and marketing of crop varieties >>> PPA and PVPA

breeding: the propagation and genetic modification of organisms for the purpose of selecting improved offspring; several techniques of hybridization and selection are applied *meth* >>> Table 35

breeding line: a group of plants with similar traits that have been selected for their special combination of traits from hybrid or other populations; it may be released or used for further breeding approaches *meth*

breeding method >>> breeding system

breeding orchard: a planting of selected trees, usually clonally or grafted propagated; it is designed to ease breeding work *meth*

breeding population: a group of individuals selected from a wild, experimental, or crossing population for use in a breeding program; usually phenotypically selected for desirable traits *meth* >>> Table 35

breeding size: the number of individuals in a population involved in reproduction during particular generations and breeding procedures *meth*

breeding system: the system by which a species reproduces; more specifically, the organization of mating that determines the degree of similarity and/or difference between gametes effective in fertilization *meth* >>> Figure 31 >>> Table 35

breeding true: producing offspring with phenotypes for particular characters that are identical to those of the parents; homozygous individuals necessarily breed true, whereas heterozygotes rarely do so *meth gene*

breeding value: the value of an individual as defined by the mean value of its progeny, either on the basis of individual traits or a selection index *meth*

breeding zone: an area within which a single population of improved trees can be planted without fear of misadaptation *fore*

brevicollate: short necked *bot*

brick grit test: a type of seedling emergence (vigor) test utilizing uniformly crushed brick gravel through which seedlings must emerge to be considered vigorous; it was originally developed by HILTNER and IHSSEN (1911) for detecting seed-borne *Fusarium* infection in cereals; with modifications, the seeds are planted on damp brick grit or in a container of sand covered with 3 cm of damp brick grit, then germinated in darkness at room temperature of a specific time *seed*

bridge parent: a parent that is sexually compatible with two reproductively isolated species and can be used to transfer genes between them *meth*

bridge-breakage-fusion-bridge cycle: a process that can arise from the formation of dicentric chromosomes; daughter cells are formed that differ in their content of genetic material due to duplications and/or deletions in the chromosomes *cyto gene*

bridging cross: a method of bypassing an incompatibility barrier between two species or genotypes by using a third species or genotype that is partly compatible with each of them in an intermediate cross *meth*

bridging species: a species used in a bridging cross in order to bring together the two incompatible species *eco*

bristle: a stiff, sharp hair *bot*

broad wing: the larger of the two parts of the glume of, for example, wheat, which are separated by the keel *bot*

broad-base terrace: a low embankment that is constructed across a slope to reduce runoff and/or erosion (e.g., in rice or grape cultivation) *agr hort*

broadcast: scattered upon the ground with the hand (e.g., in sowing seed, instead of sowing in drills or rows) *meth agr hort*

broadleaf: sometimes used to designate a broad group of nongrasslike (weedy) plants *agr bot*

broom: a symptom in which lateral branches proliferate in a dense cluster on the main branch (e.g., witch's broom) *phyt agr hort*

browning: discoloration due to phenolic oxidation of freshly cut surfaces of explant tissue; in later culture this phenomenon may indicate a nutritional or pathogenic problem, generally leading to necrosis *biot*

bruising (in potato): a gray or blue-black localized discoloration that develops in the tuber flesh as a result of physical impact *agr*

brush: the tuft of hair at the top of, for example, wheat grain *bot* >>> coma

brushing: spreading spores of fungi by brush on leaves or flowers for infection experiments *phyt;* in crossing experiments, spreading pollen on stigmata of (emasculated) flowers in order to initiate fertilization *meth*

BSA >>> bulked segregant analysis

Bt: an abbreviation for *Bacillus thuringiensis;* it is a naturally occurring soil bacterium used as a biological pesticide (biopesticide); engineered plants have a gene from *Bt* inserted into their own genetic material; this new gene produces a natural protein that kills insects after the protein is ingested; the toxins are specific to a small subset of insects; for example, cotton has been genetically altered to control the tobacco budworm, bollworm, and pink bollworm; potatoes have been altered to control the Colorado potato beetle; hybrids of "*Bt*-maize" altered to be resistant to the European corn borer; the *Bt* toxin degrades rapidly to nontoxic compounds *phyt* >>> biological control

***Bt* gene:** a gene from the bacteria *Bacillus thuringiensis* that gives resistance to lepidopterous insects; by biotechnological means it was successfully transferred to cotton, tobacco, etc. *biot*

BTT >>> balanced tertiary trisomic

bud: an immature shoot, protected by tough scale leaves, from which the stem and leaves or flowers may develop *bot*

bud dormancy >>> dormancy

bud eye: a dormant bud in the axil of a leaf; used to propagate through bud-grafting *meth hort*

bud mutation >>> bud sport

bud pollination: a procedure utilized in maintaining self-incompatible parent lines by self-pollination; hybrid seed production in plant species having a sporophytic self-incompatibility system is dependent upon the production of inbred lines homozygous for a self-incompatibility allele *S;* in these species, a protein secretion covers the stigmatic surface just prior to anthesis and acts as a barrier to penetration of the stigma by germinating pollen grains; when buds are opened and the pollen applied before the protein barrier is formed, seed set can be obtained (e.g., in rapeseed) *meth*

bud pruning: removal of lateral buds from a stem to prevent them from developing into branches *hort fore*

bud scale: a modified leaf, without lamina, protecting a bud *bot*

bud sport: a somatic mutation occurring in a bud of a plant; it results from local genetic alteration and produces a permanent modification; it is usually retained by grafting; this sort of mutation is often used in fruit tree breeding *gene meth hort*

bud union: the site of junction on a stem, usually swollen, where a graft bud has joined the stock following the process of budding; frequently found at or near soil level *bot hort*

bud-bearing >>> gemmiferous

budded: grown from a bud grafted onto a desirable understock *hort meth*

budded >>> budding

budding: formation of buds as a result of cell division in a localized area of a shoot, usually promoted by cytokinins and inhibited by auxins *bot*; in horticulture, the grafting of a bud on to a plant *hort*

budding strip: a strip of rubber or other material used to hold grafts *meth hort*

budding union >>> bud union

buffer: a solution mixed by a weak acid and a weak alkali; it prevents changes of the pH value, therefore it is a suitable medium for enzyme reactions *chem prep*

bulb: an underground storage organ, comprising a short, flattened stem with roots on its lower surface, and above its fleshy leaves or leaf base, surrounded by protective scale leaves *bot*

bulb planter: a sharp-edged, tapered cylinder used to remove a plug of soil or sod in which a bulb is placed; the plug is then returned to the hole in order to cover the bulb *meth hort*

bulbil: a small, bulblike structure, usually formed in a leaf axil, that separates from the parent and functions in vegetative reproduction *bot* >>> Figure 28

***bulbosum* technique** >>> *Hordeum bulbosum* procedure

bulbous: tuber-shaped; forming tubers *bot*

bulbous plant >>> bulbous

bulk breeding: the growing of genetically diverse populations of self-pollinated crops in a bulk plot with or without mass selection, generally followed by a single-plant selection; it is a procedure for inbreeding a segregating population until the desired level of homozygosity is achieved; the seeds to grow each generation is a sample of that harvested from plants of the previous generation; it is usually used for the development of self-pollinated crops; it is an easy way to maintain populations during inbreeding; natural selection is permitted to occur, which can increase the frequency of desired genotypes compared with an unselected population; it can be used in association with mass selection with self-pollination; disadvantages are: (a) plants of one generation are not all represented by progeny in the next generation, (b) genotypic frequencies and genetic variability cannot be clearly defined, and (c) natural selection may favor undesirable genotypes *meth* >>> Figure 16 >>> Table 24

bulk population selection: selection procedure in self-pollinating crops; segregating populations are propagated as bulks until segregation is virtually ceased, at which time selection is initiated *meth* >>> Figure 16 >>> Table 24 >>> bulk breeding

bulked segregant analysis (BSA): a rapid mapping strategy suitable for monogenic qualitative traits; when DNA of a certain number of plants is bulked into one pool, all alleles must be present; two bulked pools of segregants, differing for one trait, will differ only at the locus harboring that trait *biot*

bullet planting: setting out young trees grown in bullet-shaped rigid plastic tubes, which are injected into the ground by a spring-loaded gun, sometimes into prepared holes *fore hort*

bumblebee: any of several large, hairy social bees of the family *Apidae,* sometimes utilized for pollination of special crop plants; frequently bumblebees are used for successful seed multiplication of legume and *Brassica* crops in greenhouses, especially of genebanks *zoo seed*

bumper mill: a machine designed to clean timothy seed by a continuous bumping action on an inclined plane; the uncleaned seed is metered onto the plane, which is continuously bumped by sets of knockers; the cylindrical timothy seeds are rolled into separate grooves while non-cylindrical contaminants are jarred off the end of the inclined plane and separated *seed*

bundle sheath: a layer of cells enclosing a vascular bundle in a leaf *bot*

BUNSEN burner: a hot-flame burner using a mixture of gas and air ignited at the top of a metal tube; this device is used for sterilizing tools and container openings during aseptic transfer in vitro experiments; after R. W. BUNSEN (1811-1899) *prep*

bunt: stinking smut; a seed-borne disease of grasses caused by *Tilletia* spp.; the grain is replaced by masses of fungal spores that have a characteristic fishy smell *phyt*

bur: the rough, prickly covering of the seeds of certain plants (e.g., chestnut) *bot*

burlap: a loosely woven fabric made of jute or hemp; used to protect newly seeded lawns from wind, water, and birds *meth agr hort*

bursiculate: baglike *bot*

bushel: a dry-volume measure of varying weight for grain, fruit, etc., equal to 4 pecks or 8 gallons (2150.42 cubic inches or 36.368 liters); a bushel of wheat and soybeans each weighs ~60 pounds: a bushel of maize, rye, grain sorghum, and linseed each weighs ~56 pounds; a bushel of barley, buckwheat, and apples each weighs ~48 pounds *agr*

bushy grasses: grasses forming tufts *bot*

butyrous: butterlike *bot*

C

C >>> cytosine

C banding: a cytological staining technique for chromosomes that labels regions around the centromere with Giemsa stain; usually a bandlike and darkly stained structure appears, which consists of heterochromatin; the technique is intensively used in chromosome identification and genome characterization, including structural changes and polymorphisms *cyto meth* >>> banding >>> Giemsa staining

^{14}C dating >>> radiocarbon dating

C mitosis: mitosis in single-celled organisms is responsible for the production of new individuals (asexual reproduction); mitosis in multicellular organisms, such as plants, is responsible for the growth of the organism and the repair of damaged tissues; first, DNA must be replicated so that there is a duplicate set of genetic information to be given to each daughter cell; second, the DNA must be divided so that each daughter cell gets the same set of information; mitosis is a three-step process: (1) replication of genetic material in the mother cell, (2) separation of the replicated genetic material, and (3) formation of the two daughter cells; when polyploidy is induced with colchicine, an alkaloid of the meadow saffron *(Colchicum autumnale)* that inhibits mitosis, the development of the nuclear spindle is hampered; the mitosis that takes place after treatment with colchicines is called a C mitosis; it also enables an easier detection and identification of chromosomes than a normal mitosis does; during the prolonged metaphase of a C mitosis, the chromosomes form an X-shaped structure since the chromatids are still connected at the centromere though they may repel each other; after some time, the chromatids finally part, but they do not segregate; they become enclosed by a new nuclear membrane and proceed to their interphase state; the number of chromosomes has now doubled—a diploid nucleus has developed into a tetraploid one *cyto*

C value: the DNA quantity per genome (i.e., per chromosome set); the content of diploids is referred to as the 2C; haploid cells contain the 1C amount of DNA *cyto*

C3 pathway: most common pathway of carbon fixation in plants; this photosynthesis produces at first a 3-carbon (C3) compound (phosphoglyceric acid); in C3 plants, about 25 percent of the net carbon uptake is reevolved immediately in photorespiration *phys*

C4 pathway: a carbon fixation found in some plants that have high rates of growth and photosynthesis and that are adapted to high temperatures, strong light, low carbon dioxide levels, and low water supply; this photosynthesis produces at first a 4-carbon (C4) compound (phosphoenolpyruvate, PEP); in C4 plants, photorespiration is suppressed to a very large extent due to the presence of a very efficient C2-concentrating mechanism *phys*

CA >>> combining ability

CAAS: Chinese Academy of Agricultural Science

CAAT box >>> CAT box

cadastral gene: a plant gene that controls the expression of floral homeotic genes *biot*

cadastre: an official register of the ownership, extent, and value of real property in a given area, used as a basis of land taxation *agr*

caffeine: a white, crystalline, bitter alkaloid, $C_8H_{10}N_4O_2$, usually derived from coffee or tea *chem phys*

Cajal body (CB): a nuclear structure that is found in both plants and animals; it contains components of at least three RNA-processing pathways *biot*

calceolate: shoelike in form *bot*

calcicole: a plant that grows best in calcium-rich soils *bot agr*

calcicolous plant >>> calcicole

calciphobe: reduced growth on calcium-rich soils *bot*

calciphyte >>> calcicole

calcium (Ca): a silver-white bivalent metal, combined in limestone or chalk *chem*

calicle >>> callycle

callogenesis >>> callus

callose: hard or thick and sometimes rough organic matter *bot*

callus (calli *pl*): tissue that forms over a wound or that develops from actively dividing plant tissue in a tissue culture; usually a disorganized mass of undifferentiated cells *biot;* in botany, the thickened part of the base of (e.g., oat grain) *bot*

callus culture: the in vitro culture of callus, often as first stage in the regeneration of whole plants in culture *biot* >>> tissue culture

callus induction: undifferentiated plant tissue is produced at wound edges; callus can also be induced and grown in vitro by varying the ratio of hormones (e.g., auxin and cytokinin) in the growth medium *biot*

callycle: a protective structure around a flower formed collectively by the sepals *bot*

calycular: cuplike *bot*

calypter >>> calyptra

calyptra: a cap or hood covering a flower or fruit *bot* >>> root cap

calyx (calyces *pl*): the outer part of a flower; all the sepals of a flower *bot*

cambium: in the stem and roots of vascular plants, a layer of cells lying between the xylem and phloem *bot*

camerate: chamberlike *bot*

campanulate: bell-like in form *bot*

Canada balsam: resin distilled from the bark of *Abies balsamea* (balsam fire) and other similar species; used in cytology for mounting (e.g., chromosome spreads) *cyto micr*

candidate tree: a tree that has been tentatively selected for inclusion in a breeding program, but has not yet been measured or compared with surrounding trees *meth fore*

candidate variety: breeding strains, lines, or hybrids of high grade that by a breeder or institution are announced for official national (and international) performance testing in order to get released as a certified variety *seed*

cane: a long and slender, jointed, rigid, woody stem that is hollow or pithy (e.g., in grasses, palms, rattan, bamboo or sugarcane) *bot;* in viticulture, a mature, woody, brown shoot as it develops after leaf fall; canes were last year's fruiting or renewal shoots; the buds on the canes will produce this season's fruiting shoots *hort*

cane sugar >>> saccharose

canescent: densely covered with grayish or whitish, short, soft hairs *bot*

canker: a plant disease in which there is sharply limited necrosis of the cortical tissue (e.g., in apple); in rape, it causes leaf spotting over winter and cankers on the stem later; the latter are the more serious and appear after flowering caused by *Leptosphaeria maculans,* asexual stage *Phoma lingam phyt;* sharply defined dead area of tissue on stem *phyt*

cannabinol (THC) >>> hemp

Canola: a type of rapeseed that has been developed and grown in Canada; Canola is a registered trademark, corresponding to specified low contents in erucic acid in oil and in glucosinolates in meals equivalent to double 0 in the EU standard; it was initially obtained by conventional breeding, but in recent years herbicide-tolerant varieties have been developed *agr*

canopy: the vertical projection downward of the aerial portion of plants, usually expressed as percent of ground so occupied *bot meth*

canopy temperature depression (CTD): the cooling effect exhibited by a leaf as transpiration occurs; it gives an indirect estimate of stomatal conductance, and is a highly integrative trait being affected by several major physiological processes including photosynthetic metabolism, evapo-transpiration, and plant nutrition; it has potential for complementing early generation phenotypic selection in plants *phys*

cap: structure at the 5' end of eukaryotic mRNA introduced after transcription by linking the terminal phosphate of 5' GTP to the terminal base of the mRNA; the added G is methylated, giving a structure of the form 7MeG5' ppp5'Np *biot* >>> cap site

cap site: the probable transcription initiation site of a eukaryotic gene; the primary transcripts of most eukaryotic mRNAs have an A (adenine) in the first position and the cap is added 5' to it *gene biot* >>> cap

capillarity: the process by which moisture moves in any direction through the fine pores and as films around particles *agr*

capillarity moisture: that amount of water that is capable of movement after the soil has drained; it is held by adhesion and surface tension as films around particles and in the finer pore spaces *agr*

capitulum: flower head; an aggregation of flowers on a flat platform and edged by bracts *bot*

cappiliform: hairlike *bot*

capsid: protein coat that encloses DNA or RNA molecules of bacteriophage or virus *biot*

capsular fruit >>> capsule

capsule: a dehiscent fruit with a dry pericarp usually containing many seeds *bot*

carbohydrate: the dominating substances of the cell sap are soluble carbohydrates; they occur as disaccharides, such as saccharose and maltose or as monosaccharides, such as glucose and fructose *chem phys* >>> Table 16

carbon cycle: the biological cycle by which atmospheric carbon dioxide is converted to carbohydrates by plants and other photosynthesizers, consumed and metabolized by organisms, and returned to the atmosphere through respiration or decomposition *phys*

carbon dioxide: a gaseous compound that is formed when carbon combines with oxygen *chem phys*

carbonate: a salt or ester of carbonic acid *chem*

carboxyl group: the univalent group COOH, characteristic of organic acids *chem phys*

carboxylase: any of the class of enzymes that catalyze the release of carbon dioxide from the carboxyl group of certain organic acids *phys*

carboxylation: introduction of a carboxyl group into an organic compound *phys*

carding: the process of untangling and partially straightening fibers by passing them between two closely spaced surfaces *meth agr*

carinate: keeled, boatlike *bot*

carmin: used for preparation of carminic acid; this is a red die used for coloring or staining chromosomes and other cytological material; carmin is prepared from cohineal *meth cyto* >>> cochineal

carmine staining >>> aceto-carmine staining

carnivorous plant >>> insectivorous plant

carnous: fleshy *bot*

CARNOY's fixative: a fixator solution, which consists of 6 parts ethanol: 3 parts chloroform: 1 part acetic acid; used in chromosome analysis *micr cyto*

carotene: a yellow compound of carbon and hydrogen that occurs in plants; it is a precursor of vitamin A *chem phys*

carotin >>> carotene

carpel: one of the female reproductive organs of the flower, comprising an ovary and usually with a terminal style tipped by the stigma *bot* >>> Figure 35

carrier: a heterozygote that carries an allele (recessive) that is not expressed *gene*

carrying capacity: the density of a density-regulated population at equilibrium *gene*

cartenoid: a yellow, orange, red or brown pigment, which is located in the chloroplast and chromoplast of plants; it acts as photosynthetic accessory pigment *bot*

caruncle: a reduced aril in the form of a fleshy, often waxy or oily outgrowth near the hilum of some seeds *bot*

caryopsis: the single-seeded, dry, and nutlike fruit of the grasses in which the mature ovary wall (pericarp) and the seed coat (testa) are fused *bot*

casein: a protein precipitated from milk; sometimes used for in vitro techniques *chem biot*

cash crops: agricultural production such as grains, hay, root crops, and fiber; in contrast to vegetables and fruits *agr* >>> truck crop

castrate >>> emasculate

CAT box: a conserved nucleotide sequence within the promoter region of numerous eukaryotic structural genes *gene biot*

***cat* gene** (CAT): chloramphenicol acetyl-transferase gene (and protein); it is used as a selective marker for cloning vectors and as a reporter gene *biot* >>> chloramphenicol acetyl-transferase

catabolic: pertaining to an enzymatic reaction leading to the breakdown of a complex biological molecule into less complex components, which may either yield energy in the form of ATP or be used in subsequent anabolic reactions *chem phys*

catalase: an enzyme that catalyzes the degradation of hydrogen peroxide to water and the oxidation by hydrogen peroxide of alcohols to aldehydes during seed germination *phys*

catalyst: a substance that initiates or accelerates a chemical reaction without apparent change in its own physical or chemical properties *chem*

catalyze: to induce or accelerate a chemical reaction by a substance that remains unchanged in the process *chem phys*

cataphyll: in cycads, a scalelike modified leaf that protects the developing true leaves *bot*

cataphyllary leaf >>> cataphyll

catch crop: a method of increasing agricultural or horticultural productivity by filling in the empty spaces, for example, it is created when slower-growing vegetables are harvested with fast-growing crops *agr* >>> stubble crop >>> underplant crop

cation: a positively charged ion in solution *phy chem*

cation exchange: the exchange between cations in solution and another cation held on the exchange sites of minerals and organic matter *chem agr*

catkin: a pendulous spike, usually of simple, unisexual flowers *bot*

caudate: having a tail *bot*

caudicle: an extension of tissue derived from the anther and connected to pollinia (e.g., in orchids) *bot hort*

caulescent: becoming stalked, having a stem *bot*

cauliflorous: borne on the trunk *bot*

cauliflory >>> cauliflorous

cauliflower mosaic virus (CaMV): a virus that infects cauliflower and other Cruciferae; it is transmitted by insects; the genome size is about 8 kb; it consists of double-stranded DNA with some single-stranded segments; in molecular genetics, it is used as a vector for transformation experiments *phyt gene biot*

cauliflower mosaic virus 35S promoter (CaMV 35S): a promoter (specific sequence of DNA) that is often utilized in genetic engineering to control expression of inserted gene; in other words, synthesis of desired protein in a plant *biot*

cauline leaf: a leaf formed on the florescence stem of a rosette plant (e.g., in *Arabidopsis*) *bot*

caulocarpous: bearing fruits on the stalk or trunk *bot*

caulogenesis: shoot formation *bot;* as de novo shoot development from callus *biot*

CBP >>> CHORLEYWOOD baking process

CCC >>> chlorocholine chlorid

cccDNA: covalently closed circles of DNA; it does not show nicks; only cccDNA can be supercoiled *biot*

cDNA >>> complementary DNA

cecidium (cecidia *pl*): a plant gall generally caused by an insect but sometimes by a fungus *phyt*

cell: the basic structural and functional unit of a plant; it is a system surrounded by membranes and is compartmentalized into specific functional areas and/or organelles with special tasks *bot*

cell adhesion: the contact between cells that is involved in cell aggregation and intercellular communication *cyto*

cell culture: the growing of dispersed cells in vitro *biot* >>> biotechnology

cell cycle: the sequence of events that occurs between the formation of a cell and its division into daughter cells; it is conventionally divided into G0, G1, (G standing for gap), S (synthesis phase during which the DNA is replicated), G2 and M (mitosis) *cyto*

cell division: the reproduction of a cell by karyogenesis and cytogenesis *bot*

cell envelope: the different surface components of the cell that are present outside the cytoplasmic membrane *bot*

cell fusion: fusion of two previously separate cells occurs naturally in fertilization; it can be induced artificially by the use of fusogens such as polyethylene glycol; fusion may be restricted to cytoplasm, nuclei may fuse as well; a cell formed by the fusion of dissimilar cells is referred to as a heterokaryon *bio biot* >>> biotechnology

cell generation time: the time span between consecutive divisions of a cell *cyto* >>> cell cycle

cell heredity: inheritance of a cellular level *gene*

cell hybridization: the fusion of a somatic cell in vitro and formation of viable cell hybrids *biot* >>> cell fusion

cell line: a population of cells that derives from a primary cell culture *cyto biot*

cell manipulation >>> biotechnology

cell membrane: a component of the cell surface with a discrete structure and function *bot*

cell nucleus >>> nucleus

cell plate: the structure formed between daughter nuclei after karyokinesis *cyto*

cell population: a group of cells that is static (without mitotic activity), expanding (showing scattered mitosis) or renewing (in which mitosis is abundant) *bot*

cell recognition: the mutual recognition of cells due to antigen-antibody or enzyme-substrate reactions *phys*

cell sap: the interorganelle fluid of the cell *bot*

cell selection: selection within a population of genetically different cells in vitro by different means and different approaches *biot*

cell sorting: a procedure that uses a mechanical device in order to separate mixtures of cells by their size, DNA content, etc. *cyto meth* >>> sorting

cell strain: a population of cells derived either from a primary culture or from a single cell *biot*

cell surface: the multicomponent structure surrounding a cell *bot*

cell suspension: cells and small aggregates of cells suspended in a liquid medium *biot*

cell synchrony: when a population of cells proceeds through the stages of cell cycle with synchrony (i.e., it divides at one time) *biot phys cyto*

cell tetrad >>> tetrad

cell transformation: a stable heritable alteration in the phenotype of a cell, usually brought about by viral or bacterial infection, but also by experimental means *biot*

cell tray >>> growing tray

cell wall: the steadfast external coat that surrounds the cell *bot*

cellular endosperm: a type of endosperm in which the early development is characterized by cell wall formation accompanying each nuclear division *bot*

cellulase: an enzyme that digests cellulose; sometimes used for maceration of plant tissue in order to improve spreading of chromosomes *phys micr cyto*

cellulose: a long-chain complex carbohydrate compound (polysaccharide); in the chief substance forming cell walls and the woody parts of a plant *bot*

center of diversity: a geographical location or local region where a particular taxon exhibits greater genetic diversity than it does anywhere else; N. I. VAVILOV developed this concept; he considered that the centers of diversity are also the centers of origin of a crop species; but the centers of diversity and the center of origin have subsequently been found to be distinct phenomena; the global centers of origin of crop plants and/or centers of diversity are summarized as follows: (1) China (mountains of Central and

Western China and adjoining areas), soybean, *Brassica* spp., radish, poppy, millets, buckwheat, fruit trees, mulberry, naked oat, naked barley; (2) India and Indo-Malaya (India [without the Northwest], Burma, Indochina, Malaysia), rice, sugarcane, banana, cocos palm, pepper, jute; (3) Central Asia (northwest of India, Pakistan, Afghanistan, Tadzhikistan, Tienshan, Uzbekistan), bread wheat, broad bean, pea, lentil, carrot, onion, grape, spinach, apricot; (4) West Asia (Transcaucasia region, Iran, Turkmenistan, Asia Minor), emmer wheat, einkorn wheat, rye oats, barley, vetches, alfalfa, clovers, plums, pea, lentil, fig; (5) Mediterranean coastal and adjacent regions (regions surrounding Mediterranean Sea), vegetables, rape, lupins, beets, clovers, pea, lentil, flax, olive, broad bean, seradella; (6) South Mexico–Central America (South Mexico, Central America, Antilles), maize, *Phaseolus* beans, sweet pepper, sweet potato, cotton, sisal, cacao, tomato, cucumber, pumpkins; (7) South America (Peru, Chile, Bolivia, parts of Brazil), maize, potato, tomato, cotton, peanuts, bananas, tobacco, rubber tree; (8) North America, lupins, grape, strawberry, sunflower *gene tax evol* >>> center of origin >>> center of domestication

center of domestication: the area believed to be that in which a particular crop species was first cultivated *gene* >>> center of diversity

center of origin: an area from which a given taxonomic group of plants has originated and spread and/or where wild-type species are found in greatest genetic variation; the theory was first published by V. I. VAVILOV in 1922 *gene* >>> center of diversity >>> Table 17

Centgener method: one of the earliest established pure-line systems of plant breeding based on 100 selected plants *meth*

centimorgan: equals 1 percent crossing-over *gene* >>> MORGAN unit

central cell: the largest cell in the center of the embryo sac surrounding the egg apparatus at micropylar end *bot*

central dogma: it refers to F. CRICK's seminal concept that in nature genetic information generally flows from DNA to RNA to protein *genet*

central mother cell: a large, vacuolated subsurface cell in a shoot apical meristem *bot biot*

central nervure >>> mid rip

centric: chromosomes having a centromere as opposed to acentric (having no centromere) *cyto* >>> Figure 11

centric constriction: the visible bight along a chromosome that bears the centromere *cyto* >>> Figure 11

centric fission: a chromosomal structural change that results in two acrocentric or telocentric chromosomes from one metacentric chromosome, as opposed to centric fusion *cyto* >>> Robertsonian translocation >>> Figure 37

centric fusion: the whole-arm fusion of chromosomes by the joining together of two telocentric chromosomes to form one chromosome *cyto* >>> Robertsonian translocation >>> Figure 37

centric region: the region where the centromere is placed *cyto* >>> Figure 11

centrifugal divider: a seed separator whose mode of operation is based on centrifugal forces *seed*

centrifuge: an apparatus that is used to spin liquids in a circular motion at high rates of speed; particles that are suspended in a liquid medium can be separated according to their density, the heavier particles collecting at the outer rim of the circle and the less dense ones collecting in layers toward the center *meth prep*

centriole: in mitosis, this small spherical body forms the center of the astral rays *cyto*

centromere: the structure to which the two halves of a chromosome, the chromatids, are joined; it contains the kinetochore that attaches to the spindle during nuclear division *cyto* >>> Figure 11

centromere interference: an inhibitory influence by the centromere on crossing-over and the distribution of chiasmata in its vicinity *cyto*

centromere misdivision: a transverse instead of lengthwise division of the centromere resulting in telocentric chromosomes *cyto* >>> Figure 37

centromere orientation: the process of orientation of centromeres during prometaphase of mitosis and meiosis, which contributes to a proper segregation of chromatids or chromosomes during anaphase *cyto*

centromere repulsion: the mutual repulsion of the centromeres of paired chromosomes toward the end of the meiotic prophase *cyto*

centromere shift: the displacement of centromeres by structural changes of the chromosomes (e.g., translocations, inversions etc.) *cyto*

cephalobrachial >>> acrocentric

ceraceous: waxy or waxlike *bot*

cereal(s): member of the grass family in which the seed is the most important part used for food and feed *agr bot* >>> caryopsis >>> Table 15

cereous >>> ceraceous

certation: the competition in growth rate between pollen tubes of different genotypes resulting in unequal chances of accomplishing fertilization *bot*

certified seed: seed produced under an officially designated system of maintaining the genetic identity of, and provisions for, seed multiplication and distribution of crop varieties; also a class of certified seed that is the progeny of registered or foundation seed (or basic seed); it is grown in compliance with regulations determining standards of germination, freedom from diseases and weeds, and trueness to type *seed* >>> Table 28

certify >>> certification >>> certified seed

cesium (Cs): a rare, highly reactive, soft metallic element of the alkali metal group *chem* >>> cesium-chloride-density-gradient centrifugation

cesium-chloride-density-gradient centrifugation: a method for purification of DNA by means of centrifugation in a cesium chloride solution, developed by MESELSON and STAHL *meth*

CFU >>> colony-forming units

CGIAR: Consultative Group of International Agricultural Research; under its auspices are the institutes as follows: CIAT, CIMMYT, CIP, ICARDA, ICRISAT, IITA, IRRI

chaff: the glumes, husks, scales, or bracts found with mature inflorescences and that separate from seeds during threshing, winnowing, or processing; in general, straw or hay that has been finely cut for animal feed *agr*

chaffy >>> chaff

chaffy grass divider: a subsampling device used to divide a sample of chaffy grass seed into a working sample *seed*

chalaza: the base of an ovule bearing an embryo sac surrounded by integuments *bot*

chalky: a color descriptor characterizing kernel endosperm of cereal grains (e.g., a rice grain with a high level of chalk is generally undesirable); the chalky appearance arises from the structure of the endosperm; voids cause light to be refracted and hence the endosperm appears white to reflected light and opaque to transmitted light *meth agr*

chambered >>> camerate

chapati: a flat pancakelike bread of India, usually of whole-wheat flour, baked on a griddle *meth*

chaperone: molecules that associate with an immature protein and cause it to fold into its final and active structure *phys*

character: an attribute of a plant resulting from the interaction of a gene or genes with the environment *gene*

characteristic >>> character

charged-coupled device (CCD) camera: a camera used for digital imaging; it contains a light-sensitive silicon chip; when light is falling on that chip, it creates an electrical charge at a specific location *micr*

chartaceous: paperlike *bot*

chasmogamy: fertilization after opening of flower, as opposed to cleistogamy *bot*

check cross: the crossing of an unidentified genotype with a phenotypically similar individual of defined genotype; F2 segregation analysis serves to establish whether the phenotype resulted from the action of identical or nonidentical alleles of the same gene locus or from the action of nonallelic genes *meth*

check cultivar: a commercial cultivar or experimental strain with well-known characteristics and performance that normally is included for comparison purposes with other selections in all testing procedures *meth*

check plot: in field testing of beeding material, an experimental plot system is usually applied; since variability of the land was recognized as a serious problem, first a duplicate plot system was used and later the check plot system; in the latter, standard or check varieties are sown in every fifth, tenth, twentieth, or more plot, depending upon circumstances; a map of the field is prepared, and plots are grouped around a check plot according to their nearness or, where the soil is highly variable, according to the character of the soil; the average yield of all check plots is determined; then, additions are made to below-average check plots yields to bring them up to the average, and subtractions are made from above-average yields *meth agr*

check strip >>> check plot

chelate: a claw-structure formed as a result of the reaction of a metal ion with two or more groups on a ligand, mugeinic acid is one of the many natural chelates; it can play an important role in uptake of metal ions from the soil; some plants (e.g., rye) may exude chelates into the rhizosphere *chem*

chemical desiccation: a method sometimes used to screen for postanthesis stress tolerance as destruction of the plant's photosynthetic system; the chemical desiccants most commonly used for cereals are magnesium chlorate, potassium iodide, and sodium chlorate *meth*

chemical mutagen: a chemical capable of causing genetic mutation *gene* >>> mutagen

chemical-hybridizing agents: compound applied to plants prior to anthesis to selectively induce male sterility *meth*

chemotaxis: oriented movement toward or away from a chemical stimulus *bot*

chemotherapy: control of a plant disease with chemicals (chemotherapeutants) that are absorbed and translocated internally *phyt*

chemotrophic: any organism that oxidizes inorganic or organic compounds as its principal energy source *bot*

***chi* sequence:** an octamer, nonpalindromic, that provides a hotspot for recBCD-mediated genetic recombination in *Escherichia coli;* wild-type lamda phage lacks a *chi* sequence; a *chi* sequence has been added artificially to some lamda cloning vectors, which cannot make concatemers by sigma-type replication; the *chi* sequence stimulates the formation of lamda dimers by the host recombination functions *biot* >>> lamda phage

Chi square (chi2): a statistical procedure that enables researchers to determine how closely an experimentally obtained set of values fits a given theoretical expectation *stat*

chiasma (Xta, chiasmata *pl*): a cross-shaped structure forming the points of contact between nonsister chromatids of homologous chromosomes, first seen in the diplotene stage of meiotic prophase I *cyto*

chiasma interference: the occurrence less frequent or more frequent than expected by chance of two or more crossing-over and chiasmata in a given segment of a chromosomal pairing configuration and/or chromosome *cyto*

chiasma localization: the physical position of a chiasma in a pairing configuration and/or chromosome *cyto*

chiasma terminalization: the progressive shift of chiasmata along the arms of paired chromosomes from their points of origin toward terminal positions *cyto*

chiasmate: meiosis with normal chiasma formation *cyto*

chilling damage: damage to plants at low temperatures in the absence of freezing; common in plants of tropical or subtropical origin at temperatures < +10°C; a change in viscosity of lipids in membranes might be the reason *phys*

chilling injury >>> chilling damage

chim(a)era: a tissue containing two or more genetically distinct cell types, or an individual composed of such tissue *bot* >>> xenia >>> valence cross

chim(a)eric >>> chim(a)era

chloramphenicol (chloromycetin): an antibiotic produced by *Streptomyces venezuelae;* it is a potent inhibitor of protein synthesis *phys gene biot*

chloramphenicol acetyl-transferase: coded by a particular gene *cat;* it derives from a certain transposon of a plasmid; the resistance to chloramphenicol is used in genetic experiments and was incorporated in some cloning vectors *phys biot* >>> *cat* gene

chlorenchyma: the general term for chloroplast-containing parenchyma cells as leaf mesophyll tissues *bot biot*

chlorocholine chloride (CCC): a growth regulator used for inhibition of internode growth in cereals in order to reduce straw length and thus to increase the lodging resistance *agr*

chloromycetin >>> chloramphenicol

chlorophyll: the green photosynthetic pigment generally localized in intracellular organelles (chloroplasts) *bot*

chloroplast: a membrane-enclosed, semiautonomous, subcellular organelle containing chlorophyll; it is a site where photosynthesis takes place; it contains DNA and polysomes and it is capable of replication; chloroplasts of vascular plants contain about 100 genes, most of which encode components of the photosynthetic electron transport machinery and elements of the transcriptional and translational apparatus; although the progenitor of the chloroplast was a free-living prokaryote, the loss of genetic information to the nucleus to control plastid gene expression, has largely placed the chloroplast in a "receptor" role, where it responds to nuclear signals; nonetheless, reverse signaling also occurs, demonstrating the interdependence and need for coordination between the cellular compartments *bot*

chloroplast DNA (ctDNA) >>> chloroplast

chlorosis: yellowing or whitening of normally green plant tissue; the loss of chlorophyll and associated pigments from small lesions or from whole leaves *bot*

CHOPIN alveograph: a set of equipment to predict the baking quality of certain cereals (e.g., wheat) in the absence of baking tests; the test consists of inflating a disc of dough with air until it bursts; the maximum pressure required and the time taken is measured; from this the strength and extensibility of the dough is determined *meth*

chopping: a mechanical cut-back of young shoots of sugarbeets for seed production in order to increase the seed quantity; also called "Boyage system" *agr seed*

CHORLEYWOOD baking process (CBP): a system of monitoring bread making developed in Great Britain in the early 1960s; the system uses high-speed mixing and the use of improvers, special fats, and yeasts to reduce fermentation time; this has the effect of being able to use lower protein and lower quality wheat varieties in the grist, yet produce good bread quality *meth*

chromatic aberration: inaccurate focusing of red, blue, and green light either along or at right angles to the optical axis; axial chromatic aberration results in the red image being focused farther along the optical axis than the green image; lateral chromatic aberration results in a slightly bluer image *micr*

chromatid: one of the two daughter strands of a chromosome that has undergone division during interphase; they are joined together by a single centromere *cyto*

chromatid aberration: chromosomal changes produced in one chromatid as a consequence of spontaneous or induced mutations *cyto*

chromatid break: a discontinuity in only one chromatid of a chromosome *cyto*

chromatid bridge: a bridgelike structure caused by a dicentric chromatid with the two centromeres passing to opposite poles during anaphase; the frequency of chromatid bridges in AII of meiosis is sometimes used as a measure for the level of cytological disturbances (e.g., in induced autopolyploids and allopolyploids) *cyto*

chromatid exchange >>> sister chromatid exchange

chromatid segregation: segregation of two sister-chromatid segments of a chromosome *cyto gene*

chromatin: the deoxyribonuclein-histone complex of chromosomes; it is readily stained by basic dyes and is therefore easily identified and studied under the microscope *cyto*

chromatin domain: a region of chromatin, the exact character and size of which depends on experimental context; it can be a single nucleosome or can extend to an array of more than 100 nucleosomes *biot*

chromatin insulator: a DNA element that protects a gene from position effects *gene*

chromatin reconstitution: the reconstitution of chromatin with chromosomal constituents previously removed by chromatin dissociation *cyto*

chromatography: a technique used for separating and identifying the components from mixtures of molecules having similar chemical and physical properties; molecules are dissolved in an organic solvent miscible in water, and the solution is allowed to migrate through a stationary phase; since the molecules migrate at slightly different rates they are eventually separated *meth*

chromocenter: a central aggregation of heterochromatic chromosomal elements of the cell nucleus; the euchromatic chromosome arms extend from the chromocenter *cyto*

chromogene: a stain-producing material *bot*

chromomere: a small beadlike structure visible in a chromosome during prophase of meiosis and mitosis, when it is relatively uncoiled *cyto* >>> knob

chromomere pattern: the linear order and distribution of chromomeres along a chromosome; it was extensively studied in crops, such as tomato, maize, rye, etc. *cyto*

chromonema: the smallest light-microscopically observed strand in chromosomes or chromatids *cyto*

chromoplast: a carotinoid-containing plastid that colors ripe fruits and flowers *bot*

chromosomal: referring to the structure, constituents, and function of chromosomes *cyto*

chromosomal aberration: an abnormal chromosomal complement resulting from the loss, duplication, or rearrangement of genetic material *cyto* >>> chromosome mutation

chromosomal domain: a region of a chromosome, the exact nature and size of which depend on experimental context; a domain can be a region of chromosomal packaging, such as a loop extending from two adjacent attachments to a chromosomal axis, and can vary in size from an array or less than 100 nucleosomes (~30 kb) to potentially more than 500 nucleosomes; a chromosomal domain might also represent a functional unit of chromosomal structure defined by boundary elements or insulators at the edges of the domain *cyto*

chromosomal structural change: a change in chromosome structure spontaneously or experimentally induced *cyto* >>> translocation

chromosome: a DNA-histone protein thread, usually associated with RNA, occurring in the nucleus of a cell; it bears the genes, which constitute the hereditary material; each species has a constant number of chromosomes; in 1999, a first plant chromosome of the weed *Arabidopsis thaliana* was genetically decoded *cyto gene*

chromosome arm: one part of a chromosome apart from the centromere *cyto* >>> Figure 11

chromosome banding: the experimental production of differentially stained regions because of the distribution of different chromatin constituents along a chromosome *cyto* >>> C banding

chromosome breakage: induced or spontaneous breaks across the entire cross-section of the chromosome *cyto*

chromosome bridge: a dicentric chromosome that forms a bridge between the separating groups of anaphase chromosomes because its two centromeres are being drawn toward opposite poles *cyto* >>> chromatid bridge

chromosome coiling: the spiral or helical coiling of the chromonemata of the chromosomes during some phases of mitosis and meiosis *cyto*

chromosome complement: the group of chromosomes derived from a particular gametic or zygotic nucleus *cyto* >>> genome >>> *cf* Important List of Crop Plants

chromosome configuration: any association by chromosome pairing of chromosomes at meiosis *cyto* >>> Figure 15

chromosome conjugation: joining of homologous during meiotic prophase *cyto* >>> Figure 15

chromosome contraction: the coiling and shortening of chromosomes during mitosis and meiosis either in a natural way or experimentally by using specific chemical or cold treatment *cyto* >>> chromosome coiling

chromosome doubling: induced or spontaneous doubling of chromosome sets leading to rediploidization or to polyploids *cyto meth* >>> polyploidization >>> doubled haploid

chromosome elimination: the loss of chromosomes from nuclei during certain mitotic or meiotic stages; it is common in several artificial autopolyploids and allopolyploids *cyto*

chromosome engineering: manipulation of whole chromosome sets, individual chromosomes or even chromosome segments, by different means, for scientific analysis or improvement of performance of crop plants *meth* >>> chromosome-mediated gene transfer >>> biotechnology

chromosome length polymorphism (CLP): the phenomenon that chromosomes can substantially vary in length among individuals of a population within a species, usually due to spontaneous structural changes of chromosomes (e.g., in allogamous rye) *gene meth*

chromosome map: a map showing the location of genes on a chromosome, deduced from genetic recombination and cytological experiments *meth* >>> mapping

chromosome mosaicism: the presence of cell populations of various karyotypes in the same individual *cyto*

chromosome movement: the movement of chromosomes during mitosis and meiosis as a prerequisite for the anaphase separation of chromatids and/or chromosomes *cyto*

chromosome mutation: any structural change involving the gain, loss, or translocation of chromosome parts; it can arise spontaneously or be induced experimentally by physical or chemical mutagens; the basic types of chromosome mutations are deletions (deficiencies), duplications, inversions, and translocations *cyto*

chromosome number: the specific somatic chromosome number (2*n*) of a given species or a crop derivative of it *cyto* >>> genome >>> Table 1, 14 >>> *cf* Important List of Crop Plants

chromosome orientation >>> centromere orientation

chromosome painting: fluorescent in situ hybridization of a specifically labeled DNA probe or probes that hybridizes to the entire chromosome of the probe's origin; using different fluorescent dyes and probes a pattern of multicolored chromosomes or chromosome segments appear *cyto meth*

chromosome pairing: the highly specific side-by-side association of homologous chromosomes during meiotic prophase *cyto* >>> synaptonemal complex >>> Figure 15

chromosome polymorphism: the presence of one or more chromosomes in two or more alternative structural forms within the same population *cyto*

chromosome puffing: despiralization of the deoxyribonucleoprotein of discrete regions of a chromosome during particular cell stages *cyto* >>> polyteny >>> polytene chromosome >>> puff

chromosome pulverization: the destruction of chromosome structure, varying from an apparently total fragmentation of the chromatin to various degrees of defective condensation and erosion *cyto*

chromosome rearrangement: the structural change of the chromosome complement by chromosome mutations *cyto* >>> translocation

chromosome reduplication: the synthesis of all compounds that result in an identical copy of the original chromosome *gene*

chromosome segregation: the separation of the members of a pair of homologous chromosomes in a manner that only one member is present in any postmeiotic nucleus *gene cyto*

chromosome set: the minimum viable complement of indispensable chromosomes (each is represented once) of an individual *cyto* >>> genome

chromosome size: the physical dimensions of a chromosome *cyto*

chromosome sorting >>> sorting

chromosome staining: the pretreatment and treatment of chromosomes with different dyes in order to make them more suitable for chromosome counting or specific analyses *cyto* >>> chromosome banding >>> opuntia >>> orcein >>> aceto-carmine staining >>> FEULGEN stain >>> FISH >>> GISH

chromosome stickiness: chromosome agglutination that results in a pycnotic or sticky appearance of chromosomes; sometimes caused by gene mutations or by treatment with chemical or physical agents *cyto*

chromosome substitution: the replacement of one or more chromosomes by others from another source by spontaneous events or a crossing scheme *cyto*

chromosome theory of inheritance: it states that the chromosomes, as the carriers of the genetic information, represent the material basis of nuclear inheritance *bio gene*

chromosome-counting method: the way to determine the number of chromosomes per cell *meth*

chromosome-mediated gene transfer: the transfer of genes within and between varieties, species, or genera by means of chromosome manipulations, such as additions, substitutions, translocations, or directed recombinations utilizing specific crossing techniques, cell manipulations, or micromanipulation of chromosomes; more specifically, the use of isolated metaphase chromosomes as a vehicle for the transfer of genes between cultured cells *meth cyto* >>> biotechnology

chromosome-walking technique: a procedure that is used for the determination of a gene on a particular DNA clone of a DNA library; the total DNA of a chromosome has to be available as a series of overlapping DNA fragments; such fragments are produced either by DNA shearing or by cleavage using restriction enzymes; the fragments are used for series of hybridizations; it starts with a cloned gene, which is already identified on the same chromosome; this known gene serves as a probe for detection of clones (fragments), which contain neighboring DNA sequences; during the following hybridization that DNA sequence is used as a probe for detecting the next neighboring sequences, and so on by each hybridization one subsequently moves away from the known gene toward the unknown chromosomal site; it seems that the method is more practical for plants with small genomes than for crop plants such as wheat, because the ratio of kilobases of DNA to the genetic map units is roughly proportional to the size of the genome; the cloning of genes by this method becomes much easier if the entire genome is already represented by a contiguous array of ordered DNA clones *meth gene*

chromotype: the chromosome set *cyto* >>> genome

chronic symptoms: symptoms that appear over a long period of time *pht*

CIAT: Centro Internacional de Agricultura Tropical, Cali, Colombia; responsible for dwarf beans, cassava, and forage crops breeding and research

Cibacron blue: affinity matrix used for the purification of restriction endonucleases *biot meth*

cilia: hairs growing along margins (e.g., in sunflower) *bot*

CIMMYT: Centro Internacional de Mejoramiento de Maiz y Trigo, Mexico DF, Mexico; responsible for wheat, maize, barley, and triticale breeding and research

cincinnus >>> scorpioid cyme

CIP: Centro Internacional de al Papa, Lima, Peru; responsible for potato and sweet potato breeding and research

circadian rhythm: a type of rhythmic plant growth response that appears to be independent of external stimuli *phys*

cirrate: rolled round, curled, or becoming so *bot*

cirrose >>> cirrate

cistron: a section of the DNA or RNA molecule that specifies the formation of one polypeptide chain; the functional unit of the hereditary materials; it codes for a specific gene product, either a protein or an RNA *gene* >>> structural gene >>> gene

citrate: a salt or ester of citric acid *chem*

citric acid cycle: the metabolic sequence of enzyme-driven reactions by which carbohydrates, proteins, and fatty acids produce carbon dioxide, water, and ATP *phys*

cladode >>> cladophyll

cladogenesis: a mode of evolution (i.e., the splitting of an evolutionary line) such as a species *tax evol*

cladophyll: a leaflike flattened branch that resembles and functions as a leaf *bot*

claw >>> auricle

clay: either mineral material < 2 μm texture, a class of texture, or silicate clay minerals *agr*

clean seed: sometimes it refers to endophyte-free seed (e.g., in grasses) *seed phyt*

cleared lysate: cell extract after removal of debris by centrifugation *biot*

cleavage: the processes by which a dividing egg cell gives rise to all the cells of the organism *cyto*

cleave: to make a double-stranded cut in DNA with a restriction endonuclease *gene biot*

cleaved amplified polymorphic sequences (CAPS): PCR-amplified DNA (STS, EST, or SCAR products) that is digested with restriction endonucleases to reveal polymorphisms in restriction sites *biot*

cleft grafting >>> split grafting

cleistogamous: designating a self-pollinated plant that produces inconspicuous flowers that never open (e.g., wheat or tomato) *bot* >>> Table 18

cleistogamy >>> cleistogamous

climate: the variations of cold and heat, dryness and moisture, calm and wind in a given region or country; in general, the combined result of all the meteorological phenomena of any region, as affecting its crop or vegetable production *eco*

climate chamber >>> phytotron

climax: a successional community of plants capable of optimal development under the prevailing environment and in dynamic equilibrium with its environment *eco*

climber: a plant that clambers upward by attaching itself to other plants or objects; climbers can be distinguished as (a) stem climbers, which wind upward around an erect support and, (b) as tendril climbers, which cling to nearby objects by slender, coiling tendrils (e.g., grape) *agr hort bot*

climbing plants: a vine or other plant that readily grows up a support (climbing rose), twines up a slender support (hop, honeysuckle), or grasps the support by special organs such as adventitious aerial roots (English ivy, poison ivy, trumpet creeper), tendrils, hook-tipped leaves (gloriosa lily, rattan), or stipular thorns (catabrier); some climbing plants when not supported become trailing plants (English ivy); climbing types are to be found in nearly every group of plants, e.g., the ferns (climbing fern), palms (rattan), grasses (some bamboos), lilies (gloriosa lily), and cacti (night-blooming cereus); woody-stemmed tropical kinds—usually called lianas—are particularly abundant; a sturdy vine may strangle a supporting tree and then, as with the strangler fig, become a tree itself *bot hort*

cline: an environmental gradient and a corresponding phenotypic gradient in a population of plants; when clines are evaluated by provenance tests, they are often found to have a genetic basis *fore hort*

clipping: breaking or cutting off the shoot tips in sugarbeet seed production; it stimulates the formation of side branches and thus increases the seed quantity *meth seed* >>> chopping

clonal: genetically identical *gene* >>> Table 35

clonal expansion: the population of cells produced from a single cell; it is synonymous with clone, but is used in particular context (e.g., a cell with a particular chromosomal abnormality can, by clonal expansion, produce a population of the same type of cell within the organism) *biot*

clonal propagation: vegetative (asexual) propagation from a single cell or plant *biot* >>> Table 35

clonal seed orchard: it is established by setting out clones as grafts or cuttings for seed production *hort fore* >>> seed orchard

clonal selection: choosing the best clones from a clonal testing (e.g., in potato or forest trees) *meth hort fore* >>> Table 35

clonal test: evaluation of genotypes by comparing clones in a plantation *hort fore*

clone: a group of genetically identical cells or individuals, derived from a common ancestor by asexual mitotic division *gene;* in molecular biology, a population of genetically identical organisms or cells: sometimes it refers to cells containing a recombinant DNA molecule or to the recombinant DNA molecules themselves *biot;* in horticulture or agriculture, a group of individuals originally taken from a single specimen and maintained in cultivation by vegetative propagation; all clone specimens are exactly alike and identical to the original *hort agr* >>> Table 35

clone library >>> genomic library

clone variety: a crop variety that consists of individuals deriving from a single clonal genotype (e.g., in potato) *meth* >>> Table 35

clonic >>> clonal

cloning: the process to vegetatively propagate a certain crop and/or plant *bot hort;* in molecular genetics, the cloning of DNA molecules from prokaryotic or eukaryotic sources as part of a bacterial plasmid or phage replicon *biumes of spent medium; usually cells are separated mechanically from outflowing medium and added back to the culture biot*

cloning site: restriction site, usually unique in a vector, where DNA can be inserted (cloned) *biot* >>> cloning vector

cloning vector: a plasmid or phage suitable for insertion and propagation of DNA; many cloning vectors have special properties (e.g., for expression of cloned genes or for the detection of cloned promoters) *biot*

close breeding >>> inbreeding

closed continuous culture: in vitro culture or a bioreactor processing in which inflow of fresh medium is balanced by outflow of corresponding vollone: a group of genetically identical cells or individuals, derived from a common ancestor by asexual mitotic division *gene;* in molecular biology, a population of genetically identical organisms or cells; sometimes it refers to cells containing a recombinant DNA molecule or to the recombinant DNA molecules themselves *biot;* in horticulture or agriculture, a group of individuals originally taken from a single specimen and maintained in cultivation by vegetative propagation; all clone specimens are exactly alike and identical to the original *hort agr* >>> Table 35

closed population: a population of plants with no genetic input other than by mutation *gene*

clove: one of the small bulbs formed in the axils of the scales of a mother bulb, as in, for example, garlic *bot*

clubroot (disease of crucifers): this soilborne disease, present in many *Brassica* species, causes swollen and distorted roots by the fungus *Plasmodiophora brassicae phyt*

clump: a single plant with two or more stems coming from a root or rhizome (e.g., in sunflower) *bot;* in horticulture and forestry, the aggregate of stems issuing from the same root, rhizome system, or stool *hort fore*

cluster analysis: a technique of statistical analysis in which similar variances are grouped or clustered; the results of statistical calculations are often shown as dendrograms; particularly, in cross breeding the cluster analysis is used in order to select most diverse parents for crossing *stat*

CMS >>> cytoplasmic male sterility

coadaptation: the process of selection by which harmoniously interacting genes become accumulated in the gene pool of a population *gene*

coarse shaker (at a harvester): the straw walkers convey the straw to the rear of the straw chamber where it either falls to the ground or is fed into a straw chopper *seed*

cob: the rachis of a female maize inflorescence *agr bot*

cocaine: a bitter, white, crystalline alkaloid, $C_{17}H_{21}NO_4$, obtained from coca leaves, used as a local anesthetic *chem phys*

coccus: one of the separate divisions of a divided seedpod; it splits up into one-seeded cells *bot*

cochineal: an insect or a red dye prepared from the dried bodies of the females of the cochineal insects, *Coccus cacti* or *Dactylopius coccus,* which lives on cactuses *zoo* >>> opuntia >>> carmin

coconut milk >>> coconut water

coconut water: liquid endosperm from the center of the coconut seed; a complex, undefined addendum of variable quality and effects in some nutrient solutions (2-15 percent v/v) for plant tissue culture; it shows growth-promoting effects and cell division factors; it is replaceable in some cases by cytokinins and/or sugar *bot biot*

cocultivation: a technique for transforming protoplast, other explants, or for in vitro selection by incubating them with a low density of transformed bacteria or a certain concentration of selective substances *biot*

coculture >>> cocultivation

coding sequence: the part of a gene that determines the sequence of amino acids of a protein, as opposed to noncoding sequences, such as promoter, operator, intron, or terminator regions *gene*

coding strand: the strand of duplex DNA that is transcribed into a complementary mRNA molecule *gene*

codogenic: that strand of double-stranded DNA (sense strand) that is used for genetic transcription *gene*

codominance: the expression of both alleles in the heterozygote *gene*

codominant: a heterozygote that shows fully the phenotypic effects of both alleles at a gene locus *gene*

codon: the triplet sequence of nucleotides in mRNA that acts as a coding unit for an amino acid during protein synthesis; it binds by base pairing to a complementary sequence, the anticodon, in tRNA *gene*

coefficient of inbreeding >>> inbreeding coefficient

coefficient of parentage (COP): the probability that a random gene from one individual is identical by descent with the a random gene from another individual (e.g., two varieties having one parent in common statistically showing on 50 percent of loci the same alleles (i.e., cop = 0.5), or having one grandparent in common showing 25 percent of loci the same alleles (i.e., cop = 0.25), or no parent in common cop = 0, respectively *stat meth*

coefficient of relationship: the probability that two individuals have inherited a certain gene from a common ancestor *stat* >>> coefficient of parentage

coefficient of variation: the standard variation expressed as percentage of the mean *stat*

coenocytes: an organism or a portion thereof that is multinucleate; the nuclei are not each separate in one cell, such as in some protoplast or cell fusion products *bot.biot*

coenospecies: a group of individuals of common evolutionary origin comprising more than one taxonomic species *bot evol*

coenzyme: a nonprotein, organic substance that acts as cofactor for an enzyme *phys*

cofactor: a nonprotein component that is required by an enzyme in order for it to function, and to which it may be either tightly or loosely bound *phys*

cohesive ends >>> sticky ends

coiling: when chromosome cores first become visible in late prophase of mitosis, sister cores are adjacent to one another and run along the inner sides of sister chromatids; as prophase proceeds, the proteinaceous cores separate, and sister chromatids coil; the chromatids coil because cores actively shorten; probably the contraction of a core located to one side of a chromatid causes the chromatid to coil with the core to the inside and chromatin to the outside *cyto*

colcemid: a synthetic equivalent of colchicine *cyto meth*

colchicine: a poisonous alkaloid drug ($C_{22}H_{25}NO_6$) that is obtained from meadow saffron *(Colchicum autumnale);* it has disruptive effect on microtubular activity; thus it affects tissue metabolism generally and mitosis and meiosis in particular; it is used for induction of polyploidy *chem phys cyto* >>> C mitosis >>> polyploidization >>> doubled haploid

colchiploidy: polyploidy induced by application of colchicine *cyto*

cold frame: a bottomless box consisting of a wooden, concrete, stone or metal frame with a glass or polyethylene top; it is placed on the ground over plants to protect them from cold or frost in order to speed up germination or to get an earlier harvest *hort meth*

cold stimulus >>> stratification

cold test: a type of stress test that shows the performance of seeds in cool, moist soil in the presence of various soil microorganisms; the test is conducted by planting the seeds in moist, unsterilized field soil, exposing them to cool (+5-10°C) temperatures for about one week, then allowing them to germinate in the same soil at warmer conditions *seed*

cole crops: vegetables of the genus *Brassica hort* >>> Figure 8

coleoptilar tiller >>> axillary tiller

coleoptile: the first leaf in grasses, which ensheaths the plumula *bot*

coleorhiza: a transitory membrane covering the emerging radicle (root apex) in some species; it serves the same function for the root as the coleoptile does for the plumule *bot*

colinearity: the correspondence between the order of nucleotides in a section of DNA (cistron) and the order of amino acids in the polypeptide that the cistron specifies *gene*

collar: the structure at the top of the culm above which lies the ear in cereals; in barley, the type of lodicules, which appear to enwrap the base of the caryopsis when seen in position by removal of the lemma, is called collar *bot* >>> Table 30

collective fruit >>> multiple fruit

collenchyma: a supporting tissue composed of more or less elongated living cells with nonlignified primary walls *bot*

colloid: a substance that is composed of two homogenous phases, one of which is dispersed in the other *chem*

colony: in tissue culture, a visible mass of cells *biot;* sometimes, a group of plants where all plants arise from one root system (e.g., in sunflower) *bot*

colony hybridization: a technique for using in situ hybridization to identify bacteria carrying a specific clone; it is only suitable for DNA fragments cloned onto multicopy vectors *biot*

colony-forming units (CFU): a measure for number of viable bacteria *biot*

color separator: a machine that separates seed on the basis of their surface color; it is used for seed cleaning or type separation *seed*

colorimeter >>> colorimetry

colorimetry: the methods used to measure color and to define the result of the measurement *micr*

colter >>> coulter

columella: an elongated floral axis that supports the carpels in certain plants *bot*

column chromatography: the separation of organic compounds by percolating a liquid containing the compounds through a porous material (e.g., ion exchange resin) in a cylinder *meth*

column diagram >>> histogram

coma: a tuft of hairs attached to a seed like the brush on wheat grains *bot*

combination ability >>> combining ability

combination breeding: a breeding method that utilizes the genetic diversity of individuals or varieties in order to create and to select new phenotypes on the basis of genetic recombination of useful characters of parental material *meth* >>> Figures 5, 7, 16 >>> Tables 5, 35

combine (harvester): a self-propelled grain harvester; in one operation it combines cutting, threshing, separation, cleaning, and straw dispersal *agr*

combining ability (CA): the average performance of a strain in a series of crosses (general CA); deviation in a particular cross from performance predicted on the basis of general combining ability (specific CA) *gene* >>> Figure 19 >>> Table 35

common scab (of potato): a fungal disease *(Streptomyces scabies)* affecting the tuber skin; diseased tubers are unattractive but yield and cooking quality remain unaffected *phyt*

community: a naturally occurring group of various organisms that inhabit a common environment, interact with each other, and generally are independent of other groups *eco*

compactation: increase in bulk density of soil due to mechanical forces, such as tractor wheels or combines *agr*

companion crop: two crops grown together for the benefit of one or both; it is particularly used of the small grains with which forage crops are sown *agr* >>> catch crop

companion planting >>> companion crop

companion species: usually a weed species that grows in close proximity to a crop species and which may be ancestral to the crop; the two species may exchange genes by viable spontaneous hybrids (e.g., *Aegilops* spp. near a wheat field, as it is seen sometimes in Turkey or other countries of the Middle East or the Midwest of the United States) *eco*

comparative mapping: localization (mapping) of a common set of DNA probes onto linkage maps of different species; this approach shows synteny of markers among related species or genera; such sort of maps have been established for cereals (wheat, maize, oats, rye, rice, sorghum, millet) and Solanaceae, such as potato, tomato, and paprika *gene meth*

compartmention: the subdivision of cells into parts by different membranes *bot*

compatibility: in molecular biology, compatible plasmids have different replication functions and can coexist in the same cell; incompatible plasmids are very similar or identical in their replication functions and in the absence of selection one of the two plasmids will survive in the cells while the other is lost spontaneously from a culture; the smaller, faster-replicating plasmid is generally favored in this process *biot*

compensating trisomic: an aneuploid individual with an extra chromosome in which a missing standard chromosome is compensated by two novel interchange chromosomes; the two novel chromosomes carry two different arms of the missing chromosome *cyto* >>> Figures 14, 17, 37

competency: an ephemeral state, induced by treatment with cold cations, during which bacterial cells are capable of uptaking foreign DNA *biot*

competent: in tissue culture, able to function or to develop; in molecular biology, competent (cells) have the ability to take up DNA; *Escherichia coli* is artificially made competent by washing in the cold with $CaCl_2$; some bacteria become naturally competent during certain growth phases *biot*

competition for food >>> survival of fittest

compilospecies: a genetically aggressive species that assimilates genomes from related species *bot*

complementary DNA (cDNA): DNA complementary to a purified mRNA and produced by RNA-dependent DNA polymerase *gene biot*

complementary effect: genes that by interaction produce a phenotype qualitatively distinct from the phenotype of any of them separately *gene*

complementary genes: mutant alleles at different loci, which complement one another to give a wild-type phenotype; dominant complementarity occurs where the dominant alleles of two or more genes are required for the expression of a particular trait; recessive complementarity is the case of suppression of a particular trait by the dominant allele of either gene, so that only the homozygous double recessive displays the trait *gene* >>> Table 6

complementation: the complementary action of gene products in a common cytoplasm; the support of a function by two homologous pieces of genetic material present in the same cytoplasm, each carrying a recessive mutation and unable by itself to support that function *gene;* in molecular biology, it refers to the ability of a cloned gene to overcome a host mutation *biot*

complete block: a simple experimental design in which all testers are included in each replication of the experiment and are arranged in a random order within replication; for accuracy, it should be used with small numbers of entries only; replications may be placed end to end or opposite each other, so that the total area covered by the experiment will be as nearly square in shape as possible; entries with apparent weakness may be discarded before harvest and data may still be analyzed by an analysis of variance *meth stat* >>> Table 25

complete cross >>> diallel cross

complete diallel: a mating design and subsequent progeny test resulting from the crossing of a certain number of parents in all possible combinations including selfs and reciprocals; because of severe inbreeding depression in the selfs, these are often skipped, nevertheless the test is still called a full diallel *meth*

complete dominance >>> dominance

complete flower: a flower that has pistils, stamens, petals, and sepals *bot*

complete penetrance: the situation in which a dominant gene always produces a phenotypic effect or a recessive gene in the homozygous state always produces a detectable effect *gene*

complete randomized design: the structure of the experiment is assumed that the treatments are allocated to the experimental units completely at random *stat*

completely randomized experiment >>> complete randomized design

complex character: a trait that is not inherited by single gene or simple manner of inheritance *gene*

complex heterozygous: special type of genetic system based on the heterozygosity for multiple reciprocal translocations *cyto*

complex locus: a cluster of two or more closely linked and functionally related genes constituting a pseudoallelic series *gene*

component of variance >>> variance

composite: a plant of the immense family Compositae, regarded as comprising the most highly developed flowering plants *bot;* a mixture of genotypes from several sources, maintained by normal pollination *seed*

composite cross: a population derived from the hybridization of several parents, either by hand-pollination or by the use of male sterility *meth*

composite fruit: a seed distribution unit that includes many ovaries connected by fruit walls or other suitable tissue; if the flower basis (receptacle) or other flower components are thick and fleshy (e.g., in strawberry, apple, or fig), it is called false fruit or pseuduocarp *bot*

composite mixture: breeder seed obtained by mechanically combining seed from two or more strains; the mixture is increased through successive steps in a certified seed program and distributed as a synthetic variety *seed*

compost: plant and animal residues that are arranged into piles and allowed to decompose *agr hort*

compound cross: a combination of desirable genes from more than two inbred lines, breeding strains, or varieties *meth*

compound cyme: a determinate inflorescence where there is secondary branching, and each ultimate unit becomes a simplé cyme *bot*

concatemer: tandem repeats of identical DNA molecules; lamda phage DNA must be concatemer in order to be packaged *biot* >>> lamda phage

concatenate: interlocked circles (e.g., plasmids) *biot*

concave: shaped like the inside of an egg *bot*

condensation of chromosomes >>> chromosome contraction >>> coiling

condenser: a lens or combination of lenses that gathers and concentrates light in a specified direction, often used to direct light onto the projection lens in a projection system *micr*

condenser iris diaphragm: the substage iris diaphragm located at the front focal plane of the condenser lens of a microscope; with KOEHLER illumination, the iris lies in a plane conjugate with the rear focal plane of the objective lens *micr*

conditional mutation: a mutation that has the wild-type phenotype under certain environmental conditions (temperature, age, nutrition) and a mutant phenotype under other conditions *gene*

conditioned dominance: dominance affected by the presence of other genes or by environmental influence *gene*

conditioned storage: storage of seed under controlled conditions of temperature and relative humidity *seed meth*

conditioner: a material or substance added to a fertilizer that keeps it flowing free *meth agr*

conditioning: the term used to describe the process of cleaning seed and preparing it for market; sometimes called processing *seed* >>> Table 11

conduction: plasmid mobilization involving cointegrate formation *biot*

conductivity test: an electrical conductivity test that associates the concentration of leachates from seeds, after soaking in water, to their quality *seed*

cone: a fruit with overlapping scales in which seeds are formed *fore*

cone collection: harvesting of cones after seed maturation but before their dispersal *fore hort*

confidence belt >>> confidence limit

confidence limit: a term for a pair of numbers that predict the range of values (confidence interval) within a particular parameter *stat*

confocal optics: a microscope optical system in which the condenser and objective lenses both focus onto one single point in the specimen *micr*

congenic strain: a variant plant strain that is obtained by backcrossing a donor plant strain to an inbred parental strain for at least eight generations while maintaining by appropriate selection the presence of a small genetic region derived from the donor strain *gene*

conical divider: an inverted metal cone below a spout from a hopper; the seeds fall over the cone to be evenly dispersed; a series of bugle or riffle dividers separate the seeds into channels *seed*

conidiophore: a threadlike stalk upon which conidia (spores) are produced; a specialized hypha upon which one or more conidia may bear *bot*

conidium (conidia *pl*): any asexual spore formed on a conidiophore *bot*

conifer: a species of plant that bears it seeds in cones, such as a pine tree *bot*

coniferous tree >>> conifer

conjugation: a process whereby organisms of identical species, but opposite mating types, pair and exchange genetic material (DNA) *gene;* in molecular biology, natural process of DNA transfer between bacteria in which the DNA is never exposed; it is insensitive to externally added DNase *biot*

connective: the tissue joining the two cells of an anther *bot*

conoidal: nearly conical *bot*

consensus sequence: if a particular nucleotide sequence is always found with only minor variations, then the usual form of that sequence is called consensus sequence; the term is also used for genes that encode the same protein in different organisms *gene*

conservation: maintenance of environmental quality and resources *seed*

conservation tillage: seed bed preparation systems that have about 30 percent or more of the residue cover on the surface after planting *agr*

constitutive heterochromatin: the material basis of chromosomes or segments that exhibit heterochromatic properties under most conditions (e.g., centromeric or telomeric heterochromatin) *cyto*

constitutive mutation: causes genes that usually are regulated to be expressed without regulation *gene*

constriction: an unspiralized segment of fixed position in the metaphase chromosomes (nucleolar ~ , primary or centric ~ , secondary ~) *cyto*

containment: measures taken to prevent release of recombinant DNA molecules into the natural environment; biological and physical methods are applied *biot*

contiguous (contig) map: the alignment of sequence data from large, adjacent regions of the genome to produce a continuous nucleotide sequence across a chromosomal region *biot*

continuous culture: an in vitro suspension culture continuously supplied with nutrients by the inflow of fresh medium; the culture volume is normally constant *biot* >>> closed continuous culture

continuous scale: a scale for scoring quantitative data for which the number of potential values is not predefined and is potentially limitless (e.g., seed weight in grams) *stat*

continuous variation: variation in the expression of inherited traits in which a series of nondiscrete, intermediate types, which cannot be divided into separate categories, connect the extremes with no obvious breaks between them *gene* >>> quantitative character

contrasting genetic character: a character with marked phenotypic differences *gene*

control: an economic reduction of crop losses caused by plant diseases *phyt agr*

control pollination: in horticulture and forestry, to purposely pollinate the female flowers of a tree with pollen from a known source; usually the flowers are isolated from undesirable pollen by covering them with a pollen-tight cloth or paper bag before they are receptive; it is a way to produce full-sib families *meth hort fore*

controlling element: a mobile (autonomous or nonautonomous) genetic component capable of producing an unstable, mutant target gene *gene*

controlling gene: a gene that is involved in turning on or off the transcription of structural genes; two types of genetic elements exist in this process; a regulator and a receptor element; the receptor elements is one that can be inserted into a gene, making it a mutant, and can also exit from the gene; both of these functions are under control of the regulator element *gene*

convergence: the evolution of unrelated species occupying similar adaptive areals, resulting in structures bearing a superficial resemblance *evol*

convergence breeding: a breeding method involving the reciprocal addition to each of two inbred lines of the dominant favorable genes lacking in one line and present in the other; backcrossing and selection are performed in parallel, each of the original lines serving as the recurrent parent in one series *meth* >>> Figure 31

convergence-divergence selection: a breeding scheme in which selection of promising genotypes is made in a bulk population at different locations followed by massing of selection and allowing mating among them in a pollination field; the harvested bulk seeds constitute the basis for the next propagation cycle *meth*

convergent crossing >>> convergence breeding

convex: shaped like the outside of an egg *bot*

coorientation >>> centromere orientation

COP >>> coefficient of parentage

copper (Cu): a malleable ductile metallic element having a characteristic reddish brown color; as a trace element it is needed by plants; deficiency can cause severe problems of growth; as iron efficiency, copper efficiency is genetically controlled (e.g. on rye chromosome arm 5RL a dominant gene and/or gene complex is located, increasing Cu efficiency not only in rye but also in wheat when the gene is transferred into the recipient) *chem phys* >>> white leaf disease >>> mugeinic acid >>> chelate

coppice: natural regeneration originating from stump sprouts, stool shoots, or root suckers *fore hort*

coppice method: a method of regenerating a forest stand in which the cut trees produce sprouts, suckers, or shoots *fore*

coppice selection method: a method in which only selected shoots of usable size are cut at each felling, leading to uneven-aged stands *fore*

coppice shoot: any shoot arising from an adventitious or dormant bud near the base of a woody plant that has been cut back *fore hort*

coppice-of-two-rotations method: a coppice method in which some of the coppice shoots are reserved for the whole of the next rotation; the rest being cut *fore*

coppice-with-standards method: regenerating a forest stand by coppicing; selected trees grown from seed are left to grow to larger size than the coppice beneath them; the method is used to provide seeds for natural regeneration of standards in subsequent rotations *fore*

copulation: the fusion of sexual elements *gene*

copy error: an error in the DNA replication process giving rise to a gene mutation *gene*

copy number: the number of molecules per genome, of a plasmid or a gene, that a cell contains *gene biot*

copy-choice hypothesis: the interpretation of intrachromosomal genetic recombination that is not regarded as a physical exchange of preformed genetic strands *gene*

cordage: ropes *agr*

cordon: an extension of the grapevine trunk, usually horizontally oriented and trained along the trellis wires; it is considered permanent (or perennial) wood *hort*

core collection: the basic sample of a germplasm collection; it is designed to represent the wide range of diversity in terms of morphology, geographic range, or genes; it contains, with a minimum of repetitiveness, the genetic diversity of a crop species and its wild relatives; it is not intended to replace existing gene banks collections but to include the total range of genetic variation of a crop in a relatively small and manageable set of germplasm accessions *meth seed*

corepressor: a metabolite that in conjugation with a repressor molecule binds to the operator gene present in an operon and prevents the synthesis of a repressible enzyme *gene*

coriaceous: leathery *bot*

cork: in woody plants, a layer of protective tissue that forms below the epidermis *bot*

cork cambium >>> phellogen

cork layer: layer of dead protective tissue between the bark and cambium in woody plants *bot*

corm(us): an underground storage organ formed from a swollen stem base, bearing adventitious roots, and scale leaves; it may function as an organ of vegetative reproduction or in perennation *bot*

corn: the edible seed of cereal plants other than maize *bot* >>> caryopsis

corn: *US* maize

corneous: it refers to hard, vitreous, or horny endosperm in cereal grains *bot agr*

corn-loft >>> granary

corolla: a collective term for all the petals of a flower; a nonreproductive structure; often arranged in a whorl; encloses the reproductive organs *bot*

correlation: the degree to which statistical variables vary together; measured by the correlation coefficient, which has a value from zero (no correlation) to –1 or +1 (perfect negative or positive correlation) *stat*

correlation breaker >>> outlier

correlation coefficient: a measure for the degree of association between two or more variables in an experiment; it may range in value from –1 to +1 *stat* >>> correlation

corresponding gene pair: a pair of genes in a parasite that corresponds with a pair of genes in a host, which function together to bring about a specific outcome *phyt*

cortex >>> rind

corymb: a racemose inflorescence in which the lower pedicels are longer than the upper so that the flower lies as a dome or dish, and the outline is roundish or flattish *bot*

cos site: the site of the circular form of phage lamda or others that is cleaved by the terminase to generate the cohesive 12 bp 5' overhang ends of the linear phage as it is packaged into the capsid *biot* >>> cosmid >>> lamda phage

cosmid: a synthetic word derived from the designations **cos** and plas**mid;** a cosmid is a plasmid (e.g., pBR322) with so-called cos-sites of the DNA; they offer the chance of incorporation of alien DNA fragments of sizes between 32-45 kb *gene* >>> cos site

cosmopolite: plant of worldwide distribution *eco*

cosuppression: silencing of a gene by addition of transgenic DNA copies *biot*

cotransformation: an event of two plasmids entering the same cell by transformation *biot*

C_0t curve: graphic representation of the progress of a (liquid) hybridization experiment; used to determine the complexity of DNA mixtures (e.g., the size of the genome) *gene biot*

C_0t value: an expression for the rate of DNA renaturation (annealing-reannealing); DNA renaturing at low C_0t is composed of highly repetitive sequences and DNA renaturing at high C_0t values is minimal or nonrepetitive *gene biot*

cotyledon: the leaf-forming part of the embryo in a seed; it may function as a storage organ from which the seedling draws food, or it may absorb and pass on to the seedling nutrients stored in the endosperm; once it is exposed to light it develops chlorophyll and functions photosynthetically as the first leaf *bot* >>> Table 16

cotyledonary node: the point of attachment of the cotyledons to the embryonic axis *bot*

coulter: a sharp blade or wheel attached to the beam of a plough, used to cut the ground in advance of the ploughshare *agr*

coumarin: a white crystalline compound ($C_9H_6O_2$) with a vanilla-like odor; it gives sweetclover its distinctive odor; it is also known as a chemical growth inhibitor that has germination-inhibiting capability *phys*

couple method (of breeding): a breeding method exclusively used in breeding of allogamous plants; from an original population (e.g., of sugarbeet), single plants are selected and, subsequently, pair-wise crossed, preventing unwished pollination; the crossing partners should be as similar as possible in spite of color, growth habit, etc.; the offspring is grown in separate plots during the following year; the selection of individuals from the plots and a repeated pair-wise crossing can be realized during the fourth year; during the fifth year offspring is grown in plots and again selected for progeny testing *meth* >>> Figure 41 >>> Table 35

coupling of factors: linkage in which both dominant alleles are in the one parent *gene* >>> linkage

covariance analysis: an analysis of the mean of the product of the deviation of two variates from their individual means; it measures the interrelationship between variables *stat*

cover crop: a crop grown between orchard trees or on fields between the cropping seasons of a main crop, to protect the soil against erosion and leaching and for improvement of soil *agr*

crease: the fold on a cereal grain *bot*

criss-cross inheritance: the transmission of a gene from mother to son or father to daughter *gene* >>> criss-crossing

criss-crossing: a continuous, rotational crossbreeding system alternately using males or pollinators of two different breeds; this system is simple to manage and breeds its own replacements; it utilizes the benefit of hybrid vigor; compared to the common F1, some hybrid vigor can be lost, but that loss is more than compensated for by reduced management effort and cost *meth*

cristae >>> mitochondrion

critical difference: a value indicating least significant difference at values greater than which all the differences are significant *stat*

crop: a species expressly cultivated for use *agr* >>> Table 35 >>> crop plant

crop divider (at the harvester): separates the standing crop from the material being cut *agr*

crop evolution: the adaptation of a crop over generations of association with humans *evol*

crop plant: a plant expressly cultivated for use; the majority of crops can be classified as (1) root and tuber crops (potato, yams), (2) cereals (e.g., wheat, oats, barley, rye, rice, maize), (3) oil and protein crops (rapeseed, pulses),

(4) sugar crops (sugarbeet, sugarcane), (5) fiber crops (cotton, jute), or (6) forage crops (grasses, legumes); agronomic crops can be classified as (a) green manure crops, (b) cover crops, (c) silage crops, or (d) companion crops; about 2 percent of the 250,000 higher plant species are used in agriculture, horticulture, etc.; economically, the most important families are the legumes and the grasses, which account for more than a quarter of the total species; they are followed by Rosaceae, Compositae, Euphorbiaceae, Labiatae, and Solanaceae, all with more than 100 taxa; among the families with 50 to 100 crop species, Liliaceae, Agavaceae, and Palmae are worth mentioning, whereas more than 50 percent of the families have fewer than ten crop species *agr* >>> Tables 1, 35

crop residue: that portion of a plant left in the field after harvest (maize stalk or stover, stubble) *agr*

crop rotation: the alternation of the crop species grown on a field; usually this is done to reduce the pest and pathogen population or to prevent one-track exhaustion *agr*

cross: bringing together of genetic material from different individuals in order to achieve genetic recombination *meth*

cross back >>> backcrossing

crossability: the ability of two individuals, species, or populations to cross or hybridize *bot eco*

crossbred >>> self

crossbreeding: outbreeding or the breeding of genetically unrelated individuals; this may entail the transfer of pollen from one individual to the stigma of another of a different genotype *meth* >>> Figure 7 >>> Table 35

crossbreeding barrier: a pre- and/or postfertilization condition (i.e., progamous or postgamous incompatibility) that prevents or reduces crossbreeding or any form of gene transfer; it is caused by genetic, environmental, physical, or chemical influences >>> Table 35

cross-fertilization: the fusion of male and female gametes from different genotypes or individuals of the same species, as base of genetic recombination *bot* >>> allogamy >>> cross-pollination >>> Table 35

cross-hybridization: in biotechnology, the hydrogen bonding of a single-stranded DNA sequence that is partially but not entirely complementary to a single-stranded substrate; often, this involves hybridizing a DNA probe for a specific DNA sequence to the homologous sequences of different species *biot*

crossing barrier: any of the genetically controlled mechanisms that either entirely prevent or at least significantly reduce the ability of individuals of a population to hybridize with individuals of other populations *gene* >>> crossbreeding barrier

crossing group(s): any group of individuals that comprises a unique set of parents: (1) diallel crossing group—controlled crosses are made between each pair of parents in the group but crosses with parents outside the group are excluded, (2) factorial crossing group—a limited number of parents are used as male testers in controlled crosses with an unlimited number of female parents, (3) open-pollinated crossing group—all parents in a breeding population are included in a progeny test or series of tests *meth fore hort* >>> Table 35

crossing-over: the exchange of genetic material between homologous chromosomes by breakage and reunion; it occurs during pairing of chromosomes at prophase I of meiosis; the temporary and visible joins between chromosomes during crossing-over are called chiasmata *gene cyto* >>> Figure 24

crossing-over map: a genetic map made by utilizing crossing-over frequencies as a measure of the relative distances between genes in one linkage group (chromosome) *gene*

cross(ing)-over unit: a 1 percent crossing-over value between a pair of linked genes *gene* >>> MORGAN unit

cross-inoculation: inoculation of one legume species by the symbiotic bacteria from another *agr*

cross-pollinating crop >>> crossbreeding >>> cross-pollination >>> xenogamy >>> Table 35

cross-pollination: the transfer of pollen from the stamen of a flower to the stigma of a flower of a different genotype but usually of the same species, with subsequent growth of the pollen tube *bot* >>> allogamy >>> Table 35

cross-protection: plant protection conferred on a host by infection with one strain of, for example, a virus that prevents infection by a closely related strain *phyt*

cross-resistance: resistance associated with a change in one genetic factor that results in resistance to different chemical pesticides that were never applied *phyt*

cross-sterility: the failure of fertilization because of genetic or cytological conditions (incompatibility) in crosses between individuals >>> crossing barrier >>> Table 35

crown: the stem-root junction of a plant (e.g., the overwintering base of an herbaceous plant) *hort;* the term is also used for the treetop *bot fore*

crown gall: a common and widespread plant disease, which can affect a very wide range of woody and herbaceous plants (fruit trees, roses, etc.); it is caused by the bacterium *Agrobacterium tumefaciens;* galls are formed at the crown (stem-root junction) or, less frequently, on roots, stems, or branches *phyt*

crucifer: a plant belonging to the Brassicaceae or mustard family, a large dicotyledonous family of important crop and ornamental plants (turnip, cabbage, etc.) *bot* >>> Figure 8

crust: a surface layer of soil that becomes harder than the underlying horizon when dry *agr*

cryoability: the ability of plant material (seeds, tissue, organs) to preserve or store under very low temperatures, usually in liquid nitrogen (–196°C) *phys*

cryobank: the preservation or storage under very low temperatures, usually in liquid nitrogen (–196°C) *meth seed*

cryodamage: damage caused by exposure to cold conditions *agr*

cryopreservation: the preservation or storage in very low temperatures, usually in liquid nitrogen (–196°C) *meth*

cryoprotectant: a chemical, which is used to protect seeds, cultured material, tissue, organs or cells from the low temperature in cryopreservation (e.g., glycerol) *prep*

cryoscopy: a technique for determining the molecular weight of a substance by dissolving it and measuring the freezing point of the solution *meth*

cryostat: a device designed to provide low-temperature environments in which experiments may be carried out under controlled conditions *prep*

cryptochrome >>> photomorphogenesis

cryptogam: a plant (e.g., fern) that reproduces by means of spores rather than seeds *bot*

cryptogam(ous): reproduction by spores or gametes rather than seeds *bot*

cryptogams >>> cryptogam(ous)

cryptomeric gene >>> cryptomerism

cryptomerism: the phenomenon that a gene or an allele does not show a phenotypic effect unless it is activated by another genetic factor, which leads to a sudden change of qualities in the progeny, not recognized among the ancestors *gene*

crystallography: the determination of, for example, the protein structure; the protein is crystallized and the crystals examined using X rays; the diffraction angles of the X rays are used to compute the relative positions of components of the protein, and thus its structure *phy*

$CsCl_2$ gradient: a method used to separate DNA or phages according to buoyant density *biot*

CSSA: Crop Science Society of America

CTD >>> canopy temperature depression

cuckoo chromosome: an alien chromosome that shows a preferential transmission during generative reproduction; found in certain wheat-*Aegilops* crossing progenies *cyto*

cucullate: hood- or cowllike in form *bot*

cucumber mosaic virus (CMV): one of the most widely occurring plant viruses; infects more than 800 plant species and has a considerable negative impact on agriculture worldwide; in addition to an RNA genome, certain strains of CMV naturally harbor an RNA of 330-400 nucleotides, called sat-RNA, that depends on the virus for its entire cycle, has no significant similarity with the viral genome, and apparently does not encode protein; sat-RNAs are capable of attenuating CMV strains, resulting in a sharp decrease in virus titer in infected plants and an almost complete lack of symptoms (e.g., lethal tomato necrosis) *phyt*

culled >>> off-grade

culling: the postharvest removal of pathogen-infected or damaged fruit, seeds, or plants by screening procedures; the culled or off-graded material can later be individually analyzed or discarded *meth*

culm: the jointed stem in cereals, grasses, or sedges; filled with pith or solid *bot*

cultigen: a cultivated plant or group of plants for which there is no known wild ancestor (e.g., maize *Zea mays*) *bot tax*

cultivar: a contraction of "cultivated variety" (abbreviated cv.); refers to a crop variety produced by scientific breeding or farmer's selection methods *bot* >>> variety

cultivar identification system: a classification system based on sequence-tagged microsatellite loci analysis with fluorescent primers and suitable computer software; allows unequivocal identification of varieties, paternity testing, and duplicate identification *meth biot*

cultivar mixture: a mixture of different varieties in order to improve the environmental adaptability or the resistance to pathogens *seed*

cultivation: the art or process of agriculture *agr*

cultivator: an implement drawn between rows of growing plants to loosen the earth and destroy weeds *agr*

culture: a growth of one organism or of a group of organisms for the purpose of production, trade, and utilization or for experiments *agr*

culture collection: a collection of cultures of more or less defined or characterized viruses, bacteria, and other organisms; usually used for reference and comparison with new isolates *phyt*

culture medium: medium on or in which tissues, organs, or cells are cultured; supplies the mineral and hormonal requirements for the growth *meth biot*

culture tube: a tube in which tissue, organs, cells, or organisms are cultured *prep*

cumulative genes: polymeric nonallelic genes *gene*

cupule: a cuplike structure at the base of some fruits *bot*

cushion plants: have small, hairy, or thick leaves borne on short stems and forming a tight hummock *bot*

cut flowers: flowers that are cut off the plant and used as decoration *hort*

cut surface >>> cutin >>> cutinize

cut-and-come-again: applied to any plant that is cut or sheared after flowering and blooms again (e.g., *Petunia* spp., pansy) *hort*

cuticle: a thin, waxy, protective layer covering the surface of the leaves and stems *bot*

cutin: the complex mixture of fatty-acid derivatives with waterproofing qualities of which the cuticle is composed *bot*

cutinize: to impregnate a cell or a cell wall with cutin, a complex fatty or waxy substance, which makes the cell more or less impervious to air and moisture *bot*

cutout: the occurrence of physiologically indeterminate growth (e.g., in cotton) *agr*

cutting: a section of a plant that is removed and used for propagation; cuttings may consist of a whole or part of a stem (leafy or nonleafy), leaf, bulb, or root; a root cutting consists of root only; other cuttings have no roots at the time they are made and inserted; as opposed to division, a kind of propagation that consists of part of the crown of a plant or of its above-ground portion and roots *hort*

cv. >>> cultivar

cybrid: the hybrid formed from the fusion of a cytoplast and a whole cell; the cytoplast may transmit cytoplasmic components independently of the cell genome *bot*

cyclical parthenogenesis: a life history in which a sequence of apomictic generations is followed by amphimictic generations *bot*

cycloheximide (actidione): an antibiotic from *Streptomyces griseus;* antibacterial and antifungal *biot*

cyme: an inflorescence in which each axis ends in a flower *bot*

cysteine (Cys): an aliphatic, polar alpha-amino acid that contains a sulfydryl group *chem phys*

cytochimera: different tissues or parts of them differ in chromosome number *cyto*

cytochrome (Cyt): an iron-containing pigment that plays a major role in respiration; more detailed, one of a group of haemoproteins, which are classified into four groups designated a, b, c, and d; they function as electron carriers in a variety of redox reactions in virtually all aerobic organisms *phys*

cytodifferentiation: the sum of processes by which during the development of the individual the zygote specialized cells, tissue, and organs are formed *cyto*

cytogamy: the fusion or conjugation of cells *cyto*

cytogenetic map: a map showing the locations of genes on a chromosome *cyto gene*

cytogenetics: scientific discipline that combines cytology with genetics *gene cyto*

cytogenic male sterility >>> cytoplasmic male sterility

cytogony: the reproduction by single cells *cyto*

cytohet: a cell containing two different cytoplasmic genomes (e.g., mitochondria) that differ in one or more genes contributed by two parents; thus, the individual is cytoplasmatically heterozygous *gene*

cytokinesis: during the division of a cell, the division of the constituents of the cytoplasm; it usually begins in early telophase with the formation of a cell plate, which is assembled within the phragmoplast across the equatorial plane; the phragmoplast is a complex array of GOLGI-derived vesicles, microtubule, microfilaments, and endoplasmatic reticulum that assembles during the late anaphase and is dismantled upon completion at the new wall *cyto*

cytokinin: one of a group of hormones, including kinetin, that act synergistically with auxins to promote cell division but, unlike auxins, that promote lateral growth *phys*

cytology: the branch of biology dealing with the structure, function, and life history of the cell *cyto*

cytolysis: breaking up or solution of the cell wall *cyto*

cytomixis: the extrusion or passage of chromatin from one cell into the cytoplasm of an adjoining cell *cyto*

cytoplasm: the part of a cell that is enclosed by the plasma membrane, but excluding the nucleus *cyto*

cytoplasmic inheritance: a non-Mendelian (extra-chromosomal) inheritance via genes in cytoplasmic organelles (mitochondria, plastids) *gene*

cytoplasmic male sterility (CMS): pollen abortion due to cytoplasmic factors, which are maternally transmitted, but which act only in the absence of pollen-restoring genes; this type of sterility can also be transmitted by grafting *gene* >>> Figure 2

cytoplasmon: all cytoplasmic hereditary constituents of a cell excepting those localized in the plastids and mitochondria *cyto*

cytoplasm-restorer >>> cytoplasmic male sterility

cytoplast: the cytoplasm as a unit, as opposed to the nucleus *bot*

cytosine (C): a pyrimidine base that occurs in both DNA and RNA *chem gene*

cytosol: the water-soluble components of cell cytoplasm, constituting the fluid portion that remains after removal of the organelles and other intracellular structures *bot*

cytostatic: any physical or chemical agent capable of inhibiting cell growth and cell division *phys*

cytotaxonomy: the study of natural relationships of organisms by a combination of cytology and taxonomy *tax* >>> Table 17

cytotype: any variety of a species whose chromosome complement differs quantitatively or qualitatively from the standard complement of that species *cyto*

D

Dalton (Da): a unit equal to the mass of the hydrogen atom (1.67×10^{-24} g); the unit was named after J. DALTON, a chemist of nineteenth century *chem*

***dam* gene:** DNA adenine methylation gene of *Escherichia coli;* it methylates the sequence GATC (*Sau3A* cleaves methylated and unmethylated DNA, *MboI* cleaves only unmethylated DNA, *DpnI* cleaves only the methylated sequence) *biot*

dammar resin: a hard, lustrous resin derived from Asian trees of the monkey-puzzle family *prep*

damping-off: a disease of young seedlings in which the stems decay at ground level and the seedlings collapse; by planting in sterile soil, treating seeds with a fungicide, or soaking the soil with a fungicide it can be prevented, as avoiding overwatering and planting warm-season plants in cold soil *phyt*

dandelion weeder >>> asparagus knife

dark field microscopy: a microscope designed so that the entering center light rays are blacked out and the peripherical rays are directed against the object from the side; as the result the object being viewed appears bright upon a dark background *micr*

dark reaction: the phase of photosynthesis, not requiring light, in which carbohydrates are synthesized from carbon dioxide *phys*

dark respiration >>> dark reaction

dark room: a room in which film, photographic material, is handled or developed and from which the actinic rays of light are excluded *prep;* a dark room is also used in order to simulate short-day conditions; it is applied either to prevent flowering of long-day plants or to induce flowering of short-day plants under long-day conditions *meth*

dark-field >>> dark field microscopy

Darwinism: the theory that the mechanism of biological evolution involves natural selection of adaptive variations *bio*

data management: the organization of the recording of data, its preparation for analysis and its interpretation *prep*

data sheet: a specially prepared form on paper or computer for recording data *meth*

database: a store of a large amount of information (e.g., in a form that can be handled by a computer); recently, in breeding, numerous databanks are in use, such as field registration, nursery plans, selection data, data of international observation trials, statistical values, data recording, etc. *stat*

dauermodification >>> persistent modification

daughter cell: the cells resulting from the division of a single cell *cyto*

daughter chromosome: any of the two chromatids of which the replicated chromosome consists after mitotic metaphase or anaphase II of meiosis *cyto*

daughter nucleus: the nuclei that result from the division of a single nucleus *cyto*

daylength: the number of hours of light in each 24-hour cycle *phys*

daylength insensitivity: these plants will flower independent of daylength (e.g., tomato or cotton) *phys*

daylength response >>> daylength sensitivity

daylength sensitivity: plants that will flower only when the daily photoperiod shows a critical length *phys* >>> short-day plants >>> long-day plant >>> day-neutral plants

day-neutral plants: no daylength requirement for floral initiation *phys*

debearder: in seed precleaning procedures, it has a hammering or flailing action that removes awns, beards, or lint from seed and tends to break up seed clusters of the chaffy grasses, as well as multiple seed units of nonchaffy forms *seed* >>> Table 11

decarboxylase: an enzyme that removes the carboxyl group from an organic compound *chem phys*

decay: the destruction of plant material by fungi and bacteria *agr*

deciduous: a plant whose leaves are shed at a season or growth stage *bot*

decimal code for the growth of cereal plants: a decimal code used for describing different growth stages of cereal plants; it is applied for comparison of several morphological stages; there are several schemes of description *phys meth* >>> Table 13

declining vitality: seeds that are aged or have been subjected to unfavorable storage conditions; they usually show a slow germination; some of the essential plant parts are frequently stunted or lacking; saprophytic fungi may also interfere with the growth of the seedlings *seed*

decompose: to rot or putrefy *agr*

decondensation stage: a stage between interphase and prophase of mitosis in which heterochromatin is decondensed for a short period *cyto*

decontaminate: to free from contamination; purify *meth*

decumbent: lying on the ground with the end ascending *bot*

decurrent: describes the open type of collar in barley where the margin of the platform is incomplete, merging with the neck; in general, extending downward from the point of insertion *bot*

dedifferentiation: a loss of specialization of a cell; it can be observed when differentiated cells are placed in vitro culture *biot*

deficiency: the absence or deletion of a segment of genetic material *cyto* >>> chromosome mutation

deficiency disease: any disease caused by the lack or insufficiency of some nutrient, element, or compound (e.g., copper, zinc, or iron deficiency in cereals) *phys*

definite host: the host in which the parasite attains sexual maturity *phyt* >>> host

deflexed: bent sharply downward *bot*

defoil: to strip a plant of leaves *meth*

defoliant: a chemical or method of treatment that causes only the leaves of a plant to fall off or abscise (e.g., it is applied before harvest of potato) *agr*

defoliate: to remove leaves of a plant *meth*

defoliated >>> defoliation

defoliation: the process of leaves being removed from a plant (e.g., to make harvest easier) *bot*

degeneracy of genetic code >>> degenerated code

degenerated code: a term applied to the genetic code because a given amino acid may be encoded by more than one codon *gene*

degermed: grains from which the embryo (germ) has been removed *agr meth seed*

degradation: the progressive decrease in vigor of successive generations of plants, usually caused by unfavorable growing conditions or diseases; viruses may cause great loss of vigor; in agriculture, the change of one kind of soil to a more highly leached soil *agr*

degree of dominance >>> dominance

degree of freedom (DF): the number of items of data that are free to vary independently; in a set of quantitative data, for a specified value of the mean, only (n −1) items are free to vary, since the value of the nth item is then determined by the values assumed by the others and by the mean *stat*

degree of genetic determination: the portion of total variance that is genetically determined *gene*

dehiscence: the bursting open at maturity of a pod or capsule along a definite line or lines *bot*

dehiscent fruit >>> dehiscence

dehull: removal of outer seed coat (hull) or to remove the glumes of cereal caryopses *seed*

dehusk: to remove the husk (e.g., the leaf sheath of a maize cob or the outer layer of a coconut) *seed*

dehydration: the elimination of water from any substance *phys*

dehydrogenase: an enzyme that catalyzes the removal of hydrogen from a substrate *phys*

deletion: the loss of a chromosomal segment from a chromosome set; the size may vary from a single nucleotide to sections containing several genes *cyto* >>> deficiency >>> chromosome mutation

deletion mapping: the use of overlapping deletions to localize the position of an unknown gene on a chromosome or linkage map *gene*

deltoid: shaped like the Greek letter "delta" *bot*

demic selection: special type of intergroup selection that does not necessarily involve direct competition; it has an effect on the general genetic composition of a population if subsets of a population have different gene frequencies *meth*

denaturation: reversible or irreversible alterations in the biological activity of proteins or nucleic acids that are brought about by changes in structure other than the breaking of the primary bonds between amino acids or nucleotides in the chain *chem gene*

denatured protein: a protein whose properties have been altered by treatment with physical or chemical agents *chem*

dendrogram: a genealogical diagram that resembles a tree; an evolutionary tree diagram may order objects, individuals genes, etc., on the basis of similarity *evol meth* >>> cluster analysis

dendrology: the branch of botany dealing with trees and shrubs *bot*

denitrification: the conversion of nitrate or nitrite to gaseous products, chiefly nitrogen and/or nitrous oxide, resulting in the loss of nitrogen into the atmosphere and therefore undesirable in agriculture *chem agr eco*

density gradient centrifugation: the separation of macromolecules or subcellular particles by sedimentation through a gradient of increasing density under the influence of a centrifugal force; the density gradient may either be formed before the centrifugation run by mixing two solutions of different density (e.g., in sucrose density gradients), or it can be formed by the process of centrifugation itself (e.g., in $CsCl_2$ and Cs_2SO_4 density gradients) *meth*

dent corn: a variety of maize, *Zea mays* ssp. *indentata,* having yellow or white kernels that become indented as they ripen *agr*

deoxyribonuclease: any of several enzymes that break down the DNA molecule into its component nucleotides *phys*

deoxyribonucleic acid (DNA): a nucleic acid, characterized by the presence of a sugar deoxyribose, the pyrimidine bases cytosine and thymine, and the purine bases adenine and guanine; its sequence of paired bases constitute the genetic code *chem gene*

dephosphorylation: the removal of a phosphate group from an organic compound, as in the changing of ATP to ADP *chem phys*

derivative hybrid: a hybrid arising from a certain cross between two hybrids

dermatogen: a specialized meristem in flowering plants in which floral induction begins; gives rise also to the epidermis *bot*

descendant: an individual resulting from the sexual reproduction of one parental pair of individuals *gene*

descent: the act, process or fact of descending *gene*

descriptor: an identifiable and measurable characteristic used to facilitate data classification, storage, retrieval, and use *stat*

desiccant: a chemical applied to crops that prematurely kills their vegetative growth; often used for legume seed crops so the seed can be harvested prior to normal plant senescence *phys meth* >>> defoliation >>> chemical desiccation

desiccate >>> desiccation

desiccation: the process of drying out *phys* >>> chemical desiccation

desmosome: a plaquelike site on cell surfaces that function in maintaining cohesion with an adjacent cell *bot biot*

desynapsis: the premature separation of paired chromosomes during diplotene or diakinesis of meiotic prophase; it is often genetically controlled but can also be induced by special environmental conditions (e.g., heat) *cyto*

desyndesis >>> desynapsis

detasseling: artificially removing (cutting or pulling) the tassel of the female parent to prevent selfing during hybrid seed maize production *meth seed*

detergent: any synthetic organic cleaning agent that is liquid or water-soluble and has wetting-agent and emulsifying properties *chem*

determinate: descriptive of an inflorescence in which the terminal flower opens first, thus arresting the prolongation of the floral axis *bot*

deterministic process >>> stochastic

detoxicate >>> detoxication

detoxication: the metabolic process by which toxins are changed into less toxic or more readily excreted substances *phys*

developmental cycle: the gradual progression of phenotypic modifications of an organism during development *phys*

developmental genetics: the study of mutations that produce developmental abnormalities in order to gain understanding of how normal genes control growth, form, behavior, etc. *gene*

developmental stage >>> growth stages

devernalization: the reversion of vernalization by nonvernalizing temperatures or other means *phys* >>> vernalization

deviation: the departure of a quantity from its expected value *stat*

dextrin(e): a soluble gummy substance, formed from starch by the action of heat, acids, or enzymes *chem phys*

dextrose: an aldohexose monosaccharide that is a major intermediate compound in cellular metabolism; the dextrorotatory form of glucose, occurring in fruits and commercially obtainable from starch by acid hydrolysis *chem phys*

DF >>> degree of freedom

DFP >>> DNA fingerprint(ing)

DH lines >>> doubled-haploid lines

diadelphous: showing stamens united in two sets by their filaments (e.g., in pea and bean flowers); nine out of ten stamens are usually united while one is by itself *bot*

diakinesis: the last stage in the prophase of meiosis I, when the paired homologous chromosomes are highly contracted but before they have moved onto the metaphase plate *cyto* >>> Figure 15

diallel: in either the complete or incomplete diallel, identities of both seed and pollen parents are maintained for each family *meth* >>> diallel cross >>> complete diallel >>> incomplete diallel

diallel cross: the crossing in all possible combinations of a series of genotypes *meth* >>> Table 24

diallel crossing group >>> crossing group(s)

diallel mating >>> diallel cross

diallelic >>> diallel cross

diapause: a period of dormancy *phys;* in insects, a state during which growth and development is temporarily arrested *zoo*

dibber >>> dibble

dibble: a pointed tool used to make holes in the ground for seeds and seedlings *prep hort* >>> pricking-out peg

dibble planting >>> dibbling

dibbling (seed): sowing in holes made by a pointed tool *meth hort fore*

dicentric: a chromosome or chromatid with two centromeres *cyto* >>> Figure 11

2,4-dichlorophenoxy acetic acid (2,4-D): a crystalline powder, $C_8H_6O_3Cl_2$, used for killing weeds and as a growth hormone *phys biot*

dichogamy: the condition in which male and female parts of a flower mature at different times *bot*

dichophase: the phase of the mitotic cycle in which a cell is determined for further mitotic differentiation or for special cell functions *cyto*

dichotomous ramification: branching, frequently successively, into two more or less equal arms *bot*

diclinous species: species having pistils and stamens on different flowers *bot* >>> Table 32

dicot: an abbreviated name for dicotyledon, which refers to plants having two seed leaves *bot* >>> dicotyledons

dicotyledonous >>> dicotyledons

dicotyledons: plant species having two cotyledons; flower parts arranged in fours or fives or multiples thereof; net-veined leaves and vascular bundles in the stem, arranged in a ring *bot* >>> Table 32

didiploid *syn* **allotetraploid:** two different diploid chromosome sets present in one cell or organism *cyto*

dieback: the death of tips or shoots due to damage or disease *phys*

differential centrifugation: a method(s) of separating subcellular particles by centrifugation of cell extracts at successively higher speeds; it is based on differences in sedimentation coefficients that are roughly proportional to particle size; in other words, large particles (nuclei, chloroplasts, or mitochondria) are sedimented at lower speeds than small particles (ribosomes) *prep meth biot*

differential chromosome staining: differential staining of euchromatic and heterochromatic chromosome regions by specific pretreatment and dyes; it leads to characteristic banding patterns, which often allow the identification of single chromosomes, genomes, or chromosome segments *cyto micr* >>> banding

differential host: the special species of varieties of host plants whose reactions are used for determining physiologic races *phyt* >>> host

differential medium: an in vitro cultural medium with an indicator (e.g., a special dye), which allows various chemical reactions to be distinguished during plant growth *meth*

differential selection: the difference between a selected plant, family, or clone and the average of the population from which it is taken *meth*

differential staining: in microbiology, staining procedures that divide bacteria into separate groups based on staining properties; in cytology, staining procedures that divide chromosomes or genomes into separate segments based on structural and biochemical properties *meth cyto*

differential variety: a host variety, part of a set differing in disease reaction, used to identify physiologically specialized forms of pathogen *phyt* >>> differential host

differentiation of cells: the development of specialized kinds of cells from nonspecialized cells in a growing tissue *bot phys*

diffuse stage: a meiotic prophase stage in which the chromosomes become reorganized; they may almost disappear *cyto*

dig: to break up and turn over piecemeal the soil or ground *meth agr*

digametic >>> heterogam(et)ic

digenic: it refers to an inheritance that is determined by two genes *gene*

digestibility: the attribute of forage biomass to be digested by grazing animals—an important selection criterion in forage crop breeding *agr* >>> Table 33

digitate: fingerlike; or a compound, with the members arising together at the apex of the support *bot*

digoxigenin (DIG): antigenic alkaloid from *Digitalis* spp. (foxglove), which is used to label DNA *biot* >>> in situ hybridization

digynoid: in parthenogenesis, the progeny derives from an unreduced egg cell *bot*

dihaploid: a haploid cell or individual containing two haploid chromosome sets—not to be confused with doubled-haploid *cyto*

dihybrid: a cross between individuals that differ with respect to two specified gene pairs *gene*

diisosomic: a cell or an individual, which has a pair of homologous isochromosomes for one arm of a particular chromosome *cyto* >>> aneuploid >>> Figure 37

diisotrisomic: a cell or an individual that lacks one chromosome but carries two homologous isochromosomes of one arm of a particular chromosome *cyto* >>> aneuploids >>> Figure 37

dikaryon: a dinucleate cell *cyto*

dikaryotic: a cell showing two nuclei *bot*

dimer: a protein that is made up of two polypeptide chains or subunits paired together *chem*

dimerization >>> dimer

dimethyl sulfoxide (DMSO): a liquid solvent, C_2H_6OS, approved for better penetration of specific substances through the cell wall *chem meth*

dimorph >>> dimorphism

dimorphism: the occurrence of two forms of individuals within one population or other taxa (e.g., sexual dimorphism or the presence of one or more morphological differences that divide a species into two groups) *bot*

dioecious: possessing male and female flowers or other reproductive organs on separate, unisexual, individual plants (e.g., in hemp or spinach) *bot*

dioecism: the phenomenon of plants showing either male or female sex organs *bot*

diphasic: chromosomes that show both euchromatic and heterochromatic segments *bot*

diphyletic: a group of species that share two ancestries *evol* >>> monophyletic >>> polyphyletic

diplandroid: in parthenogenesis, the progeny derives from an unreduced sperm cell *bot*

diploid: a cell with two chromosome sets or an individual with two chromosome sets in each cell; a diploid state is written as "2n" to distinguish it from the haploid state "n" *gene* >>> Table 1 >>> Figure 8

diploid parthenogenesis: a type of gametophyte apomixis by which a diploid embryo sac cell results in a diploid embryo; the diploid condition is the result of cytogenetic mechanisms occurring in the egg stage *bot* >>> Figure 28

diploidization: in polyploids, a natural or induced mechanism in which the chromosomes pair completely or partially as bivalents, although polyploid sets of chromosomes are present; it may be caused by a structural differentiation of homologous chromosome sets or by genetic control; for example, in bread wheat three homoeologous genomes are available (AABBDD); they do not pair as hexavalents but exclusively as bivalents *cyto*

diploidizing mechanism: a mechanism whereby the chromosomes of a polyploid sometimes form exclusively or partly bivalents instead of multivalents during meiosis; usually it is under strictly genetic control (e.g., the

Ph [pairing homologous] locus on chromosome 5B of hexaploid wheat) *cyto*

diploidy: the presence of two homologous sets of chromosomes in somatic cells *cyto* >>> Table 1 >>> Figure 8

diplonema >>> diplotene

diplontic >>> diploid

diplophase: the diploid generation phase after fertilization to meiosis *cyto*

diplospory: a type of agamospermy (apomixis) in which a diploid embryo sac is formed from archesporial origin *bot* >>> Figure 28

diplotene: the stage in the prophase of first meiosis, when the paired homologous chromosomes separate except where they are held together by chiasmata *cyto*

dipping: the immersion of seedling roots in a solution or water prior to planting *fore hort meth*

direct embryogensis: embryoid formation directly on the surface of zygotic or somatic embryos or on seedling plant tissues in culture without an intervening callus phase *biot*

direct organogenesis: organ formation directly on the surface of relatively large intact explants without an intervening callus phase *biot*

directed dominance >>> directional dominance >>> dominance

directional cloning: in biotechnology, DNA inserts and vector molecules are digested with two different restriction enzymes to create noncomplementary sticky ends at either end of each restriction fragment; it allows the insert to be ligated to the vector in a specific orientation and prevents the vector from recircularizing *biot*

directional dominance: a type of dominance in which the majority of dominant alleles have positive effects in one direction *gene*

directional mutation: a genetic change that favors a certain genotype or population *gene*

directional selection: selection resulting in a shift in the population mean in the direction desired by the breeder *meth*

dirty seed: endophyte-infected seeds (e.g., in grass) *seed* >>> clean seed

disaccharide: any of a group of carbohydrates, as sucrose, that yield monosaccharides on hydrolysis *chem phys*

disassortative mating: occurs if the plants mating resemble each other less than plants belonging to pairs of random plants, with regard to some trait *gene* >>> assortative mating

disbud: removing buds, shoots, or growing tips (with finger and thumb) of a tree, vine, or flowering plant to encourage production of sideshoots or high-quality flowers and fruits; pinching out is also used when small side shoots are completely removed; it is done when single stems are desired, especially when training to form the "trunks" of standard (tree-form) specimens *hort*

disc floret: a small flower, usually one of a dense cluster (e.g., sunflower) *bot*

disc flower: the tubular flowers of the head, as distinct from the ray (e.g., in Asteraceae or sunflower) *bot*

disc grain-grader: discs revolve through a seed mass and a certain size of seeds are lifted and discharged, while the other size (e.g., the longer ones) is rejected by the disc indents *seed*

disc plough >>> disk plough

discontinuous character: variation in which discrete classes can easily be recognized (e.g., flower color, straw length) *gene* >>> qualitative character

disease: a condition in which the use or structure of any part of a living organism is not normal; harmful deviation from normal functioning of physiological processes; six types of causal agents can be considered: (1) fungi, (2) bacteria, (3) viruses, (4) nematodes, (5) insects, and (6) plant parasites *phyt* >>> seed-borne pathogens

disease avoidance: avoiding the disease by, for example, growing the crop sufficiently early so that the vulnerable part of the plant's growing cycle is over before the disease-causing organism arrives in the area; this stops the disease from starting; it is sometimes called "passive resistance" *phyt*

disease control: several types of disease control can be classified: (1) disease resistance, (2) protection, (3) avoidance, (4) exclusion, (5) eradication, (6) therapy *phyt*

disease cycle: a cyclical sequence of host and parasite development and interaction that result in disease and in reproduction of the pathogen *phyt*

disease eradication: this control measure is applied to a situation in which the disease is present in the area; it involves removing the disease by, for example, burning all stubble of the diseased crop, in order to prevent transfer of the disease from the previous crop to the next season *phyt*

disease protection: involves protecting the plant with a chemical; this is applied before the disease starts and prevents the beginning of problems; systemic fungicides can penetrate and move inside the plant; they therefore have a greater exposure to the pathogenic organism; some are taken up by roots, other by leaves *phyt* >>> systemic pesticide

disease resistance: the ability to resist disease or the agent of disease and to remain healthy *phyt*

disease therapy: this applies to removing the particular part of the plant that is diseased; often applied to large and valuable plants, such as fruit trees, where a diseased branch may be removed by a tree surgeon *phyt*

disinfectant: a chemical treatment used to disinfect seed for planting; it is commonly useful for surface-borne pathogens *seed meth*

disjunction (of daughter chromosomes): the separation of homologous chromosomes at the anaphase stage of mitosis and meiosis, and movement toward the poles of nuclear spindle *cyto*

disjunctional separation >>> alternative disjunction

disk: enlarged growth of the head made up of a circular arrangement of fused petals (e.g., in sunflower) *bot* >>> disc flower

disk flower >>> disc flower

disk plough: a plough with saucer-shaped units for breaking the soil *agr*

dislocation: the displacement of a chromosome segment away from its original position in the chromosome *cyto*

disome: a cell or an individual showing the two homologous sets of chromosomes *cyto* >>> aneuploid >>> Figure 37

disomic >>> disome

disomy >>> disome

dispermy: the entering of two sperm cells into one egg cell *bot*

dispersal: the spread of a pathogen within an area of its graphical range *phyt*

dispersing agent: a chemical added to a pesticide formulation to aid the efficient distribution of particles of the active ingredient *phyt prep*

disporic: having two spores *bot*

disruptive selection: a selection that changes the frequency of alleles in a divergent manner, leading to the fixation of alternative alleles in members of the population; the result after several generations of selection should be two divergent phenotypic extremes within the population *meth*

dissecting microscope: usually, a low-power microscope (50x magnification) used to facilitate dissection, examination, or excision of small plant parts; however, in recent biotechnology high-power microscopes also are applied for dissections *prep*

dissepiment: a partition within an organ of a plant (e.g., the membrane that separates sections of the orange and other citrus fruits) *bot*

dissemination: the spread of seeds or spores *eco* >>> dispersal

dissimilation >>> assimilation

distal: farthest from the point by which it is attached to the starting point *gene cyto*

distant hybridization: the crossing and/or hybridization of members of different genera *meth*

distichous: in two vertical ranks *bot*

distinguishable hybrid: a type of hybrid in which intermediate inheritance is phenotypically expressed (i.e., the heterozygous gene constitution is visible by the phenotype) *gene*

distyly: the presence of either pin or thrum flowers; pin x pin and thrum x thrum crosses are incompatible due to alleles at a single locus; the thrum morphology is controlled by a dominant allele *S* and the pin morphology by the recessive alleles *s bot* >>> self-incompatibility

disulfide bridge: a covalent bond formed between two sulfur atoms; it is a particular feature of peptides and proteins, where it is formed between the sulfydryl groups of two cysteine residues, helping to stabilize the tertiary structure of these compounds *chem*

ditelocentric >>> ditelosomic

ditelomonotelosomic: a cell or an individual that has a pair of telocentric chromosomes for one arm and a single telocentric chromosome from the other arm *cyto* >>> aneuploid >>> Figure 37

ditelosomic: a cell or an individual that has two telocentrics of one chromosome arm *cyto* >>> aneuploid >>> Figure 37

ditelotrisomic: a cell or an individual that has two telocentric chromosomes of one arm plus one complete chromosome of the homologue *cyto* >>> aneuploid >>> Figure 37

ditertiary compensating trisomic: a cell or an individual with a compensating trisomic chromosome in which a missing chromosome is compensated by two tertiary chromosomes *cyto* >>> Figure 14

diurnal rhythm >>> circadian rhythm

divalent >>> bivalent

divergent: set at an angle to one another herring-bone fashion, as in the lateral spikelets of the two-row barley spike *bot*

diversifying selection: selection in which two or more genotypes show optimal adaptation under different environments *meth*

division: a method of propagation by which a plant clump is lifted and divided into separate pieces; it includes roots and a growing point *meth hort*

DNA >>> deoxyribonucleic acid

DNA clone: a section of DNA that has been inserted into a bacterium, phage, or plasmid vector and has been replicated many times *gene*

DNA content: usually, the total DNA amount per nucleus, given as picograms, *cf* Important List of Crop Plants *cyto*

DNA fingerprint(ing) (DFP): the unique pattern of DNA fragments identified by Southern hybridization (using a probe that binds to a polymorphic region of DNA) or by polymerase chain reaction (PCR) using primers flanking the polymorphic region *meth gene* >>> DNA-amplified fingerprinting

DNA hybridization: base pairing of DNA from two different sources; in biotechnology, a technique for selectively binding specific segments of single-stranded (ss) DNA or RNA by base pairing to complementary sequences of ssDNA molecules that are trapped on a nitrocellulose membrane *gene biot*

DNA library >>> clone library

DNA ligase (polynucleotide ligase): an enzyme that creates a phosphodiester bond between the 5'-PO_4 end of one polynucleotide and the 3'-OH end of another, thereby producing a single, larger polynucleotide *biot*

DNA methylation: the methylation of DNA bases by endogenic methylases *gene chem*

DNA polymerase: any enzyme that catalyzes the synthesis of DNA strands from deoxyribonucleoside triphosphates using single-stranded DNA as a template *gene phys*

DNA polymerase I (KORNBERG enzyme): used for nick translation and for the production of the KLENOW fragment *biot*

DNA polymorphism: one of two or more alternate forms (alleles) of a chromosomal locus that differ in nucleotide sequence or have variable numbers of repeated nucleotide units *biot*

DNA probe: a more or less defined piece of DNA that is used for DNA-DNA or DNA-RNA hybridization experiments *gene meth*

DNA repair: the reconstruction of DNA molecule after different sorts of DNA-strand damages by endogenic enzymes; it is involved in the recombination process and promotes the survival of an organism after partial DNA damage *gene*

DNA replication: the process whereby a copy of a DNA molecule is made, and thus the genetic information it contains is duplicated; the parental double-stranded DNA molecule is replicated semiconservatively (i.e., each copy contains one of the original strands paired with a newly synthesized strand that is complementary in terms of AT and GC base pairing) *gene*

DNA sequencing: methods and procedures for determining the nucleotide sequence of a DNA fragment and/or chromosome *biot gene*

DNA-amplified fingerprinting: a technology based on amplification of random genomic DNA sequences achieved by a single short (5-8 bases) oligonucleotide primer of arbitrary sequence; it produces a characteristic spectrum of short DNA pieces of varying complexity that are resolved on polyacrylamide gel (PAGE) following silver staining; it is used to detect genetic differences between genotypes as well as for detecting polymorphism even between organisms that are closely related, such as near isogenic lines *biot meth* >>> DNA fingerprint(ing)

DNA-DNA hybridization: the annealing of two complementary DNA strands to produce hybrid nucleic acid molecules; it is used to identify the base sequences in two polynucleotide chains from different sources *gene meth*

DNase: nuclease specific for DNA *biot* >>> deoxyribonuclease

DNase I: an endonuclease that makes random single-stranded nicks in DNA; used for nick translation *biot*

dome: in cereals, the zone of cells at the tip of the apical meristem (shoot apex), which, by cell division, forms the site for production of leaf and spikelet primordia *bot* >>> meristematic tip

domesticated populations >>> domestication

domestication: the selective breeding by humans of species in order to accommodate human needs *evol*

dominance: the quality of one of a pair of alleles that completely suppresses the expression of the other member of the pair when both are present; the degree of dominance is expressed by the ratio of additive genetic variance to total phenotypic variance; in case the ratio equals 1, the trait shows complete dominance, if the ratio is greater than 1, the trait shows overdominance, if the ratio is less than 1, the trait shows incomplete dominance *gene* >>> Tables 2, 6, 9, 20, 21

dominance hypothesis: in hybrid breeding, dominant alleles of genes that should have stimulating effects on heterosis, while recessive ones should show inhibitory effects *gene* >>> dominance >>> Figure 18

dominance variance: that portion of the genetic variance attributable to dominant gene effects *stat* >>> dominance

dominance-recessiveness relation >>> dominant

dominant: in diploid organisms, a gene that produces the same phenotypic character when its alleles are present in a single dose (heterozygous) per nucleus, as it does in a double dose (homozygous); a gene that is masked in the presence of its dominant allele in the heterozygous state is called to be recessive to that dominant *gene* >>> dominance >>> Figures 6, 18 >>> Tables 6, 20, 21

dominant epistasis: one dominant factor *A* is epistatic of another factor *B*, or *B* is hypostatic to *A* *gene* >>> Table 6

donator >>> donor

donor: an individual, line, population, or variety from whom pollen or genetic material is used for transfer to another *meth*

donor parent: the parent from which one or a few genes are transferred to the recurrent parent in backcross breeding *meth*

donor plant: the source plant used for propagation, crossing, etc., whether a simple individual, an explant, graft, or cutting *meth*

dormancy: a resting condition with reduced metabolic rate found in nongerminating seeds and nongrowing buds *phys*

dormant >>> dormancy

dormant bud >>> dormancy

dormant seeding: sowing during late autumn or early winter after temperatures become too low for seed germination to occur until the following spring *meth agr*

dormant spray: a pesticide applied to dormant, leafless plants to control insects and diseases *phyt*

dorsal: in general, upon or relating to the back or outer surface of an organ (abaxial); the side of the caryopsis on which the embryo is situated *bot* >>> adaxial

dosage compensation: a genetic process that compensates for genes that exist in two doses in the homozygous dominants, so that the heterozygotes produce the same amount of gene product as the homozygotes *gene*

dosage effect: the influence upon a phenotype of the number of times a genetic element is present *gene*

dose effect >>> dosage effect

dot-blot analysis: a variant of the Southern transfer; different concentrations of RNA or DNA may be determined; nonradioactive DNA will be denatured and with different concentrations transferred to nitrocellulose filters; only small dots are transferred; that DNA may hybridize with radioactively labeled probe; after autoradiography, the intensity of blackness is used as a simple measure of DNA concentration or DNA homology *meth biot*

double cropping: the more or less contemporary growing of two crops on the same field; for example, it might be to harvest a wheat crop by early summer and then plant maize or soybeans on that acreage for harvest in autumn; this practice is only possible in regions with long growing seasons *agr*

double cross: a cross between two F1 hybrids; the method used for producing hybrid seed; four different lines (A, B, C, D) are used; A x B → AB hybrid and C x D → CD hybrid; the single-cross hybrids (AB and CD) are then crossed and the double-cross hybrid (ABCD) seed is used for the commercial crop *meth seed* >>> Figure 31

double crossing-over: the situation in which two crossing-overs take place within a tetrad *cyto gene*

double digestion: cleavage of a DNA molecule with two different restriction enzymes *biot*

double ditelocentric: the phenomenon in which both arms of a certain chromosome are present as telocentrics, and each telocentric with its homologue *cyto* >>> Figure 37

double fertilization: the union of one sperm nucleus with the egg nucleus to form the diploid zygote and of the other sperm nucleus with the two polar nuclei to form a triploid endosperm nucleus; the male gametophyte, pollen grain plus pollen tube, actually contains three sperm nuclei, but one, the vegetative nucleus, degenerates once double fertilization has been accomplished *cyto* >>> Figure 35

double flower: flowers that have more than one row of petals; stamens and sometimes the pistils are transformed into petals or sometimes the petals split to form several more; completely double flowers have usually lost their reproductive organs and are therefore unable to produce seeds; these varieties of plants are bred to stay fresh longer than single flowers when cut; commonly, plants do not show double flowers; they can only be formed when three specific genes are simultaneously mutating; these genes are the main regulators for flower formation; their DNA sequences are almost identical; by induced mutations and subsequent combination of those genes double flowers can be induced (e.g., in *Arabidopsis*) *bot gene hort*

double helix: a structure of DNA consisting of two helices around a common axis *chem gene*

double hybrid >>> double cross

double monoisosomic: the presence of two isochromosomes, one for each arm; it can derive from a double monotelosomic individual *cyto* >>> Figure 37

double monotelosomic: the presence of two telocentrics, one for each arm *cyto*

double recessive: an individual that is homozygous for a recessive allele (e.g., *aa*), as opposed to homozygous for a dominant allele (e.g., *AA*) *gene*

double reduction: the genetic outcome of chromatid segregation, as opposed to chromosome segregation, whereby two sister-chromatid segments are included in the same meiotic product *cyto*

double telotrisomic: the presence of two telocentrics, one for each arm of a missing whole chromosome, but together with the complete homologue *cyto* >>> aneuploid >>> Figure 37

double-cross hybrids: hybrids resulting from crossing two single cross hybrids *seed* >>> double cross >>> Figures 22, 31

doubled haploid: a diploid plant that results from spontaneous or induced chromosome doubling of a haploid cell or plant, usually after anther or microspore culure by using different means *biot* >>> Figures 17, 26 >>> Table 7

doubled haploid (DH) method: a method used to speed up the production of homozygotes and to decrease the population size for selection; in other words, generating haploid plants by parthenogenesis or by anther culture followed by doubling of the number of their chromosomes (spontaneously or induced) *biot meth* >>> doubled-haploid lines (DH lines) >>> Figures 17, 26 >>> Table 7

doubled-haploid lines (DH lines): homozygous lines derived from haploidization and doubling again the chromosome number *biot meth* >>> doubled haploid (DH) method >>> Figures 17, 26 >>> Table 7

doubling time: the average time required to double the number of individuals of a population *gene*

dough quality >>> "strong" flour

dough strength >>> "strong" flour

downstream: a term used for description of the position of a DNA sequence within a DNA or protein molecule; it means that the position of the sequence lies toward the direction of the synthesis of a DNA or protein molecule *gene*

downy: covered with soft hairs or down *bot*

downy mildew (of rape): the most common rapeseed disease *(Peronospora parasitica);* it causes yellow discoloration of the upper leaf surface and white fungal growth on the lower surface; though most severe in the autumn, the disease can reach high levels in cool wet springs; severe infections limit the ability of small plants to respond to good growing conditions *phyt*

drain: in order to remove water from the soil by artificial means (e.g., drainage ditches, buried perforated plastic pipes, or a gravel sump) *meth agr*

drainage: excess of water can be harmful to crop production; wet soils are usually low in temperature and low in oxygen content; drainage can be facilitated by open ditches or different subsoil drains *agr*

dressing: manure, compost, or other fertilizers *agr*

drift: changes in gene and genotypic frequencies in small populations due to random processes *gene*

drill: a machine for seeding crops by dropping them in rows and covering them with earth; the term is also used for a row of seeds deposited in the earth (i.e., the trench or channel in which the seeds are deposited); in general, to sow seeds in rows (i.e., the field was drilled, not sown broadcast) *agr*

drill-row >>> drill

drip irrigation: a system of watering by which moisture running through a porous hose; the water is slowly released through tiny holes or emitters to the plant roots; it is one of the most efficient of irrigation technologies *agr*

drought hardening: adapting plants to survive periods of time with little or no water by stepwise reducing water supply or germinating and/or growing under insufficient moisture conditions *meth* >>> Table 33

drought stress >>> stress protein(s)

drought-tolerant: plants that can survive periods of time with little or no water *agr* >>> Table 33

drupe: a fleshy fruit, such as a plum, cherry, coconut, walnut, peach, or olive, containing one or more seeds, each enclosed in a stony layer that is part of the fruit wall (hard endocarp) *bot*

dry farming: a method of farming in arid and semiarid areas receiving less than 500 mm rainfall the year without using irrigation; the land being treated so as to conserve moisture; the technique consists of cultivating a given area in alternate years, allowing moisture to be stored in the fallow year; moisture losses are reduced by producing a mulch and removal of weeds *agr*

dry matter (DM): the substance in a plant or plant material remaining after oven drying to a constant weight at a temperature slightly above the boiling point of water *meth*

dry rot: a disease characterized by the formation of dry, shriveled lesions, often caused by the fungus *Phoma lingam phyt*

dry season: a period each year during which there is little precipitation *eco*

dry weight: moisture-free weight *meth agr*

dsDNA: double-stranded DNA *biot*

dummy trial >>> blindfold (trial)

duplex type: a polyploid plant that shows two dominant alleles for a given locus *gene* >>> nulliplex type >>> autotetraploid >>> Table 3

duplicate genes: two or more pairs of genes in one diploid individual, which alone or together produce identical effects *gene* >>> Figure 36

duplication: a chromosomal aberration in which more than one copy of a particular chromosomal segment is produced within a chromosome set *gene* >>> Figure 36

duplication of germplasm: a duplicated seed sample that is prepared for safety reasons; usually the two samples are kept at different locations *seed*

durable resistance: resistance that remains effective and stable during the agronomic life of a crop; this type of resistance is usually determined by several genes and is a main target of resistance breeding *phyt*

DUS testing: the methods and standards of the identification and description of varieties are elaborated by different teams of the international organization UPOV established for the protection of plant varieties; the so-called TG/01/2/1/ contains the most important prescriptions; these procedures, prescriptions, and methods are called DUS testing (D = distinctness, U = uniformity, S = stability); for the legal protection of varieties, the candidates (1) have to have names that have not yet been used for the registration of recognized varieties, (2) must have been discovered distinct of the other "known" varieties (D), (3) have to be uniform and homogeneous (U), and (4) have to be stable (S); the characteristics involved in DUS testing can be divided into two main types: (a) measured characteristics and (b) visually observed (bonitated, qualitative) characteristics; the measured characteristics have values on the continuous scale; visually observed data have values in the interval 1, 2 to 9; in other words, visually observed traits are of the so-called ordinal type, while measured data are on the so-called interval or ratio scale; phenological data must also be mentioned as a special type of measured characteristic (e.g., number of days until flowering, etc.) *seed*

dust: a method of pesticide control in which a dry substance is applied by spraying seeds, tubers, or bulbs *phyt*

dust separator >>> aspirator

duster: a device consisting of a bin, a wand with a nozzle, and a crank mechanism; it is used for applying dust or powder to plants *phyt meth*

dusting >>> dust

dwarf fruit trees: a small fruit tree reaching a height at maturity of 150-200 cm; bred for convenient harvest technology as well as bearing early and normal fruits *meth hort* >>> Table 33

dwarfing gene: one category of genes that control the height of the plant *gene* >>> Table 33

dyad: a pair of cells (i.e., one of the products of the disjunction of the tetrads at the first meiotic division, contained in the nuclei of secondary gametocytes) *cyto*

dyeing flowers: blossoms capable of producing dyes *hort*

dysgenic >>> eugenic

dysploid: a plant or species in which the chromosome number is more or less than the expected normal euploid number *cyto*

dysploidy: abnormal ploidy (e.g., the appearance of diploid or triploid individuals in a normally tetraploid population or of triploid and tetraploid ones in a normally diploid population) *cyto* >>> anisoploid

E

ear >>> spike

ear lifter (on a harvester): improves cutting efficiency with laid crops or overhanging ears *agr*

early generation test: selection schemes in which poor recombinants are discarded already in F2 and F3 generations *meth*

earsh >>> stubble

ear-to-row planting >>> ear-to-row selection

ear-to-row selection: a separate growing of progenies (i.e., the separate sowing of lines or families); a procedure developed by the German breeder T. ROEMER (*syn* Ohio method) *meth* >>> Illinois method

echinate: having sharply pointed spines *bot* >>> prickly

ecological dominance: the state in plant communities in which one or more species, by their size, number, or coverage, exert considerable influence or control over the other species *eco*

ecological niche: the position occupied by a plant in its community with reference both to its utilization of its environment and its required associations with other organisms *eco*

ecology: the study of the interrelationships between individual organisms and between organisms and their environment *eco*

economic trait loci (ECL): sites in the genome that determine characteristics of economic importance *gene*

ecophene: the range of different phenotypes produced by one genotype within a certain environment *eco*

ecospecies: a locally adapted species; it shows minor changes of morphology and physiology to another species, which are related to a habitat and are genetically determined *eco*

ecosystem: the complex of an ecological community together with a biological component of the environment, which function together as a stable system *eco*

ecotype: a locally adapted population of a widespread species; it shows minor changes of morphology and physiology that are related to a habitat and are genetically determined; the individuals of an ecotype are only uniform in regard to the traits that provide them with special adaptation to specific environments; all the other characters may vary; ecotypes may be found in perennial clover, alfalfa, grasses, or other forage crops *eco*

ecovalence: this parameter is a quantitative measure for the evaluation of ecological adaptability *eco*

ectoparasite: an external parasite *phyt* >>> opuntia

ectosite >>> ectoparasite

ectozoon >>> ectoparasite

edaphon: all (micro)organisms living in the soil close to the plant *agr*

eddish >>> stubble

edging: a row of plants set along the border of a plot or flower bed *agr meth*

EDTA >>> ethylenediamine tetra acetic acid

eelworm: in cereals, there are two species of eelworm attacking plants, the cereal cyst nematode *(Heterodera avenae),* which can infest cereal crops, and the oat stem eelworm *(Ditylenchus dipsaci),* which attacks oats and rye, as well as several other crops and weeds *phyt* >>> nematode

effective breeding population (size): in general, the size of population, which is adjusted mathematically to permit comparisons with others; more specifically, it is the size of an equivalent ideal population, which is expected to experience the same increase in homozygosity over time (i.e., drift) as the population in question; the ideal population is one in which mating is at random in the absence of selection and in which all individuals have the same expected contribution to the next generation *meth*

effective seedling: any seedling that has survived in reasonable vigor for some arbitrary time and is so sited that it should make an effective contribution to the crop *agr fore*

efficiency of plating: number of plaques formed by a phage lysate; it can be different for different hosts *biot*

efflorescence: the time or state of flowering; in general, blooming or flowering *bot* >>> anthesis

egg: a female gamete or germ (egg) cell *bot* >>> Figure 25

egg apparatus: the structure containing the embryo sac, within which is the egg cell *bot* >>> Figure 25

egg cell >>> egg

egg mother cell: a megasporocyte from which an egg cell is derived during megasporogenesis *bot*

electrical field fusion >>> cell fusion

electrofusion >>> cell fusion

electromorphs: allozymes that can be distinguished by electrophoresis *phys chem*

electron microscope: a microscope that permits magnification of particles up to 200,000 diameters; instead of having the specimen exposed to a light source, a stream of electrons is directed on the object; the higher resolving power of the electron microscope is largely the result of the shorter wavelength associated with electrons; the electrons are accelerated in a high vacuum through electromagnetic lenses and focused on the specimen; they are projected on a fluorescent screen where the image of the particle may be viewed or onto a photographic plate and/or film *micr*

electropermeabilization >>> electroporation

electrophoresis: the migration of charged particles under the influence of an electric field within a stationary liquid; the latter may be a normal solution or held upon a porous medium such as starch acrylamide gel or cellulose acetate *meth*

electroporation: the application of a short sharp electrical shock to protoplasts in order to force the incorporation or uptake of DNA by producing transient holes in the cellulose membrane *biot*

electrostatic separator: a machine that separates seed on the basis of their ability to accept and retain an electrical charge *seed*

elicitor: a molecule produced by the pathogen host (or pathogen) that induces a response by the pathogen (or host) *phyt*

elimination >>> culling

ELISA >>> enzyme-linked immunosorbent test

elite: high-grade seed used for both seed and ware production; or, an agronomically superior and high-performing local cultivar *seed* >>> base seed

elite germplasm: germplasm that is adapted (i.e., selectively bred and optimized to new environment); for example, maize, which is native to Mexico, has been adapted to many locations in the world *gene seed*

elite hybrid variety >>> elite

elite plants >>> elite

elite seed: the class of pedigreed seed corresponding to the foundation seed class in some countries *seed* >>> base seed

elite strains >>> elite

elite tree: a tree that has been shown by progeny testing to produce superior offspring *meth hort fore*

eluate: a liquid solution resulting from eluting *prep*

emasculate: to remove the anthers from a bud or flower before pollen is shed; a normal preliminary step in crossing to prevent self-pollination; there are basically two methods used for emasculation: individual emasculation and mass emasculation; mass emasculation is applied, for example, in monoecious maize by mechanical detasseling, in hemp by removing male plants from the dioecious population, in rice by spike treatment with high temperatures (warm water of about +42°C), in wheat by using gametocides, or in tobacco by dark-phase treatment during anthesis *meth*

emasculation >>> emasculate

EMBL3 vector: a replacement vector for the cloning of large (~20 kb) DNA fragments; it derives from phage lamda *biot* >>> lamda phage

embryo: the rudimentary plant within a seed that arises from the zygote, sometimes from an unfertilized egg cell, or from progressive differentiation in cell culture *bot* >>> Table 8

embryo culture: a method of inducing the artificial growth of embryos by excising the young embryos under septic conditions and placing them on suitable nutrient media; the method is often applied when postgamous incompatibility exists (e.g., in wide crosses) *biot*

embryo percent: the amount of embryo compared with endosperm and other seed parts, and/or the percent of embryo in the whole seed *agr meth*

embryo rescue >>> embryo culture

embryo sac: the female gametophyte is formed by the division of the haploid megaspore nucleus—the site of fertilization of the egg and development of the embryo *bot* >>> Figures 25, 28, 35

embryo sac development >>> embryo sac

embryo sac mother cell >>> embryo sac

embryogenesis: the formation of an embryo; after double fertilization, seed development begins with endosperm, embryo growth, and differentiation; during subsequent seed maturation the seed storage proteins, starch, and lipids accumulate before the seed undergoes desiccation and dormancy; during late embryogenesis endosperm and mature embryo express seed storage reserve genes; many other hydrophyllic proteins with certain function that share similar sequence repeats also are accumulated; they are called late embryogenesis-abundant proteins, protecting cells during seed desiccation (e.g., abscisic acid) *bot* >>> Table 8

embryogenic: related to or like an embryo *bot*

embryoid: an embryolike structure *bot biot*

embryonic leaves: initial leaves, which are easily discernible in the germ of a mature grain *bot*

emergence date: the time or date when plants start to emerge spikes, buds, flowers, or showing coleoptiles, first leaves, etc. *agr meth*

empty fruits: in seed production, a fruit without a seed, for example, it happens in the composite family (sunflower, chicory, lettuce, etc.); the so-called seed is a fruit (achene) of which the outer structures are comparable to the pod in the legumes; the true seed may or may not form inside the fruit, which is not (clearly) visible from outside *seed*

empty glume >>> spikelet glume

enantiomer: mirror image form of a molecule; there may be more than two, as several parts of the molecule may have a mirror image form *chem phys*

enation: an outgrowth from the surface of a leaf or other plant part *bot*

encapsidation (of viruses): the process when a virus forms the protein shell that surrounds the virus nucleic acid *phyt*

encapsulation: the process of enclosing fragile organic material in a protective casing, sometimes of a semisolid nature; it is used for planting or moving somatic embryos *seed meth biot*

endangered species: a species on the verge of extinction *eco*

endemic: confined to a given geographic region; diseases continuously occurring in a particular area *eco phyt*

endemic species >>> endemic

end-filling: conversion of a sticky end (5' overhang only) to a blunt end by enzymatic synthesis of the complement to the single-stranded overhang *biot*

end-labeling: attaching a radioactive or nonradioactive label specifically to the ends of a DNA molecule *biot*

endocarp: inner layer of the fruit wall (pericarp); it may be hard and stony (peach pit), membranous (apple core), or fleshy (orange pulp) *bot*

endocytosis: the uptake of cellular material through the cell membrane by formation of a vacuole *bot*

endodermis: a specialized tissue in the roots and stems of vascular plants, composed of a single layer of modified parenchyma cells forming the inner boundary of the cortex *bot*

endoduplication: the doubling of the haploid chromosome complement owing to the failure of cell-wall formation *cyto*

endogamy: sexual reproduction in which the mating partners are more or less closely related *bot* >>> inbreeding >>> intrabreeding

endogenous rhythm: a type of rhythmic plant response or growth capacity that is not affected by external stimuli *phys*

endomitosis: a doubling of the chromosomes within a nucleus that does not divide, thus producing a polyploid; the doubling may be repeated a number of times, which leads to endopolyploidy *cyto*

endonuclease: any enzyme that cuts DNA at specific sites corresponding to specific base sequences within a polynucleotide chain; each endonuclease cuts at its own specific site; it is used to identify genomes, genotypes, and chromosomes, and it acts as a genetic marker *chem phys*

endoplasm: the granular central material of the cytoplasm *bot*

endoplasmic reticulum: a system of minute tubules within the cytoplasm; two types are recognized, rough and smooth; these are particularly concerned with pathways of protein and steroid synthesis *bot*

endopolyploid: diploid individuals whose cells containing 4C, 8C, 16C, 32C, etc. amounts of DNA in their nuclei *gene cyto* >>> endomitosis >>> endoduplication

endoreduplication: chromosome reduplication in the interphase of the mitotic cell cycle *gene cyto*

endosperm: in grasses, the reserve food material in the caryopsis lying outside the embryo; it is a starchy tissue that is formed during grain development; it provides nourishment for the developing embryo and for the seedling after germination until it can establish itself; usually the endosperm is triploid and originates during the double fertilization when one of the two sperm nuclei is fused with the two polar nuclei *bot* >>> Tables 8, 16

endosperm texture: the tendency of, for example, wheat endosperm to fracture either along the outlines of the cells (hard) or across the cells in a more irregular way (soft); it can be assessed by "grinding time" or by inspection of the grain cross-section *bot meth*

endotoxin: a poison produced within a cell and released only when the cell disintegrates *phys*

engraft >>> ingraft

enhancer: a modifier gene that may enhance the action of another gene *gene*

enneaploid: a polyploid plant with nine chromosome sets *cyto*

entire leaf: a leaf margin that has no teeth (i.e., smooth) *bot*

entomologist: an insect specialist *phyt*

entomology: the branch of zoology dealing with insects *phyt*

entomophilous: insect-borne pollen; entomophilous pollen is usually characterized by a sticky, mainly liquid layer that fills in the interstices of the sculpted pollen wall, or exine, and is responsible for its adhesion to insect and other vectors; this liquid pollen coat, the tryphine, is also responsible for pollen adhesion to the stigmatic surface in many species *bot* >>> Table 35

enucleate (of cells): removing the nucleus from a cell *cyto biot*

environment: the sum of biotic and abiotic factors that surround and influence an organism *eco*

environmental mutagen: a substance that may act as a mutagen in the environment of an organism *gene eco*

environmental resistance: all characteristics that protect the reproduction of plants against negative influences of the environment *eco*

environmental variance (VE): that portion of the phenotypic variance caused by differences in the environments to which the individuals in a population have been exposed *stat*

enzyme: a molecule, wholly or largely protein, produced by a living cell that acts as a biological catalyst *phys*

enzyme-linked immunosorbent test (ELISA): a highly sensitive immuno assay based on enzyme reactions; usually two antibodies are used; the first antibody binds to an antigen and mutates itself toward the second antibody, the antiglobin; the antiglobin is linked to an enzyme (e.g., horseradish peroxidase); the enzyme activity has to be easily detected (e.g., by a staining reaction); the degree of enzyme reaction quantitatively shows how many antibodies of the first type (i.e., of the original antigen) is present *meth*

e.o.p. >>> efficiency of plating

epiallele(s): alternative states of a gene that have an identical DNA sequence but differ in methylation or chromatin structure and, hence, level of expression *gene*

epiblast: the primordial outer layer of a young embryo *bot*

epicarp: the outermost layer of a pericarp, as the rind or peel of certain fruits *bot*

epicotyl: the portion of the embryo or seedling above the cotyledons *bot*

epidemic: a rapid increase in disease over time and in a defined space *phyt*

epidemic potential: the biological capacity of a pathogen to cause disease in a particular environment *phyt*

epidemic rate: the amount of increase of disease in a plant population per unit of time *phyt*

epidemiology: the study of disease epidemics, with an effort to tracing down the cause *phyt*

epidermis: the outermost layer of cells; it protects them against drying, mechanical injury, or pathogens *bot*

epigeal: seed germination in which the cotyledons are raised above the ground by elongation of the hypocotyl (e.g., bean) *bot*

epigeal germination: it characterizes a type of germination in which the cotyledons are raised above the ground by elongation of the hypocotyl *bot*

epigenesis: the concept that an organism develops by the new appearance of structure and functions, as opposed to the hypothesis that an organism develops by the unfolding and growth of entities already present in the egg at the beginning of development *bio*

epigenetic: reversible, nonhereditary variation that may be the result of changes in gene expression *gene*

epigenetic regulation: regulation of gene activity mediated by a reversible change in DNA modification or chromatin structure *gene*

epigenetics: the study of the mechanisms by which genes bring about their phenotypic effects *gene*

epigenotype >>> epimutation

epimutation: a heritable change in gene activity that is not due to a change in base sequence; generally applied to an abnormal change in gene activity (e.g., in a tumor or a cell culture); in development, normal changes in gene activity occur and these are often somatically inherited; such changes are not epimutations; the pattern of epimutations in an individual or in a cell is referred to as the epigenotype *gene*

epiphytic: growing on other plants for its physical support but not drawing nourishment from them *bot eco*

episome: a bacterial plasmid that can integrate reversibly with the bacterial chromosome and replicate it *bot biot*

episperm >>> seed coat

epistasis: the nonreciprocal interaction of nonallelic genes; the situation in which one gene makes the expression of another; more specifically, gene interaction in which one gene interferes with the phenotypic expression of another nonallelic gene so that the phenotype is determined effectively by the former; the latter is described as hypostatic *gene* >>> Tables 6, 20, 21

epistasy >>> epistasis

epistatic >>> epistasis

epistatic variance >>> variance >>> epistasis

epitope: a site on an antigen at which an antibody can bind, the molecular arrangement of the site determining the specific combining antibody *chem sero*

equational division: the division of each chromosome during the metaphase of mitosis or meiosis into two equal longitudinal halves (i.e., sister chromatids), which are distributed to the two cell pols, and then incorporated into separate daughter nuclei *cyto*

equational separation >>> equational division

equatorial plane: the plane between the two daughter nuclei of a dividing cell *cyto*

equatorial plate: an arrangement of the chromosomes in which they lie approximately in one plane, at the equator of the spindle; it is seen during metaphase of mitosis and meiosis *cyto*

equilateral: the type of panicle (e.g., in oats) in which the branches appear to spread equally on all sides *bot*

eradicant: a chemical substance that destroys a pathogen at its source *phyt*

eradicate: to remove entirely; to pull up by the roots *meth hort agr* >>> disease eradication

eradication: control of plant disease by eliminating the pathogen after it is established or by eliminating the plants that carry the pathogen *phyt agr* >>> disease eradication

erect: upright *bot*

erectoides mutant: in cereals, a mutant form showing upright tillers and/or leaves *gene*

ergot: a disease that affects many grasses, including cereals; the conspicuous, hard, black sclerotia of the fungus *(Claviceps purpurea)* replace the ovaries (i.e., the grain in the spikelet of an infected plant); the germination of the sclerotia takes place after overwintering, the ascospores, from perithecia in stalked capitula (stromata), causing infection of the stigmas of susceptible plants; in the sphacelia stage, which comes after infection, there is a sweet liquid or nectar with an attraction for the insects, which take the conidia from plant to plant; later in the year the life cycle is completed by the development of sclerotia; the sclerotium contain alkaloids, which can cause severe poisoning or even death if ingested by animals or humans; most ergot for medicine comes from naturally or artificially diseased rye; it is chiefly of use in gynecology and obstetrics; its properties are dependent on a number of alkaloids of which ergotamine and ergotoline are the most important physiologically *phyt* >>> rye

error variance: variance arising from unrecognized or uncontrolled factors in an experiment with which the variance of recognized factors is compared in tests of significance *stat*

ersh >>> stubble

escape: applied to a plant that has escaped from cultivation and naturalized more or less permanently *eco;* in plant pathology, the failure of inherently susceptible plants to become diseased, even though disease is prevalent *phyt*

Escherichia coli (E. coli): an aerobic bacterium; because of its rapid multiplication, it is extensively used in molecular genetics and genetic engineering *biot*

espalier: a trelliswork of various forms on which the branches of fruit trees, grapevine, etc. are extended horizontally, in fan or other shape, in a single plane in order to provide better air circulation and sun exposure for the plants *hort meth*

essential oils: the volatile, aromatic oils obtained by steam or hydrodistillation of botanicals; most of them are primarily composed of terpenes and their oxygenated derivatives; different parts of the plants can be used to obtain essential oils, including the flowers, leaves, seeds, roots, stems, bark, and wood; certain cold-pressed oils, such as the oils from various citrus peels, are also considered to be essential oils, but these are not to be confused with cold-pressed fixed or carrier oils, such as olive, grapeseed, and apricot kernel, which are nonvolatile oils composed mainly of fatty acid triglycerides *chem phys*

EST(s) >>> expressed sequence tag(s)

establishment: the process of developing a crop to the stage at which the young plant may be considered established (i.e., safe from juvenile mortality and no longer in need of special protection) *agr fore*

establishment period: the time elapsing between the initiation of a new crop and its establishment *agr*

ester: a compound that is formed as the condensation product; water is removed, of an acid and an alcohol, while water is formed from the OH of the acid and H of the alcohol *chem phys*

esterase: a member of hydrolases that may catalyze esters; esterases were frequently used as biochemical markers in breeding experiments *chem phys*

estimation error: the amount by which an estimate differs from a true value; this error includes the error from all sources (e.g., sampling error and measurement error) *stat*

estivation: stagnating or otherwise nonfunctional during the summer period *bot agr hort*

étagère: a series of open shelves for growing and displaying plants or in vitro cultures *hort biot meth*

ethanoic acid >>> acetic acid

ethanol: a liquid solvent *chem*

ethene >>> ethylene

ether extract: fats, waxes, oils, etc., that are extracted with warm ether in chemical analysis *prep meth*

ethereal oil >>> essential oil

ethidium bromide (EB): fluorescent molecule that intercalates between base pairs of DNA and RNA *chem biot*

ethyl alcohol >>> ethanol

ethyl methane sulfonate (EMS): a chemical compound that acts as a mutagen in plants *chem meth* >>> mutagen

ethylene: a volatile plant hormone; synthesis is promoted by auxin or damage in seedlings, in shoot apex, and various organs; it is known as a ripening hormone; it stimulates flowering and fruit ripening; it inhibits elongation of stems, roots, and leaves *chem phys*

ethylenediamine tetra acetic acid: known as edetic acid or tetra acetic acid, this chelator is a synthetic amino acid having a molecular weight of 292.25 and a molecular formula of $C_{10}H_{16}N_2O_8$; it is used in a wide range of biochemical and chemical procedures *chem biot*

etiolation: a plant syndrome caused by suboptimal light, consisting of small, yellow leaves and abnormally long internodes *phys meth*

etiology: the study of causes of diseases *phyt*

E-type: in sugarbeet breeding the high-yielding (E = Ertrag = yield) varieties with average sugar content *seed*

euapogamy: a form of apomixis in which the sporophyte develops from a gametophyte without fertilization and formation of zygotes *bot*

EUCARPIA: European Association for Research on Plant Breeding

eucell: a eukaryotic cell showing a nucleus, nuclear envelopes, chromosomes, and nuclear divisions *bot*

eucentric: a chromosomal interchange by which the translocated segment does not change the relative position to the centromere *cyto*

euchromatin: chromatin that shows the staining behavior of the majority of the chromosome complement; it is uncoiled during interphase and condenses during mitosis, reaching a maximum density at metaphase *cyto*

euchromatization: the induced or spontaneous change of heterochromatin into euchromatin *cyto*

euchromosome: a chromosome showing the typical features of the standard complement of a given species *cyto* >>> autosome

eugenic: favorable to the genetic quality of a population, as opposed to dysgenic *gene*

euhaploid: a haploid genome showing no deviating number of chromosomes compared with the standard genome of the species *cyto*

eukaryon: the highly organized nucleus of a eukaryote *bot*

eukaryote: an organism whose cells have a distinct nucleus *bot*

eukaryotic >>> eukaryote

euploid: a cell and/or plant having any number of complete chromosome sets *cyto*

euploidy >>> euploid

eupycnotic: normally coiled and normally stainable chromosomes *cyto*

eusexual: showing regular alternation of karyogamy *bot*

eusom: a plant showing each member of the chromosome complement with the same copy number *cyto*

evaluation: the recording of those characters whose expression is often influenced by environmental factors *meth*

everbloomer: a plant that blooms continuously throughout the growing season, for example, busy lizzie *(Impatiens walleriana) hort*

evergreen: vascular plants that produce and shed leaves throughout the year; their branches are never bare like those of deciduous trees (e.g. citrus trees) *bot*

everlasting >>> evergreen

eversporting: the situation in which a certain trait is permanently varying in each generation due to unstable or highly mutable genes *gene*

evolution: the process by which new species are formed from preexisting species over a period of time *evol* >>> Tables 12, 32

evolution pressure: the joint action of mutation, immigration, hybridization, and selection pressure *evol*

evolutionary >>> evolution

evolutionary breeding: breeding procedure in which the variety is developed from an unselected progeny of a cross or multiple crosses, that have undergone evolutionary changes *meth*

evolutionary divergence: the mode of evolutionary change whereby an ancestral population is split into different genotypic populations or phyletic lines *evol*

evolutionary plasticity: the degree of genetic adaptability of a species or other taxa *evol*

ex situ: out of place or not in the *original* environment (e.g., seeds stored in a genebank) *seed meth*

ex situ conservation: a conservation method that entails the removal of seed, pollen, sperm, or individual organisms from their original habitat, keeping these resources of biodiversity alive outside of their natural environment *meth seed*

ex vito: describes plants or organs that are transplanted from culture to soil or to pots *meth*

exchange pairing: the type of pairing of homologous chromosomes that allows genetic crossing-over to take place *cyto gene*

excised embryo test: a quick method for evaluating the growth potential of a root-shoot axis that has been detached from the remainder of the seed *seed meth*

excision repair: the repair of DNA defects by excision of defective olignucleotides in one of the two DNA strands and the subsequent resynthesis of the excised nucleotide sequence, utilizing the complementary base pair code in the intact strand *gene*

exhauster >>> aspirator

exhaustion test: a type of vigor test that measures the ability of seeds to grow rapidly under rigidly controlled conditions of high temperatures, relative humidity, and moisture content in continuous darkness *seed meth*

exine: the outer, decay-resistant coat of a pollen grain or spore; it shows different characteristics on the surface, which allows taxonomic differentiation between genera and even species; there are several scanning electron microscopy studies made on crop plants, also for comparisons in crop plant evolution *bot*

exocarp: the pericarp or ovary wall of angiosperm fruits is composed of three different layers; the outer layer is the exocarp *bot*

exocytosis: the extrusion of cellular material from a cell, as opposed to endocytosis *bot*

exogamy: outbreeding; all forms of sexual reproduction in which the mating of unrelated and/or more distantly related partners is dominating *bot*

exon: the portion of a gene that is transcribed into mRNA and is translated into protein (i.e., a DNA sequence in the encoding part of a gene) *gene*

exonuclease: a DNase enzyme that digests DNA beginning at the ends of the strands (e.g., exonuclease III degrades dsDNA starting from the 3' end, lamda exonuclease degrades ssDNA and dsDNA) *gene phys biot*

exotic species: a species that is not native to a region *agr hort*

exotoxin: a poison excreted by a plant into the surrounding medium *phys*

experimental alteration of germplasm >>> mutation

experimental design: the planning of a process of data collection; also used to refer to the information necessary to describe the interrelationships within a set of data; it involves considerations, such as number of cases, sampling methods, identification of variables and their scale-types, identification of repeated measures, and replications *stat*

explant: an excised fragment of a tissue or an organ used to initiate an in vitro culture *biot*

explantation >>> explant

expressed sequence tags (ESTs): DNA sequences derived by sequencing an end of a random cDNA clone from a library of interest; usually, tens of thousands of such ESTs are generated as part of a given genome project; these ESTs provide a rapid way of identifying cDNAs of interest, based on their sequence "tag" *biot*

expression variegation: a type of variation in gene expression during development that causes streaks or patches of cells to have different phenotypes *gene*

expression vector: vector constructs with a promoter sequence showing a highly efficient transcription of an inserted gene; the cell shows a high concentration of the gene product (i.e., protein) *biot*

expressivity of a gene: the degree to which a particular genotype is expressed in the phenotype *gene*

extension: 5' and 3' extensions are ssDNA regions at the ends of dsDNA; sometimes they are called overhangs or sticky ends *biot*

extensograph: a standard equipment for evaluating dough extensibility of cereal flour; dough extensibility is an important quality parameter for biscuit making and thus a breeding target *meth*

exterminate >>> eradication

extine >>> exine

extra chromosome >>> B chromosome

extrachromosomal: structures or processes outside the chromosomes *gene*

extrachromosomal element: all genetic elements that are not at all times part of a chromosome, for example, plasmids, phages, transposons, insertion sequences, plastid (mitochondrial and chloroplast) genomes *gene*

extrachromosomal inheritance: inheritance that is not controlled by chromosomal determinants but by cytoplasmic components *gene*

extranuclear: structures and processes outside the nucleus *gene*

eye: the center of a flower when it is a different color from the petals; the term is also used for an undeveloped bud on a tuber (e.g., potato), or for a cutting with a single bud *bot hort*

eye depth (in potato): common scaling: 1 = very deep; 3 = deep; 5 = medium; 7 = shallow; 9 = very shallow *meth prep*

eyepiece: the lens or combination of lenses in an optical instrument (e.g., microscope) through which the eye views the image formed by the objective lens or lenses *micr*

eyespot disease (of wheat, barley, rye, *Pseudocercosporella syn Cercosporella herptotrichoides*, of maize, *Kabatiella zeae*): common in autumn-sown wheat and barley; early symptoms appear as brown smudges on the leaf sheath below the first node; the lesions develop to become eye-shaped with diffuse brown margins and paler centers, which may carry black dots in their center; severe infections girdling the stem may result in premature ripening and white heads and weaken the stem to cause straggling and lodging *phyt* >>> *Tapesia yallundae*

F

F1 >>> filial generation

F2: the progeny produced by intercrossing or self-fertilization of F1 individuals *meth*

F3: progeny obtained by self-fertilizing F2 individuals *meth*

facilitated recurrent selection: a type of recurrent selection in which genetic male sterility is maintained in the population to maintain heterozygosity and genetic diversity and to permit the recombination and shifting of gene frequencies *meth*

factor: synonymous with gene *gene*

factor(ial) analysis: a multivariate statistical analysis in which the independent variables are grouped into factors describing the variance of the dependent variables; it is used in cluster analysis *stat*

factorial crossing group >>> crossing group(s)

factorial design >>> factorial trial

factorial trial: experimental design in which the effects of a number of different factors are investigated simultaneously; the factorial set of treatments consists of all combinations that can be formed from the different factors; the trial can include randomized complete blocks and Latin squares (e.g., treatments A and B at levels x and y); the sources of variance are replicates A, B, A x B, and error *stat*

facultative apomict: apomicts that retain sexuality so that aberrant offspring may occur *bot*

facultative growth habit >>> winter-and-spring wheat

facultative heterochromatin: heterochromatin that is present in only one of a pair of homologues or not permanently present *cyto*

facultative parasite: a mainly saprophytic organism with weakly pathogenic properties *phyt*

facultative-type (of growth habit): can be winter-and-spring type *phys*

fading >>> photobleaching

falcate >>> falciform

falciform: curved or sickle shaped (e.g., leaves or leaf hairs) *bot*

falling number (after HAGBERG): the test provides an indication of alpha amylase activity and depends on the action of this enzyme in reducing the viscosity of a heated flour-and-water slurry; alpha amylase is an enzyme involved in the degradation of starch to sugars and is usually associated with germination; sprouted grains give low falling numbers; high number is required for better bread-making quality (<200 low; 230 medium; >250 high; about 290 very high; >310 extremely high) *meth*

fallow (ground or cropland): leaving the land uncropped for a period of time; it may contribute to moisture accumulation, to improvement of soil structure, or to mineralization of nutrients; it may be tilled or sprayed to control weeds and conserve moisture in the soil *agr*

falls: the drooping or horizontal petals of irises *bot hort*

false color: representation in colors differing from the original scene *micr*

false fruit >>> pseudocarp

false node: an abnormal node that occurs in some varieties of, for example, oats; a true node, appearing devoid of branches, occurs a little distance below it but the branches remain fused with the rachis and appear at the false node *bot*

family: a group of individuals directly related by descent from a common ancestor *tax* >>> *cf* Important List of Crop Plants

family selection: the selection of progeny families on their mean performance; in addition, the best individuals are usually selected in the best families *meth*

fangy root (of beets): forked roots (in beets) caused by diseases, soil, or weather conditions *agr phyt*

fanning machine >>> winnower

fanning mill: the air-screen machine that utilizes airflow and sieving action in separating and cleaning seeds *seed* >>> air screen cleaner >>> winnower

fan-shaped >>> flabellate

fan-training >>> espalier

FAO/IAEA: Joint FAO/IAEA Division of Nuclear Techniques in Food and Agriculture, Vienna, Austria; in a broad sense, responsible for the utilization of nuclear energy in agriculture and food industry

farinogram: a curve on a kymograph chart; it provides accessory information on dough properties, such as absorption, optimum mixing time, and mixing tolerance; it may be used to estimate absorption and mixing time of a flour to be backed; specific correlations between those traits have been established *meth*

FARMER's fixative: 3 parts of anhydrous ethanol : 1 part of glacial acetic acid; a fixing and dehydrating agent used in histology; also used in conjunction with FEULGEN stain and carbolfuchsin stain for chromosome analysis *prep cyto*

fasciata-type of pea: a leaf mutant in peas; the leaves mutated into tendrils; it became a breeding target; several varieties are commercially used *gene meth agr*

fascicle: a bundle of needles on a pine tree *bot*

fasciculated root: fibrous roots, in which some of the branches are thickened *bot agr*

fat: used by most plants as a energy-rich storage substance, usually present in the seeds; in some seeds it may amount about 70 percent of the dry matter; plant fats are usually liquid at room temperature and they are mixtures of glycerine ester of many fatty acids *bot phys* >>> Table 16

fat body >>> fat

father plant: the individual or species from which pollen was obtained to create a hybrid *meth*

fatoid: a mutation that arises spontaneously in cultivated oats; the plant and grain closely resemble the variety in which it occurs but the grain shows to a varying degree certain characteristics of wild oats *(Avena fatua),* for example, a strong geniculate awn, a "horseshoe" base and dense hairs on the callus and the rachilla *bot gene*

fatty acid: a long-chained, predominantly unbranched, carboxylic acid, in which a side-chain of carbon atoms is attached to the carboxyl group, and hydrogen atoms to some or all of the carbon atoms in the side chain *chem phys*

fecundity: the potential number of offspring produced during a unit of time *bot* >>> resilience

fecundity selection: the forces acting to cause one genotype to be more fertile than another genotype *evol*

feed grain: any of several grains most commonly used for livestock feed, including maize, grain sorghum, soya, lupins, oats, rye, barley, etc. *agr*

feedback mechanism: a control device in a system; homoeostatic systems have numerous negative-feedback mechanisms, which tend to counterbalance positive changes and so maintain stability *phys*

fen: low land covered wholly or partially with water *agr*

ferment >>> enzyme

fermentation: anaerobic respiration; usually applied to the formation of ethanol or lactate from carbohydrate *phys* >>> biotechnology

ferredoxin: a nonhaem iron protein with a low redox potential that functions as an electron carrier in both photosynthesis and nitrogen fixation *phys*

fertile: a plant produces seed capable of germination or which produces viable gametes *cyto gene*

fertility: fruitfulness, ability to produce viable offspring *bot gene*

fertilization: the union of two gametes to produce a zygote that occurs during sexual reproduction *bot*

fertilize: bringing together the two gametes to produce a zygote *meth*

fertilizer: a material that is added to the soil to supply one or more plant nutrients in a readily available form *agr*

fertilizer grade: an expression that indicates the percentage of plant nutrients in a fertilizer *agr phys*

festulolium: an artificial grass hybrid between *Festuca pratensis* x *Lolium perenne,* which is used in agriculture as a forage crop *bot agr*

FEULGEN method >>> FEULGEN reaction

FEULGEN reaction: a cytochemical test that utilizes SCHIFF's reagent as a stain and DNA hydrolysis; it is highly specific for DNA detection; the method allows a wide range of chromosome studies and quantitative determination of DNA contents of nuclei applying the so-called cytophotometry; it was discovered in 1912 by FEULGEN *micr*

FEULGEN staining: a histochemical and/or cytochemical reagent for quantifying nuclear DNA content or for staining chromosomes in cells; it it prepared as follows: 900 mL distilled water is boiled in 2-liter flask, then 5.0 g basic fuchsin is slowly added; it is swirled for one minute and then vacuum filtered through two layers of filter paper in a BUCHNER funnel into a 1-liter flask; it is then cooled to 50°C; while swirling 100 mL hydrochloric acid, 1.0 N 10.0 g potassium metabisulfite ($K_2S_2O_5$) is added and swirled for two minutes; the cloudy red solution will become clear blood red; it is stored in the refrigerator for twenty-four hours, then it is removed, warmed to room temperature, and supplemented by 3.75 g activated charcoal; it is shaken vigorously for one minute and then vacuum filtered; the reagent should be clear and colorless; if not, the charcoal step must be repeated; the reagent has to be stored in the refrigerator and used within forty-eight hours; it remains stable for months *meth cyto*

few-seeded >>> oligospermous

fiber: an elongated, thick-walled, often lignified cell (sclerenchyma) present in various plant tissues, usually providing mechanical support *bot*

fiber crop: a crop plant mainly used for production of fiber (e.g., flax, abaca, hemp, etc.) *agr*

fibril: a small thread or very fine fiber; normally a fiber is constituted of a bundle of fibrils *bot*

fibrillar >>> fibril

fibrillarin: a nucleolar protein *bot*

fibrous: resembling or having fibers *bot*

fibrous root: a fine, densely branching root that absorbs moisture and nutrients from the soil *bot agr*

Ficoll: the brand name for an inert, synthetic, highly soluble polymer used as an osmotic agent; sometimes used for suspending protoplasts *prep biot*

field burning: burning plant residue after harvest (1) to aid in insect, disease, and weed control, (2) reduce cultivation problems, and (3) stimulate subsequent regrowth and tillering of perennial crops *agr meth phyt*

field capacity: water that remains in soil after excess moisture has drained freely from the soil; usually expressed as a percentage of oven-dry weight of soil *agr*

field crops >>> cash crops

field diaphragm: a variable diaphragm located in the illumination pathway *micr*

field experiment(ation): an evaluative test whereby the field performance of experimental plants is assessed in comparison to controls *meth*

field gene bank: a collection of accessions kept as plants in the field (e.g., perennial entries) *meth*

field germination: a measure of the percentage of seeds in a given sample that germinate and produce a seedling under field conditions *meth agr* >>> ground germination rate

field grafting: grafting a new variety on to an established rootstock already growing in the orchard *hort*

field laboratory >>> field test

field moisture capacity: the water that soil contains under field conditions *agr* >>> field capacity

field plane: the set of plane(s) that are conjugate with the focused specimen; in a microscope adjusted for KOEHLER illumination it includes the planes of the specimen, the field diaphragm, the intermediate image plane, and the image on the retina *micr*

field plot size >>> plot size

field resistance: synonymous to general resistance; it is under polygenic control (i.e., controlled by many genes with minor individual effects); in general, field resistance is longer lasting than race-specific resistance; field resistance slows down the rate at which disease increases in the field *phyt*

field test: an experiment conducted under regular field conditions (i.e., less subject to control than a precise contained experiment) *meth agr* >>> field experiment(ation)

field trial: experiments carried out in the field *meth* >>> field test

filament: the stalk of a stamen, which bears the anther *bot* >>> Figure 35

filamentous: threadlike *bot* >>> filiform

filial generation (F1): the offspring resulting from first experimental crossing of the plants; the parental generation with which the genetic experiment starts is referred to as P1 *gene* >>> F1

filiation >>> descent

filiform: threadlike, long, and slender *bot*

fill planting: the planting of plants in areas of inadequate stocking to achieve the desired level of stocking (density), either in plantations, areas of natural regeneration, or other trials *meth fore*

filter hybridization: hybridization of nucleic acid fragments on a filter (e.g., nitrocellulose) as a carrier *meth gene* >>> SOUTHERN transfer

final host >>> definite host

fingerprint: the characteristic spot pattern produced by electrophoresis of the polypeptide or DNA fragments obtained through denaturation of a particular protein or DNA with a proteolytic enzyme or other means *biot* >>> fingerprinting

fingerprinting: the method for combining electrophoresis and chromatography to separate the components of a protein or DNA; the protein is denatured by means of a proteolytic enzyme and the resulting polypeptide fragments produce a characteristic spot pattern, referred to as a fingerprint, after electrophoresis; it is also used with hydrolyzed fragments of nucleic acids; it became a common and very sensitive method for identification of different plant genotypes *gene*

fire-blight: a disease of fruit trees, especially of pears and apples, that blackens the foliage and is caused by bacterium *Erwinia amylovora phyt*

firm seeds: sometimes applied to grass caryopses that are dormant due to seed coats that are impervious to water or gases *bot*

FIS: International Federation of Seed Trade

FISH >>> fluorescence in situ hybridization

fishtail weeder >>> asparagus knife

fissable >>> fissile

fissile >>> fission

fission: in genetics, the division of one cell by cleavage into two daughter cells, or a chromosome into two arms *cyto*

fitness: the relative ability of a plant to survive and transmit its genes to the next generation *eco* >>> resilience

5' end: describes the different, complementary ends of a DNA single strand; ends with a phosphate group *biot gene*

fixation: the first step in making permanent preparations of tissue, etc., for microscopic study; the procedure aims at killing cells and preventing subsequent decay with the least distortion of structure *prep cyto*

fixation agent: a solution used for the preparation of tissue for cytological or histological studies; it precipitates the proteinaceous enzymes and prevents autolysis, destroys bacteria, etc. (e.g. acetic acid, formalin, FARMER's fixative, etc.) *prep*

fixative >>> fixation agent

fixed effects model: an effect of a treatment, in any experiment, which is concerned only with a certain, particular set of treatments rather than with the whole range of possible treatment effects *stat*

flabellate: fan-shaped *bot*

flabelliform >>> flabellate

flag leaf: the uppermost leaf on the grass stem and the last to emerge before the spike *bot*

flanking region: the DNA sequences extending on either side of a specific locus or gene *biot gene*

flash tape: a metalized plastic tape that produces bursts of light in response to breezes; it is suspended over crops to scare away birds *agr*

flat: a container for holding packs of plant starter cells *hort*

flexuous (spike neck): wavy or in a more or less zigzag line *bot*

flint maize: a variety of maize, *Zea mays* ssp. *indurata,* with very hard-skinned kernels *agr*

floating leaf: a leaf swimming on the surface of water (e.g., in lotus) *bot*

floating row cover: a fiber sheet, water and air penetrable; it is placed over a row or bed of plants for protection from heat, cold, or insects *hort meth*

flora: the plants of a particular region or period, listed by species and considered as a whole; in general, plants, as distinguished from fauna *bot*

floral induction: the morphological changes in the development of a reproductive meristem from a vegetative meristem; it is the morphological expression of the induced state and usually occurs inside the meristem *phys*

floral initiation >>> floral induction

floral meristem: a meristem that gives rise to a flower *bot*

floral primordium >>> floral meristem

florescence: anthesis or flowering time or the state of being in bloom *bot*

floret: a single flower consisting of the ovary, stamens, and lodicules together with its enveloping lemma and palea in grasses *bot*

floret initial >>> floral primordium

floriculture: the cultivation of flowers or flowering plants *hort*

floriferous: bearing flowers *bot*

florigen: the universal hormone that supposedly causes plants to change from the vegetative to the reproductive state *phys*

flour: the finely ground meal of grain (e.g., from cereals) separated by bolting *agr prep*

floury >>> chalky

flow cytometry >>> sorting

flow densitometry >>> sorting

flow sorting >>> sorting

flower: a typical flower of angiosperms or plants whose seeds are enclosed in an ovary; it is composed of petals, sepals, stamens, and a pistil; the flower morphology contributes to the relative importance of self- and cross-pollination; the two structures directly involved in sexual reproduction are the male stamina and the female pistil; a stamen consists of an anther, which contains the pollen grains, and a filament on which the anther is borne *bot* >>> self-incompatibility >>> Table 18

flower bud initiation >>> floral primordium

floweret: one of the segments of a cauliflower head *hort*

flowering hormone >>> florigen

fluid drilling: a mechanical procedure for planting seed; pregerminated seeds are suspended in a gel and sowed through a fluid drill seeder; this technology is potentially adaptive for sowing artificial seed, such as somatic embryos or embryoids *agr biot*

fluorescence: property of certain molecules to absorb energy in the form of light and then release this energy at a longer wavelength than the wavelength of absorption (i.e., at a lower energy level) *micr phy*

fluorescence dye: several dyes are critical for specific light specters, for example, Hoechst, DAPI, AMCA, Cascade Blue, Fura, Dansylchloride, Fluorescine FITC for green light, Lucifer Yellow, Quinacrin, Chromomycin A3, FITC, NBD Chloride for yellow light, Phycoerythrin, Propdium Iodide, Feulgen, Auramine, DiOC, Ethidium Bromide for orange light or Rhodamine or Texas Red for red light *micr*

fluorescence in situ hybridization (FISH): a technique for visual detection in the microscope of specific DNA sequences on cytological fixed chromosomes, after hybridization with DNA probes labeled with a fluorochrome; it can be done on both interphase and metaphase chromosomes *cyto meth* >>> chromosome painting

fluorescence microscopy: the common method of microscopic examination base on observing the specimen in the light transmitted or reflected by it; fluorescence preparations are self-luminous; the tissue is stained by fluorochromes, dyes that emit light of longer wavelength when exposed to blue or ultraviolet light; the fluorescing parts of the stained object then appear bright against a dark background; the staining technique is extremely sensitive *micr* >>> fluorescence dye >>> chromosome painting

fluorescence staining: very few biological samples are inherently fluorescent such that they can be imaged directly; instead, fluorescence microscopy essentially always uses a fluorochrome that is introduced through some form of staining procedure; fluorescence staining for flow cytometric, fluorescence or light microscopic analysis; the fluorescence can be directly applied, such as 4'-6-diamidino-2-phenylindole (DAPI); it is known to form fluorescent complexes with natural double-stranded DNA, showing a fluorescence specificity for AT, AU, and IC clusters; because of this property, DAPI is a useful tool in various cytochemical investigations; when DAPI binds to DNA, its fluorescence is strongly enhanced, what has been interpreted in terms of a highly energetic and intercalative type of interaction, but there is also evidence that DAPI binds to the minor groove, stabilized by hydrogen bonds between DAPI and acceptor groups of AT, AU, and IC base pairs; other applications use single-stranded DNA sequences with a fluorescent label to hybridize with its complementary target sequence in the chromosomes, allowing it to be visualized under ultraviolet light; fluorescence staining methods offer several advantages, such as high resolution, live cell staining, and the possibility of dual or multiple labeling *cyto meth* >>> fluorescence microscopy >>> fluorescence dye

fluorescent: the color exhibited when the grain or glumes of certain oat varieties are viewed under ultraviolet light or other radiation *phys*

fluorescin: a red crystalline compound, $C_{20}H_{12}O_5$, which in alkaline solutions produces an intense green *chem*

fluorescin diacetate (FDA) staining: living cells stained with FDA fluorescence in the presence of UV light; the stain is used to assess cell viability of cell cultures, etc. *meth*

fluorochrome >>> fluorescence dye

flush cut: a pruning cut to remove a tree limb in which the cut is completely flush with the tree; the resulting scar is usually too large to heal efficiently *meth hort*

flush ends >>> blunt ends

flush season: the plant growth that is produced during a short period *agr*

flux: a flowing or flow *phys*

focal plane: a plane through a focal point and perpendicular to the axis of a lens, mirror, or other optical system *micr*

focus: the ability of a lens to converge light rays to a single point *micr*

fodder: harvested grass or other crop parts for animal feed *agr*

foliaceous: leaflike shape *bot*

foliage: the leaves of plants *bot*

foliage blight >>> late blight

foliar feeding: feeding plants by spraying liquid fertilizer on the leaves *agr hort*

foliar nutrient: any liquid substance applied directly to the foliage of a growing plant for the purpose of delivering an essential nutrient in an immediately available form *agr*

foliar treatment: treating plants by spraying liquid or dry insecticides, pesticides, or herbicides on their leaves *agr hort phyt*

foliation: leafing *bot*

foliole: leaflet *bot*

foliose: leafy *bot*

follicle: a fruit with a simple pistil that at maturity splits open along one suture *bot*

follicular fruit >>> follicle

food grain: cereal seeds most commonly used for human food, chiefly wheat and rice *agr*

food legume: legume plants with nutritive value for humans, directly and indirectly consumed *agr*

food plant: plants with nutritive value for humans, directly and indirectly consumed *agr*

food species >>> food plant

foot (ft): equals 30.48 cm

foot rot >>> eyespot disease

footprinting: a method used to determine the length of nucleotide chains that are close to a protein (which bind to DNA); for example, certain types of drugs act by binding tightly to certain DNA molecules in specific locations *gene meth* >>> fingerprinting

forage: feed from plants for livestock such as hay, pasturage, straw, silage, or browse *agr*

forage crop: crop plants for feeding of livestock (e.g., alfalfa, clover, maize) *agr* >>> forage

forage shrub >>> forage

forceps >>> pincers

forcing: the practice of bringing a plant into growth or flower (usually by artificial heat or controlling daylight) at a season earlier than its natural one; it is sometimes applied in order to synchronize flowering dates of parental plants for crosses in the greenhouse *meth*

forecrop: the crop grown during the season of the respectively present cropping *agr*

foreign DNA: DNA that is not found in the normal genome concerned; usually it is directly or indirectly introduced into a recipient cell by several experimental means *biot meth*

forest tree breeding: the genetic manipulation of trees, usually involving crossing, selection, testing, and controlled mating, to solve some specific problem or to produce a specially desired product *fore gene*

forked: sometimes a morphological deviation of the common root in beets (e.g., sugarbeet) *agr* >>> fangy root

form: a botanical category ranking below a variety and differing only trivially from other related forms (e.g., in waxiness of leaves) *tax*

forma specials (f.sp.): a taxon characterized from a physiological standpoint (especially host adaptation); in general, biotypes of pathogen species that can infect only plants that are within a certain host genus or species *phyt tax* >>> Table 12

formaldehyde: a colorless gas readily soluble in water *chem phys*

formalin: an aqueous solution of formaldehyde commonly used as a fixative that functions through cross-linking protein molecules *chem prep*

formazan: colorless when dissolved in water; the chemical 2,3,5-triphenyl tetrazolium chloride is reduced to the red-colored chemical triphenyl formazan on contact with living, respiring tissue; the amount of formazan formed is used as a measure of seed viability, as it reflects oxidative metabolism *chem seed* >>> tetrazolium test

forward mutation: a mutation that alters (usually inactivates) a wild-type allele of a gene *gene* >>> back mutation

fossil: markedly outdated *evol* >>> Table 32

foundation seed: seed stocks increased from breeder seed; handled as to closely maintain the genetic identity and purity of the variety; it is a sort of certified seed, either directly or through registered seed *seed* >>> Table 28 >>> base seed

foundation single cross: it refers to a single cross in the production of a double, three-way, or top cross *meth*

founder effect: genetic drift due to the founding of a population by a small number of individuals *gene*

four-way cross >>> double cross

frame shift mutation: a mutation that is caused by a shift of the reading frame of the mRNA (usually by the insertion of a nucleotide) synthesized from the altered DNA template *gene*

free-threshing: spikes with brittle (fragile) rachis; for example, in wheat, it is controlled by a single recessive allele; the fragility of the spike reveals the main difference between wild and cultivated forms *bot gene agr*

freeze preservation >>> cryopreservation

freezing injury: a type of winter injury caused by the combined effects of low temperature, wind, and insufficient soil moisture; the low temperature injury is associated with ice formation in the extracellular spaces resulting in freeze-induced dehydration and metabolic changes; the plasma membrane remains attached to the cell wall, causing the cell to collapse *phys agr*

frego-bract mutant: a mutant bract type in cotton in which the bracts curl outward, exposing flower buds and bolls *gene*

frequency histogram: a step-curve in which the frequencies of various arbitrarily bounded classes are presented as a graph *stat*

frequency table: a way of summarizing a set of data; a record of how often each value (or set of values) of the variable in question occurs; may be enhanced by the addition of percentages that fall into each category; used to summarize categorical, nominal, and ordinal data; may also be used to summarize continuous data once the data set has been divided up into sensible groups *stat*

frequency-dependent selection: selection where the fitness of a type varies with its frequency (i.e., whereby a genotype is at an advantage when rare and at a disadvantage when common) *gene meth*

friabilin(s): proteins that determine the adhesion of the starch granules; for example, in wheat they determine the difference between hard and soft wheat; soft wheats show strong friabilins, which bind the granules and hence the endosperm fractionates into large fragments, whilst hard wheats contain weaker friabilins and hence fracture into small fragments; latest molecular and biochemical evidences associate them with differences in the structure of proteins (puroindolines) *phys meth*

frond: the leaves of ferns and other cryptogams; includes both stipe and blade; commonly used to designate any fernlike or featherlike foliage *bot*

frost damage >>> killing frost

frost killing >>> killing frost

frost mold: certain strains of at least three species of epiphytic bacteria (*Pseudomonas syringae, P. fluorescens, Erwinia herbicola*), are present on many plants; they serve as ice-nucleation-active catalysts for ice formation at temperatures as high as –1°C *phyt*

frost resistance: the capacity to survive temperatures below zero degrees Celsius *phys*

frost tolerance: the ability of plants to survive very harsh winter conditions (i.e., to withstand subzero temperatures) *agr*

frost-lifting of seedlings >>> heaving

fructiferous: bearing fruits *bot*

fructification: the process of forming a fruit body or the fruit body itself *bot*

fructose: a sugar that occurs abundantly in nature as the free form, but also with glucose in the form of the disaccharide sucrose *chem phys*

fruit: strictly, the ripened ovary; more loosely, the term is extended to the ripened ovary and seeds together with the structure with which they are combined *bot*

fruit drop: the premature abscission of fruit before it is fully ripe; it is a common process; in many fruits there are certain peak periods of fruit drop; for example, apple fruits are lost immediately following pollination (post blossom drop), when the embryos are developing rapidly (June drop), and during ripening (preharvest drop); as in leaf fall, fruit drop is associated with low auxin levels; auxin sprays have been used to prevent excessive fruit drop *hort phys*

fruit-bearing branchlet: in some fruit trees (e.g., in apple) the main type of fruit-bearing wood *hort*

fruiting body: a specialized organ for producing spores (e.g., in mushrooms) *bot*

frutex >>> shrub

fruticous >>> shrubby *bot*

f.sp. >>> *forma specials*

fuchsin(e): a greenish, water-soluble, solid, coal-tar derivative, obtained by the oxidation of a mixture of aniline and the toluidines, that forms deep-red solutions; used chiefly as a dye *micr cyto* >>> FEULGEN staining

full bloom: the developmental stage at which essentially all florets in the inflorescence are in anthesis *bot*

full diallel >>> complete diallel

full pedigree: all parents, grandparents, etc., of a particular genotype are known back to natural population *gene*

full sibs: individuals that have both parents in common; the mating of full sibs is the most extreme form of inbreeding that can occur in bisexual, diploid plants *gene meth*

fumigant: vaporized pesticide used to control pests in soil, storage rooms, and greenhouses (e.g., methyl bromide) *phyt seed*

fumigate >>> fumigation

fumigation: to expose to smoke or fumes, as in disinfecting or in exterminating vermin, fungi, etc. *seed meth*

functional hemizygosity: an autosomal gene is functionally hemizygous if only one of the two copies in a diploid cell is expressed *gene*

fungal pathogen >>> pathogen

fungicide: a chemical, physical, or biological agent that inhibits fungal growth *phyt*

fungicide control >>> fungicide

fungus (fungi *pl*): any member of the kingdom Fungi (or division Thallophyta of the kingdom Plantae), comprising single-celled or multinucleate organisms that live by decomposing and absorbing the organic material in which they grow (e.g., mushrooms, molds, mildews, smuts, rusts, and yeasts) *bot phyt*

funicle: the stalk connecting the ovule with the placenta on the ovary wall *bot*

funiculus >>> funicle

furrow: a trench in the earth, made by a plough *agr*

furrow irrigation: small, shallow channels guide water across the surface of a leveled field; crops are typically grown on a ridge or raised bed between the furrows *agr*

furrow weed: a weed that grows on ploughed land *phyt agr*

6-fururylaminopurine >>> kinetin

***Fusarium* ear blight:** all cereals may be infected during wet warm weather at flowering; the disease is most common in wheat; it starts with small brown spots on the glumes that develop to leave florets or whole spikelets prematurely bleached; pink or orange colonies of the fungus can often be seen at the base on the spikelet *phyt*

fusiform: shaped like a spindle or cigar, tapering at both ends *bot*

fusion gene: a hybrid gene created by joining portions of two different genes (to produce a new protein) or by joining a gene to a different promoter (to alter or regulate gene transcription) *biot*

fusogen(ic): a fusion-inducing agent used for protoplast agglutination in somatic hybridization studies (e.g., polyethylene glycol [PEG]) *biot*

fuzz: beard of grain *bot* >>> lint (linters)

G

G banding: a special staining technique for chromosomes that results in a longitudinal differentiation by Giemsa stain, which is a complex of stains specific for the phosphate groups of DNA; the characteristic bands produced are called G bands; these bands are generally produced in AT-rich heterochromatic regions *cyto meth* >>> differential staining

galactose (Gal): a component of milk sugar (lactose) *chem phys*

gall: a localized proliferation of plant or parasite tissue that produces an abnormal growth or swelling, usually caused by pathogenic organisms, nematodes, or insects *phyt*

gall midge: any midge of the family Cecidomyidae, the larvae of which form characteristic galls on plants *zoo phyt*

gamet precision divider: a type of mechanical halving device for subdividing a large seed sample to obtain a smaller working sample for germination or purity analysis; it has an electrically operated rotating cup into which the seed is funneled to be spun out and into one of two spouts *seed*

gamete: a specialized haploid cell whose nucleus and often cytoplasm fuses with that of another gamete in the process of fertilization, thus forming a diploid zygote *bot* >>> Tables 2, 9

gametic selection: the influences acting to cause differential reproductive success of one allele over another in a heterozygote *gene*

gametocidal >>> gametocide

gametocidal gene: a gene encoding a product that destroys cells that divide to produce the gametes *gene*

gametocide: a chemical agent used to selectively kill either male or female gametes; it is used in hybrid seed production of autogamous crops (e.g., barley or wheat) *meth*

gametoclonal: regenerated from a tissue culture originating from gametic cells or tissue *biot*

gametoclonal variation: variation among regenerants obtained from pollen and/or anther culture *biot*

gametoclone: plants regenerated from cell culture derived from meiocytes or gametes *biot*

gametocyte: a cell that will undergo meiosis to form gametes *bot*

gametogamy: the fusion of the sexual gametes and the formation of the zygotic nucleus *bot*

gametogenesis: the formation of gametes from gametocytes *bot*

gametophyte: a haploid phase of the life cycle of plants during which gametes are produced by mitosis; it arises from a haploid spore produced by meiosis from a diploid sporophyte *bot* >>> Figure 28 >>> Table 16

gametophytic apomixis: agamic seed formation in which the embryo sac arises from an unreduced initial; it includes both diplospory and apospory *bot*

gametophytic self-incompatibility: self-incompatibility is based on the genotypic and phenotypic relationship between the female and male reproductive system; alleles in cells of the pistil determine its receptivity to pollen; the phenotype of the pollen, expressed as its inability to effect fertilization, may be determined by its own alleles, referred to as gametophytic incompatibility *gene* >>> self-incompatibility

gamopetalous: having the petals of the corolla more or less united *bot*

gangrene of potato: necrosis or death of soft tissue due to obstructed circulation, usually followed by decomposition and putrefaction (*Phoma exigua* var. *foveata*) *phyt*

gap: single-stranded region in dsDNA *biot*

gas chromatograph: an analytical technique for identifying the molecular composition and concentrations of various chemicals in agricultural products, water, or soil samples *meth prep*

GAUSS distribution >>> normal distribution

Gaussian curve >>> normal distribution

GCA: >>> combining ability (CA) general combining ability

geitonogamy: when neighboring flowers of the same plant can achieve pollination, as opposed to xenogamy *bot*

gel: a two-phase colloidal system consisting of a solid and liquid that have combined to form a viscous, jellylike substance; some gels are used as inert matrices in the separation of nucleic acids or proteins by electrophoresis *prep*

gel electrophoresis: in gel electrophoresis, proteins migrate in an electric field through a porous matrix, typically composed of polyacrylamide or starch; mobilities are directly related to charge and inversely related to size; in isoelectric focusing, proteins migrate in a pH gradient to positions of electric neutrality that are independent of molecular weight *meth*

gelatin(e): a nearly transparent, glutinous substance, obtained by boiling the bones, ligaments, etc. of animals; used in making jellies *chem prep*

GEM: genetically engineered microorganism *biot*

geminivirus: a single-stranded DNA virus that causes serious diseases in cereals, vegetables, and fiber crops worldwide; like nanoviruses, it is transmitted by either whiteflies or leafhoppers; whitefly-transmitted geminiviruses are all in the genus *Begomovirus,* which typically have bipartite genomes (DNA A and DNA B) comprising circular DNAs of ~2,700 nucleotides, although some have been shown to be monopartite, having only a DNA A component; in contrast, the nanoviruses are a recently established group of plant viruses that are transmitted by either aphids or planthoppers, and have multipartite genomes comprising circular DNAs of ~1,000 nucleotides *phyt*

gemmiferous: bearing buds *bot*

gene: the hereditary unit that occupies a fixed position on the chromosome, which through transcription has a specific effect upon phenotype *gene;* it may mutate to various allelic forms; in molecular biology, a segment of DNA including regulatory sequences (promoter[s], operator[s], terminator) that encodes an RNA and/or protein molecule *biot* >>> cistron >>> Table 9

gene action: the expression of the gene based on the transcription into complementary RNA sequences, the subsequent translation of mRNA into polypeptides, which may form a specific protein *gene*

gene activation: different mechanisms of repressing and activation of genes *gene*

gene amplification: the more or less specific production of multiple copies of a gene *gene*

gene bank: an establishment in which both somatic and hereditary genetic material are conserved (seeds, pollen, whole plants, extracted DNA); it stores in a viable form, material from plants that are in danger of extinction in the wild and cultivars that are not currently in popular use; the stored genetic material can be called up when required; the normal method of storage is to reduce the water content of seed material to around 4 percent and keep it at 0°C or less (–20°C); all stored stocks are periodically checked by germination tests *meth* >>> gene library

gene center: it refers to the center of origin of a given crop plant *evol gene* >>> center of diversity >>> center of origin

gene cloning: insertion of a DNA fragment carrying a gene into a cloning vector; subsequent propagation of the recombinant DNA molecule in a host organism results in many identical copies of the gene (clones) in a form that is more easily accessible than the original chromosomal copy *gene biot*

gene construction: an experimentally engineered gene with functional and nonfunctional properties *gene biot*

gene conversion: a process whereby one member of a gene family acts as a blueprint for the correction of the other; this can result in either the suppression of a new mutation or its lateral spread in the genome *gene*

gene dosage (dose): the number of times a given gene is present in the nucleus of a cell and/or individual *gene*

gene duplication: a process in evolution in which a gene is copied twice; the two copies lie side by side along the same chromosome *gene*

gene expression: the phenotypic manifestation of a gene depending on the different levels of gene activation, or the process by which the information in a gene is used to produce a protein; in molecular genetics, the full use of the information in a gene via transcription and translation leading to production of a protein and hence the appearance of the phenotype determined by that gene; gene expression is assumed to be controlled at various points in the sequence leading to protein synthesis; this control is thought to be the major determinant of cellular differentiation *gene* >>> Table 9

gene family: a group of similar or identical genes, usually along the same chromosome, that originate by gene duplication of a single original gene; some members of the family may work in concert, others may be silenced and become pseudogenes *gene*

gene flow: the spread of new genes, which takes place within an interbreeding group as a result of crossing with immigrants *evol*

gene frequency: the number of loci at which a particular allele is found divided by the total number of loci at which it could occur for a given population, expressed as a proportion or percentage *gene*

gene insertion: the addition of one or more copies of a normal gene into a defective chromosome *biot*

gene interaction: modification of gene action by a nonallelic gene or genes, generally the interaction between products of nonallelic genes *gene*

gene library: in molecular genetics, a random collection of cloned DNA fragments in a number of vectors that ideally includes all genetic information of that species *biot*

gene linkage >>> linkage

gene location: determination of physical or relative distances of a gene on a particular chromosome *gene*

gene locus: the fixed position that a gene occupies on a chromosome *gene*

gene machine: in common literature, an idiomatic description of an automated oligonucleotide synthesizer *biot*

gene map: a graphic presentation of the linear arrangement of a chromosome or segment; it shows the relative distance between loci gained in linkage experiments *gene*

gene mapping: determination of the position of genes on a DNA molecule *biot*

gene mutation: a heritable change of gene revealed by phenotypic modifications *gene*

gene pairs: the two copies of a gene present in a diploid, one on each homologous chromosome *gene*

gene patenting: protection provided by governmental or nongovernmental institutions to the discoverer of new genes, genotypes, strains, or testing procedures so that the detailed information can be declared publicly; recently, a synthetic gene but not a natural gene itself can be patented, however its sequenced functional unit or specific utilization cân be a matter of patent *biot*

gene pool: the reservoir of different genes of a certain plant species or lower and higher taxa available for crossing and selection; it may be differentiated between (1) primary gene pools (consists of those species that readily hybridize, produce viable hybrids and have chromosomes that may freely recombine), (2) secondary gene pools (consists of those species with a certain degree of hybridization barrier due to ploidy differences, chromosome alterations, or incompatibility genes), and (3) tertiary gene pools (consists of distinct species or higher taxa with strong crossing barriers); in general, the total number of genes or the amount of genetic information that is possessed by all the reproductive members of a population of sexually reproducing organisms *gene* >>> primary gene pool >>> secondary gene pool

gene recombination >>> recombination

gene redundancy: the presence of a gene(s) in multiple copies due to polyploidy, polytenic chromosomes, gene amplification, or chromosomal duplications *gene*

gene silencing: inactivation of gene expression, usually with the implication that inactivation is not due to mutation, but rather to epigenetic modification such as methylation or change in chromatin structure; the inactive state can be heritable under most conditions yet reversible under special conditions, such as treatment with inhibitors of DNA methylation *gene*

gene splicing: combining genes from different organisms into one organism *biot*

gene stacking: the insertion of two or more (possibly synthetic) genes into the genome (e.g., the *bat* gene from *Bacillus thuringiensis* and a gene for resistance to a specific herbicide) *biot*

gene substitution: the replacement of one allele by another mutant allele in a population by natural or directed selection *gene*

gene symbol: designating a gene, usually by an abbreviation of the name or description of a given gene; in the past, genes have been described by Latin names, and subsequently the one- to three-letter abbreviations; currently, several systems of naming and symbolization are in use, although a comparable, uniform symbolization is sought *gene* >>> genetic nomenclature

gene tagging: the labeling of a gene by a marker gene or specific DNA sequence closely linked with the gene in question *gene biot*

gene targeting: the insertion of antisense DNA molecules in vivo into selected cells of the body in order to block the activity of undesirable genes; these genes might include oncogenes or genes crucial to the life cycle of parasites *biot*

gene technology: in a broad sense, it is the artificial transfer of genes between cells or individuals by means of molecular and in vitro techniques; prerequisites are (1) the presence of a gene, which is available as a DNA fragment, (2) a clonable DNA, (3) the fragment has to be transferable by different systems, (4) the incorporation of the DNA fragment into a recipient cellular genome has to be feasible, (5) the transformed cells have to be regenerable into a normal plant, and (6) the gene that was transferred must be expressed in the alien genetic background *biot*

gene transfer: the physical transfer of a gene by crossing, chromosomal manipulation, and molecular means; in biotechnology, different methods are described, such as (1) microinjection, (2) insertion via microprojectiles (particle gun, particle bombardment) using silicon fibers as carriers of the DNA, (3) direct transfer, or (4) a vector-mediated transfer *biot* >>> Figure 27

gene translocation: the transfer or movement of a gene or gene fragment from one chromosomal location to another; often it alters or abolishes expression *gene cyto biot* >>> position (positional) effect

genealogy: descent from an original form or progenitor; ancestry *evol*

gene-for-gene theory: in certain plant-pathogen interactions; a gene for resistance in the host corresponds to and is directed against a gene for virulence in the pathogen *phyt*

general combining ability >>> combining ability (CA)

general resistance: resistance against all biotypes of a pathogen; nonspecific host-plant resistance *phyt*

general seed blower: a precision seed blower used to aid in separating light seed and inert matter from heavy seed *seed*

generalized resistance >>> general resistance

generation time: the time required for a culture to double its cell number *biot*

generative: sexual processes *bot*

generative meristem: gives rise to parts, such as floral organs that ultimately produce fruits and seeds *bot*

generative nucleus: a haploid nucleus of a pollen grain that produces two sperm nuclei by mitosis (pollen grain mitosis) *cyto* >>> sperm nucleus >>> Figures 25, 35

generic: referring to the genus *tax*

generic diversity: the differences between individuals of different genera *evol eco tax*

genetic: pertaining to the origin or common ancestor or ancestral type *gene*

genetic advance: the expected gain in the mean of a population for a particular quantitative character by one generation of selection of a specified percent of the highest-ranking plants *gene*

genetic assimilation: possible extinction of a natural species as massive pollen flow occurs from another related species; or in breeding, the older crop becomes more like the new crop *gene evol*

genetic background: the remaining genetic constitution when a particular locus or allele of a given individual or taxon is studied *gene*

genetic balance: the optimal interaction of coadapted genes, alleles, or genetic systems within a given individual *gene*

genetic block: the reduction or stop of an enzyme activity caused by a specific gene mutation *gene*

genetic code: the set of correspondences between base triplets in DNA and amino acids in protein; these base triplets carry the genetic information for protein synthesis *gene*

genetic complement >>> genome

genetic complementation: the complementary action of homologous sets of genomes *gene*

genetic correlation: the correlation between the genotypic values of two characters with respect to the genetic character *stat gene*

genetic death: death of an individual without reproducing; caused by mutationally arisen alleles that reduce the fitness of a genotype and/or taxon *gene evol*

genetic distance: a measure of gene differences between individuals or populations measured by the differences of several characters; such distance may be based on phenotypic traits, allele frequencies, or DNA sequences (e.g., genetic distance between two populations having the same allele frequencies at a particular locus and based solely on that locus is zero); the distance for one locus is maximum when the two populations are fixed for different alleles; when allele frequencies are estimated for many loci, the genetic distance is calculated by averaging over these loci *gene*

genetic drift: the random fluctuations of gene frequencies in a population such that the genes amongst offspring are not a perfectly representative sampling of the parental genes *gene*

genetic engineering: the manipulation of DNA using restriction enzymes, which can split the DNA molecule and then rejoin it to form a hybrid molecule—a new combination of nonhomologous DNA; the technique allows the bypassing of all the biological restraints to genetic exchange and mixing, and even permits the combination of genes from widely differing species *biot*

genetic equilibrium: an equilibrium in which the frequencies of two alleles at a given locus are maintained at the same values generation after generation; a tendency for the population to equilibrate its genetic composition and resist sudden change is called genetic homoeostasis *gene* >>> homoeostasis

genetic erosion: the loss of genetic information that occurs when highly adaptable cultivars are developed and threaten the survival of their more locally adapted ancestors, which form the genetic base of the crop *gene evol*

genetic gain: the change achieved by artificial selection in a specific trait; the gain is usually expressed as the change per generation or the change per year; it is influenced by selection intensity, parental variation, and heritability *gene*

genetic homoeostasis >>> genetic equilibrium

genetic homology: the identity or near identity of DNA sequences, genes, or alleles *gene*

genetic information: the information contained in a sequence of nucleotide bases in a nucleic acid molecule *gene*

genetic instability: different mechanisms that give rise phenotypic variation *gene*

genetic interaction: the interaction between genes resulting in different phenotypic expressions *gene*

genetic load: the average number of lethal mutations per individual in a population *gene evol*

genetic (linkage) map: the linear arrangement of gene loci on a chromosome, deduced from genetic recombination experiments; a genetic map unit is defined as the distance between gene pairs for which one product of meiosis out of a hundred is recombinant (i.e., it equals a recombination frequency of 1 percent) *gene*

genetic mapping: the process of determination of a genetic map *gene*

genetic marker: any phenotypic difference, controlled by genes, that can be used for studying recombination processes or selection of a more or less closely associated target gene *gene*

genetic material: all single- or double-stranded DNA carrying genetic information or that is a substantial part of the genetic information *gene*

genetic nomenclature: the designation of genes by abbreviated gene descriptions or symbols, usually a beginning capital letter of the abbreviation represents a dominant allele, while a small beginning letter refers to a recessive allele *gene* >>> gene symbol

genetic polymorphism: an occurrence in a population of two or more genotypes in frequencies that cannot be accounted for by recurrent mutation *gene*

genetic recombination: a number of interacting processes that lead to new linkage relationships of genes *gene*

genetic resistance: resistance against pathogens or pests due to specific or general gene action *gene* >>> passive resistance

genetic resources: the gene pool in natural and cultivated stocks of organisms that are available for human exploitation *gene*

genetic segregation >>> segregation

genetic sterility: a type of male sterility conditioned by nuclear genes, as opposed to cytoplasmic sterility *gene*

genetic stock: a variety or strain known to carry specific genes, alleles, or linkage groups *gene*

genetic system: the organization of genetic material in a given species and its method of transmission from the parental generation to its filial generations *gene*

genetic targeting >>> gene targeting

genetic variability: individuals differing in their genotypes due to mutational, recombinational, or selective mechanisms *gene*

genetic variance (V_{gen}): a portion of phenotypic variance that results from the varying genotypes of the individuals in a population; together with the environmental variance, it adds up to the total phenotypic variance observed amongst individuals in a population; it is divided into additive (resulting from differences between homozygotes, V_{add}) and dominance variance (resulting from specific effects of various alleles in heterozygotes, V_{dom}); V_{gen}

$\doteq V_{add} + V_{dom}$; the quotient V_{add} / V_{dom} is termed heritability in a narrow sense *stat* >> heritability

genetically modified organisms (GMO): a term, currently used most often in official discussions, that designates crops that carry new traits which have been inserted through advanced genetic engineering methods (e.g., flavor-saver tomato, roundup-ready soybeans, Bt cotton, or Bt maize) *agr*

geneticist: a specialist in genetics *gene*

genetics: the scientific study of genes and heredity *gene*

geniculate: bent abruptly like a knee *bot*

genome: the total genetic information carried by a single set of chromosomes in a haploid nucleus; for example, genome sizes: lamda phage 48.5 kb, *Escherichia coli* 4,500 kb, yeast 1.6×10^4 kb, *Drosophila* 1.2×10^5 kb; in 2000, the total nucleotide sequence of a plant genome, *Arabidopsis thaliana,* was for the first time identified; the sequenced genome contains about 120 kb *gene* >>> Tables 8, 14 >>> lamda phage

genome allopolyploids >>> allopolyploid

genome analysis: the study of the genome by combination of cytogenetics, karyotyping, and crossing *cyto gene* >>> Table 8

genome doubling >>> autopolyploidization

genome formula >>> genome symbol

genome mutation: spontaneous or induced changes in the number of complete chromosomes that result either in polyploids or aneuploids *cyto gene* >>> Figure 37

genome symbol: the description of specific genome by a symbol, usually a capital letter with or without a specification *gene* >>> *cf* Important List of Crop Plants

genomic in situ hybridization (GISH): an in situ hybridization technique that uses total genomic DNA of a given species as a probe and total genomic DNA of another species as a blocking DNA; it is based on fluorescence in situ hybridization; it is a useful method to detect interspecific or intergeneric genome differentiation, chromosome rearrangements (translocations), and substitutions or additions *cyto meth*

genomic inprinting: the phenomenon whereby genes function differently depending on whether they are inherited from the maternal or paternal parent; this is thought to be caused by information superimposed on DNA sequences, which is different in male and female gametes; such information is transmitted, or inherited, in somatic cells but usually erased and reset in the germ line *gene*

genomic library: a type of DNA library in which the cloned DNA is from a genomic DNA of the plant and/or organism; since genome sizes are relatively large compared to individual cDNAs, a different set of vectors is usually employed in addition to plasmid and phage *biot* >>> bacterial artificial chromosomes (BAC) >>> yeast artificial chromosomes (YAC) >>> cosmid

genomic selection (GS): a method in which the number of polymorphic bands resembling the recurrent parent is used as the selection criterion *biot*

genomics: the scientific study of genes and their role in a structure, growth, health, and disease of a plant (e.g., how a certain number of genes contributes to the shape, function, and the development of the organism) *biot*

genotoxic: it refers to substances and circumstances inducing mutants and damage of the heritable material *gene*

genotype: the genetic constitution of an organism, as opposed to its physical appearance (phenotype); usually, it refers to the specific allelic composition of a particular gene or set of genes in each cell of an organism, but it may also refer to the entire genome *gene* >>> Tables 2, 9

genotype-environment interaction: a third type of interaction (besides allelic and nonallelic interactions), namely between genes and their environment *gene*

genotypic: phenomenons and processes that are associated with the genotype *gene*

genotypic variance >>> genetic variance (V_{gen})

genus (genera *pl*): a taxonomic grouping of similar species *tax* >>> *cf* Important List of Crop Plants

genus name: name of a taxonomic group *tax* >>> *cf* Important List of Crop Plants

geometric mean: the square root of the product of two numbers *stat*

geophyte: a land plant that survives an unfavorable period by means of underground food-storage organs (rhizomes, tubers, bulbs, e.g., onions, tulips, potato, asparagus, Jerusalem artichoke, etc.) *bot*

geotaxis: oriented movement of a motile organism toward or away from a gravitational force *bot*

germ: the embryo; or a collective name given to the embryonic roots and shoot, and the scutellum tissue of the grain *bot*

germ cell >>> gamete

germ line: the lineage of cells from which the gametes are derived and which therefore bridge the gaps between generations, unlike somatic cells in the body of an organism *bot*

germ pore: an area, or hollow, in a spore wall through which a germ tube may come out *bot*

germ tube: the filament that emerges when a spore germinates *bot phyt* >>> pollen tube

germability: the degree of potential for germination *bot* >>> germination test

germicidal: it refers to any substance or condition that kills the embryo *seed*

germinal: referring to the germ or germination *bot*

germinal cell >>> gamete

germinal selection: the selection during gametogenesis against induced mutations that retard the spread of mutant cells *gene*

germination: the beginning of growth of a seed, spore, or other structure, usually following a period of dormancy and generally in response to the return of favorable external conditions; when it takes place a root is produced that grows down into the soil; at the same time, a stem and leaves are growing upward *bot*

germination test: a procedure to determine the proportion of seeds that are capable of germinating under particular conditions *seed* >>> germinator

germinative cell >>> gamete

germinator: an apparatus with which seed germination is realized under more or less controlled conditions *seed prep*

germplasm: the hereditary material transmitted to offspring through the germ cells and giving rise in each individual to the cells *gene*

germplasm bank >>> gene bank

germplasm collection >>> gene bank

gibberellic acid (GA): a group of growth-promoting substances; they regulate many growth responses and appear to be a universal component of seeds and plants *phys* >>> short-straw mutant >>> aleuron(e) layer

gibberellin: the generic name of a group of plant hormones that stimulate the growth of leaves and shoots; they tend to affect the whole plant and do not induce localized bending movements; they are thought to act either at a transcriptional level or as inducers of enzymes; first isolated from the fungus *Gibberella fujikuroi,* which causes the Bakanae disease in rice *phys phyt* >>> aleuron(e) layer >>> short-straw mutant

Giemsa stain >>> banding

gigantism: abnormal overdevelopment due to an increase in cell size (hypertrophy); for example, in roots of crucifers infected with club root *(Plasmodiophora brassicae),* or abnormal overdevelopment as a result of an increase in the number of cells in response to a disease-production agent (hyperplasia), for example, witches broom, cankers, galls, leaf cure, or scab *phys phyt*

ginned lint: cotton fibers after they have been removed from the seed *agr*

GISH >>> genomic in situ hybridization

glabrous: without hair or smooth *bot*

gland: organs or swellings that usually secrete a watery or characteristic substance; many oily and aromatic products are glandular in origin *bot*

glandular: having or showing glands *bot*

glass slide >>> heating

glasshouse >>> greenhouse

glassy grain >>> hyaline grain

glaucous: with waxy bloom present on the surface of the plant structure; a whitish, grayish, or bluish appearance is often imparted *bot*

gliadin: any prolamin and a simple protein of cereal grains that imparts elastic properties to flour; it is a monomeric molecule between 30,000 and 75,000 kDa; it is divided in alpha-, gamma-, and omega-gliadins; it may form large polymeric structures as a result of intermolecular disulfide bonds *chem phys* >>> Table 15

globulin: one of a group of globular, simple proteins, which are insoluble or only sparingly soluble in water, but soluble in salt solutions; they occur in plant seeds (mainly in dicots), where they have a variety of functions (e.g., legumin, vignin, glycinin, vacilin, or arachin) *chem phys* >>> Table 15

glomerule: a very compact cyme; a cluster of flowers *bot*

glucide >>> carbohydrate

glucose >>> dextrose

glucoside: glucosides are soluble in water and alcohol; some of them are highly poisonous (e.g., saponin, from tung tree, *cf* Important List of Crop Plants); they are found in vegetative organs and some in seeds (e.g., salicin in bark and leaves of willows; amygdalin in seeds of almonds, peaches, or plums; sinigrin in black mustard; aesculin in horse chestnut seeds; quereitron in the bark of oaks) *phys*

glume(s): the outermost pair of bractlike structures of each spikelet; a chafflike bract *bot* >>> Table 34

glume surface: the upper external surface of the broad wing of the glume (e.g., in wheat, which is described as being rough and/or smooth when scratched with a needle point) *bot*

glutamate (Glu): a salt or ester of glutamic acid *chem phys*

glutamic acid: an amino acid involved in purine biosynthesis, occasionally added to plant tissue culture media; it may replace ammonium ions as the nitrogen source; it is of key importance in pollen growth in vitro *chem phys*

glutathione: a tripeptide containing glutamic acid, cystein, and glycine capable of being alternately oxidized and reduced; it plays an important role in the cellular oxidation *chem phys*

gluten: a term that is utilized to refer to a naturally occurring mixture of two different proteins (glutenin and gliadin) in the seeds of, for example, wheat; it is the principal protein in cereal seeds; it consists of a long polypeptide chain; in wheat, it possesses particular elasticity, which allows production of high-quality bread (strength and elasticity of the flour); for example, more of the high-molecular-weight glutenin (which is "stretchy" and imparts physical strength to a dough) results in a flour that is better suited to manufacture high-quality yeast-raised bread products *chem phys* >>> gliadin >>> glutenin >>> Table 15

glutenin: it is soluble in aqueous or saline solutions or ethyl alcohol; it can also be extracted with strong acid or alkaline solutions; it is found in cereal seeds (e.g., glutenin in wheat or oryzinin in rice); it is divided in a low-molecular-weight glutenin subunit (LMW-GS) and in a high-molecular-weight glutenin subunit (HMW-GS) of about 65,000-90,000 kDa *chem phys* >>> Table 15

glycerin(e): a three-carbon trihydroxy alcohol that combines with fatty acids to produce esters, which are fats and oils; it may serve as a cryoprotectant *chem phys* >>> cryoprotectant >>> refraction index

glycerol >>> glycerin(e)

glycine (Gly): an amino acetic acid; the simplest alpha amino acid *chem phys*

glycinin >>> globulin

glycoll >>> glycine

glycoprotein: a conjugated protein that consists of a carbohydrate covalently lined to a protein *chem phys*

glycoside >>> glucoside

glycosylation: the attachment of a carbohydrate to another molecule *phys*

glyoxaline >>> imidazole

glyphosate: a chemical compound used as a herbicide; glyphosate resistance is a subject of biotechnological approaches (e.g., tobacco and tomato transformants may show an overexpression of 5-enolpyruvyl shikimate-e-phosphate synthase [EPSPS], which is usually blocked at normal concentrations by the herbicide); transformed cells, callus, or individuals proved to be tolerant to high glyphosate concentrations *phyt biot*

gm food >>> genetically modified organisms

GMO >>> genetically modified organisms

GOLGI apparatus: a system of flattened, smooth-surfaced, membranaceous cisternae, arranged in parallel 20-30 nm apart and surrounded by numerous vesicles; a feature of almost all eukaryotic cells; this structure is involved in the packaging of many products of cell metabolism *bot*

goodness of fit: methods to test the conformity of an observed empirical distribution function of data with a posited theoretical distribution function (e.g., chi-square test) by comparing observed and expected frequency counts; the KOLMOGOROV-SMIRNOV test calculates the maximum vertical distance between the empirical and posited distribution functions *stat*

gossypol: a dark pigment, $C_{30}H_{30}O_8$, derived from cottonseed oil *chem phys*

gradation: the successive increase of organisms in a more or less cyclic or spontaneous pattern *evol eco*

gradient: a gradual change in a quantitative property over a specific distance or time *prep*

graft: to transfer a part (a small piece of tissue or organs) of an organism from its normal position to another position on the same organism (autograft) or to a different organism or species (heterograft); the stem or shoot that is inserted into a rooted plant is called the scion; the plant or part of a plant into which the scion is inserted is called the stock or understock; there are many different methods of grafting (e.g., flat grafting, split grafting or side grafting); sometimes the term is used in order to describe the point where a scion is inserted in the stock *hort* >>> transplant >>> scion

graft chimera >>> graft hybrid

graft hybrid: a plant made up of two genetically distinct tissues due to fusion of host and donor tissues after grafting *hort*

grafting: the joining together of parts of plants; the united parts will continue their growth as one plant *hort*

grafting tape: tape backed with biodegradable cloth; used in budding and grafting operations and in banding tree wounds *hort*

grafting thread: a fine-waxed string used in budding and grafting operations *hort prep*

grafting wax: a wax or related substance that is used to cover all injured parts of the rootstock and the scion after grafting and thus prevent infection by fungi or bacteria *prep hort*

grain: a cereal caryopsis that may or may not be enclosed by the lemma an palea *bot* >>> corn

grain grade: a market standard established to describe the amount of contamination, grain damage, immaturity, test weight, and marketable traits *seed meth*

Gramineae: any of the monocotyledonous, mostly herbaceous plants, having jointed stems, slender sheathing leaves, and flowers borne in spikelets of bracts; the species in which the cereals are included appeared during the Cretaceous period (136-65 million years ago) *bot evol;* generally, in grassland agriculture the term does not include cereals when grown for grain but does include forage species of legumes often grown in association with grasses *agr* >>> Table 32 >>> *cf* Important List of Crop Plants

gramineous >>> Gramineae

GRAM's stain: an important bacteriological staining procedure discovered empirically in 1884 by the Danish scientist C. GRAM; a technique used to distinguish between two major bacterial groups based on stain retention by their cell walls; bacteria are heat-fixed, stained with crystal violet, a basic dye, then with iodine solution; this is followed by an alcohol or acetone rinse; GRAM-positive bacteria are stained bright purple; GRAM-negative bacteria are decolorized; safranin is used to stain them *meth*

granary: a storehouse or repository for grain *agr*

granum (grana *pl*): stacks of circular thylakoids, composed of lamellae in higher green plant chloroplasts, containing pigments and other essential components of photosynthetic light reactions *bot*

grape sugar >>> dextrose

grasses >>> Gramineae

gravitational water: the water that flows freely through the soil in response to gravity *agr*

gravity separator: a machine utilizing a vibrating porous deck and air flow to separate seed on the basis of their different specific gravities *seed*

gray level: the brightness of pixels in a digitized video and/or computer image; for an 8-bit signal, this ranges from 0 (black) to 225 (white) *micr*

graze: the eating of crops by animals in the field *agr*

green chop: green plants cut into small sections for animal feed *agr*

green manure crops: crops that are grown for the purpose of being ploughed into the soil to improve soil fertility and organic content; phacelia may be a green manure crop in some regions, as are various legumes, such as lupins; the crops are ploughed into the soil while they are still green *agr*

green manuring >>> green manure crops

green revolution: advances in genetics, petroindustry, and machinery that culminated in a dramatic increase in crop productivity during the third quarter of the twentieth century *agr*

greenhouse: a building, room, or area, usually of glass, in which the temperature is maintained within a desired range; used for cultivating tender plants or growing plants out of season *prep hort*

grid design: for the grid design, plants or variants are divided into blocks and the best ones chosen from each *stat* >>> Figure 32

grist: a mixture of grain (e.g., wheat or barley) utilized by, for example, the miller for grinding or the maltster for producing malt; grist may contain a mixture of several varieties *meth prep agr*

grits >>> semolina

groat: the caryopsis of oats after the husk has been removed *agr seed*

ground germination rate >>> field germination

group-selection method: a method of regenerating and maintaining uneven-aged stands in which trees are removed in small groups *meth fore*

growing tray: a tray having compartments like an ice-cube tray, used for starting seeds *hort prep*

growth: the increase in size of a cell, organ, or organism; it may occur by cell enlargement or by cell division *phys*

growth analysis: a mathematical analysis of crop or plant growth using relative growth rate, net assimilation rate, leaf area growth rate, and crop growth rate *agr phys*

growth curve: a curve showing the change in the number of cells in a growing culture as a function of time *phys*

growth form: the form of a plant; the habit in which a plant grows (e.g., shrubby plant, climbing plant, leaf plant, etc.) *bot* >>> growth habit >>> habit(us)

growth habit: the mode of growth of a plant; crops may be classified as (1) annuals (e.g., barley), (2) biennials (e.g., sugarbeet), or (3) perennials (e.g., alfalfa) *phys*

growth inhibitor: any substance that retards the growth of a plant or plant part; almost any substance will inhibit growth when concentrations are high enough; common inhibitors are abscisic acid and ethylene; other inhibitors, such as phenolics, quinones, terpens, fatty acids, and amino acids affect plants at very low concentrations *phys* >>> growth promoter

growth promoter: a growth substance that stimulates cell division (e.g., cytokinin) or cell elongation (e.g., gibberellin) *phys* >>> growth inhibitor

growth rate (of crop): the crop growth rate is a specific plant growth analysis term denoting the absolute growth rate of mass per unit land, etc. *phys agr*

growth regulator: despite natural growth regulators, a synthetic compound that, when applied to a plant, promotes, inhibits, or otherwise modifies the growth of that plant *phys* >>> brassinosteroid >>> cytokinin >>> gibberellin

growth stages *syn* **developmental stage:** the discrete portion of the life cycle of a plant, such as vegetative growth, reproduction, or senescence; there are systems to various crops in order to subdivide the broad physiological and/or morphological stages *phys*

growth substance: a naturally occurring compound, other than a nutrient, that promotes, inhibits, or otherwise modifies the growth of a plant *phys*

GS >>> genomic selection

guanidine: a crystalline, alkaline, water-soluble solid, CH_5N_3, used in making resins *chem phys*

guanine (G): a purine base that occurs in both DNA and RNA *chem gene*

guanylic acid: guanosine monophosphate; a ribonucleotide constituent of ribonucleic acid that is the phosphoric acid ester of nucleoside guanosine *chem*

guard cell: a specialized type of plant epidermal cells; two of which surround each stoma; changes in their turgidity cause stomatal opening and closing *bot phys* >>> photosynthesis >>> respiration

guidepost >>> landmark

GUS gene: a gene that codes for production of beta-glucuronidase (GUS protein) in *Escherichia coli* bacteria *biot*

guttation: the tearlike extrusion of water and sometimes salts from the aerial parts of plants, particularly at night when transpiration rates are low; it is the process of water being exuded from hydathodes at the enlarged terminations of veins around the margins of the leaves *bot;* in biotechnology, used for exudation of specific proteins made by artificially inserted genes *biot*

gymnosperm(ous): a kind of plant that produces seeds but not fruits; the seeds are not borne within an ovary and are called naked *bot*

gymnosperms >>> gymnosperm(ous)

gynaecium >>> gynoecium

gynandromorph: an individual exhibiting both male and female sexual differentiation *gene*

gynic: female *bot*

gynodioecy: plant species or population in which female plants as well as hermaphroditic plants occur *bot*

gynoecious >>> gynoecium

gynoecium: the collective term for the female reproductive organs of a flower, comprising one or more carpels *bot*

gynophore: the stalk that pushes pollinated peanut flowers into the soil *bot*

H

habit(us): the general appearance of a plant *bot*

habitat: the living place of an organism or community characterized by its physical or biotic properties *eco*

habituation: the diminishing requirement of some tissue cultures for growth-regulatory substances, possibly due to endogenous production *biot*

hair: a slender outgrowth of the epidermis, common on certain leaf or stalk structures *bot*

hairy root disease: a disease in some dicots; rootlike tissue is proliferated along segments of the stem; it is caused by *Agrobacterium rhizogenes*, if it carries a Ri plasmid *phyt biot*

half sibs: progeny with only one common parent *gene*

half-diallel cross: the crossing of a series of genotypes in all combination except reciprocal combinations *meth*

half-grain method: a method in cereals separating the embryo from the endosperm by transversal dissection; it is applied when the endosperm is used for biochemical and molecular studies while the embryo can be grown for seed production; in this way preselected individuals and/or genotypes can be multiplied, it can reduce the breeding population and experimental costs *meth biot*

half-hardy annuals: plants that will survive some frost, but not a long freeze *phys*

half-life (time): the time required for one-half the atoms of a given amount of a radioactive substance to decay *phy biot* >>> radio carbon dating >>> isotope

half-shrub >>> suffrutex

half-sib progeny selection >>> method of overstored seeds (*syn* remnant seed procedure) >>> Figure 4

halomorphic: a soil that has high levels of salt *agr*

halophilic >>> halophilous

halophilous: a salt-loving organism, adapted to a high-salt environment (e.g., in a salt marsh) *bot*

halophytes: plants that grow in saline soil *bot*

HANAHAN transformation procedure: an optimized procedure for the transformation of *Escherichia coli* with plasmid DNA using $CaCl_2$ *biot*

hand weeding: manually removing the undesirable species inhibiting the growth of valued species *meth agr*

hand-dibbed: sowing individual seeds by hand according to a special plot design; it is mainly used for F1 seeds when they are rare or seeds that need special care *meth*

hanging-drop culture (technique): a method of microscopic examination of organisms or particles suspended in a drop on a special concave microscopic slide *meth micr*

haplodiploidy: sex differentiation in which males are haploids and females diploids *bot*

haploid: applied to a cell nucleus that contains one of each type of chromosome (i.e., one set of chromosomes); designated "n" *gene* >>> monoploid

haploidization: the process whereby diploid somatic cells become haploid during a parasexual cycle or by experimental means *cyto meth*

haplontic: organisms in which meiosis occurs in the zygote resulting in four haploid cells *bot*

haplophase: that part of the life cycle in which the gametic chromosome number is found in reproduction cells *bot*

haplosis: the meiotic division resulting in haploid cells and/or gametes *bot*

haplosomic: the situation in which the homologue of a pair of chromosomes is missing in somatic cells *gene* >>> monosomic >>> Figure 37

haplotype: a combination of alleles of closely linked loci found in a single chromosome; sometimes, a combination of particular nucleotide variants within a given DNA sequence *gene biot*

hard seed: a seed that is dormant due to the nature of its seedcoat, which is impervious to either water or oxygen *seed*

hard wheat: tetraploid wheat, *Triticum durum,* used for high-quality noodles, bread, and pastas *agr* >>> *cf* List of Important Crop Plants >>> Table 1

hardening (off): the gradual process of acclimating plants started indoors to outside conditions (e.g., placing them in a sheltered location outdoors for increasing lengths of time over a period of days) *meth phys hort agr* >>> acclimatization

hardiness: the capability of a plant to withstand environmental stress *phys agr*

hard-leafed >>> sclerophyllous

hardseedness >>> hard seed

hard-shelled seed >>> hard seed

HARDY-WEINBERG law (equilibrium): states that in an infinitely large, interbreeding population in which mating is random and in which there is no selection, migration, or mutation, gene and genotype frequencies will remain constant from generation to generation *gene bio*

harlequin chromosome >>> sister chromatid exchange (SCE)

harrow: an agricultural implement with spikelike teeth or upright disks, for leveling and breaking up clods in ploughed land or experimental plots; to draw a harrow over soil *agr*

harvest index (HI): the proportion of the biological yield to economic yield (i.e., grain) *phys* >>> Table 33

haulm: stems or stalks collectively, as of peas, beans, or hops, or a single stem or stalk *bot*

haustorium (haustoria *pl*): in certain parasitic fungi, an outgrowth from hypha that penetrates a host cell in order to absorb nutrients from it; in some parasitic angiosperms, outgrowth of the roots; in endosperm development, nutrient-gathering outgrowths toward surrounding tissue of the developing endosperm *bot phyt phys*

hay: herbage, as grass, clover, or alfalfa, cut and dried for use as forage *agr*

head: an inflorescence in which the floral units on the peduncle are tightly clustered, surrounded by a group of flowerlike bracts called an involucre (e.g., in sunflower) *bot*

head components: generally, all components of the inflorescence of grain and grass crops *agr*

head shattering: a preharvest lost of kernels in cereals caused by loose seeds inside the spikelets and mechanical shattering (wind, etc.) *agr*

heading: emerging spikes (i.e., from initial emergence of the inflorescence from the boot until the inflorescence is fully exerted) *phys;* in viticulture, to shorten or prune the trunk when it reaches the desired height; done in an effort to focus growth on the lower shoots *hort*

heat filter: absorption glass filter that attenuates infrared radiation, but transmits light in the visible wavelength range *micr*

heat shock protein (HSP): when certain plants are exposed to high temperature, heat shock proteins are synthesized; they provide thermal protection to subsequent heat stress *phys* >>> stress protein(s)

heating: gentle heating of the slide over a spirit or other flame flattens the cells, sticks them to the glass slide and cover slip and spreads the chromosomes, whether in prophase or metaphase *prep cyto*

heaving: lifting effect of the soil due to alternate freezing and thawing; it may result in the lifting up of plants and may tear them loose from the soil or may shear off roots *agr*

heavy feeder: a plant that requires a great amount of nitrogen because of its speedy growth (e.g., squash, potato, tomato, etc.) *agr*

heavy soil: a soil that has a high content of clay and is difficult to cultivate *agr*

hectare: equals 10,000 square meters equals 2.471 acres *agr*

hedgerow: a row of bushes, trees, or special plants forming a hedge *agr*

heel: a piece of the old branch or shoot that is detached from the old branch or shoot along with a cutting made *hort meth*

heeling in: temporarily covering the base of a plant with soil for a short time (e.g., stecklings of sugarbeet during the winter) in order to prevent frost damage of cold-sensitive genotypes *agr hort meth*

heirloom plant: a plant that was developed and in cultivation sometime in the past *agr evol* >>> heirloom variety

heirloom variety >>> heritage seeds

hemialloploid: not a normal full alloploid but a segmental alloploid, in which some parts of the unified genomes show some degree of structural conformity *cyto*

hemiautoploid: autopolyploids with a certain degree of differentiation between the diploid sets of chromosomes, either by a subsequent differentiation of previously homologous sets of chromosomes or by spontaneous or induced intervarietal or subspecies hybridization *cyto*

hemicellulose: a heterogeneous group of compounds that in plant cell walls form part of the matrix within which cellulose fibers are embedded *chem phys*

hemichromosome: a chromosome split into chromatids without previous reduplication at interphase *cyto*

hemigamy: where gametes fuse but the nuclei do not, forming a dikaryon *bot*

hemihaploid: individuals with half of the normal haploid chromosome number *cyto*

hemimethylation: the state in which a DNA duplex is methylated in one strand but unmethylated in the other *gene biot*

hemiploid: individuals with half of the somatic chromosome number *cyto* >>> haploid

hemizygous: an individual, generally diploid, having a given gene and/or allele present once (e.g., in monosomics or haploids) *gene* >>> Figure 37

herb(s): a small, nonwoody, seed-bearing plant in which all the aerial parts die back at the end of each growing season *bot*

herbaceous: nonwoody, as applied to kinds of plant growth *bot* >>> herb(s)

herbage: plant material used for animal feed *agr*

herbarium: a collection of dried, preserved, and systematically classified plants *bot tax*

herbarium beetle: *Cartodere filum;* eats the spores of certain fungi (e.g., *Lycoperdon,* smuts, etc.) that are attached to the plant material of a herbarium *phyt*

herbarium glue: an adhesive that minimizes cracking, discoloration, and shattering with age; it is used in fastening plant specimens to the herbarium sheet *meth bot*

herbarium paste >>> herbarium glue

herbicide: a chemical substance that suppresses or eliminates plant growth; it may be a nonselective or selective weed killer *phyt*

herbicide tolerance: the ability of some plants to tolerate herbicides; it is a task of genetic engineering to modify crop plants for this trait in order to apply herbicides against weeds in the field *phyt biot*

hercogamy >>> herkogamy

hereditary: transmissible from parent to offspring or progeny *gene*

hereditary determinant: any genetically acting unit of an organism that is replicated and conserved transferred from generation to generation *gene*

hereditary factor >>> hereditary determinant

heredity: the transmission of genetic characters from one generation to the next generation; it operates primarily by the germ cells in sexually reproducing species *gene*

heritability: a measure of the degree to which a phenotype is genetically influenced and can be modified by selection; it is represented by the symbol h^2; this equals V_{gen}/V_{phe} where V_{gen} is the variance due to genes with additive effects and V_{phe} is the phenotypic variance; there are two types of heritability: (1) broad-sense heritability, $h_b{}^2 = V_{gen}/\ V_{phe}$ and (2) narrow-sense heritability, $h_n{}^2 = V_{add}/\ V_{dom}$ *gene stat* >>> genetic variance (V_{gen}) >>> Figure 38 >>> Table 34

heritable >>> hereditary

heritage seeds: nonhybrid seeds of old varieties that have been passed from generation to generation *agr*

herkogamy: pollination by the neighbor individual, population, or species *bot*

hermaphrodite: a plant having both female and male reproductive organs in the same flower *bot* >>> Table 18

hermaphroditic: reproductive organs of both sexes present in the same individual or in the same flower in higher plants *bot* >>> bisexual >>> Table 18

hesperidia: a berrylike fruit with papery internal separations or septa and a leathery, separable rind (e.g., orange, lemon, lime, grapefruit) *bot*

heteroallele: an allele that differs from other alleles of the same gene by nucleotide differences at different sites within the gene; in contrast with "true" alleles, of which only four are possible at each site within the gene *gene*

heteroallelic >>> heteroallele

heteroauxin: an obsolete term for the auxin 1H-indole-3-acetic acid (IAA) *chem phys* >>> indole-3-acetic acid

heterochromatic: of chromosome regions or whole chromosomes that have a dense, compact structure in telophase, interphase, and early prophase *cyto*

heterochromatin: the chromosome material that accepts stains in the interphase nucleus (unlike euchromatin); such regions, particularly those containing the centromeric and nucleolus organizers, may adhere to form a chromocenter; some chromosomes are composed primarily of heterochromatin; these are termed heterochromosomes, such as Y chromosome in some species *cyto*

heterochromosome: any chromosome that differs from the autosomes in size, shape, and behavior *cyto*

heteroduplex DNA: a double-stranded DNA molecule formed by the annealing of strands from two different sources, as opposed to homoduplex, which has homologous strands; as a result, there are regions noncomplementary and showing abnormalities in the form of extra loops *gene*

heteroecious: a species that produces male and female gametes on different individuals *bot* >>> dioecious; the requirement of a pathogen for two host species to complete its life cycle *phyt*

heterofertilization: fertilizing of the nuclei of endosperm and embryo-forming cells by genetically different gametes *bot*

heterogamete >>> anisogamete

heterogam(et)ic: a species that sexually reproduces by two types of gametes *bot*

heterogamy: reproduction involving two types of gametes *bot*

heterogeneity index: a measure for genetic differences within populations *gene*

heterogenetic: chromosome pairing between more or less different genomes in amphiploids *cyto*

heterogenic: gametes or populations differing in alleles or genes *gene*

heterograft: heterologous graft; the scion and rootstock derive from different species *hort*

heterohistont: an individual or cell aggregate that is composed of tissues of genetically different origin *bot*

heterokaryon: cells with two or more nuclei that are genetically non-identical *bot*

heterokaryotype: a chromosome complement that is heterozygous for any sort of chromosome mutations *cyto*

heterolabeling: a chromosomal labeling pattern due to induced or spontaneous exchange of labeled and nonlabeled half-chromatids *cyto*

heterologous gene expression: expression of a gene in another host *biot*

heterologous probing: probing at low stringency with a DNA fragment that originates from another organism and thus does not have an identical counterpart in the target DNA; often it gives significant signals because of sequence conservation *biot*

heteromeric: genes that control the determination of a trait by joint gene action, but each of them shows a definitely different contribution to the final product *gene*

heteromorphic: chromosomes that differ in size and/or shape; the term is also used for meiotic pairing configuration, which is composed of different chromosomes and/or chromosome segments *gene*

heteromorphic bivalent: a bivalent consisting of nonhomologous chromosomes or segments *cyto*

heteromorphous >>> heteromorphic

heteromorphy >>> heteromorphic

heterophylly: the production of more than one leaf form on a plant species; in developmental heterophylly, juvenile leaves may differ from adult ones *bot*

heterophyte: a plant that is dependent upon another, obtaining its nourishment from other living or dead organisms, such as parasites or saprophytes *bot*

heteroplasmic >>> alloplasmic

heteroplasmonic >>> alloplasmic

heteroplastidic: cells whose plastids are different in shape *bot*

heteroploid: deviating chromosome numbers from the standard chromosome set *cyto*

heteropycnotic: chromosomes or chromosomal segments that show a different coiling or staining pattern *cyto*

heterosis: the increased vigor of growth, survival, and fertility of hybrids, as composed with the two homozygotes; it usually results from crosses between two genetically different, highly inbred lines; it is always associated with increased heterozygosity; in breeding, three types of heterosis are distinguished: (1) F1 yielding more than the mean of the parents, (2) F1 yielding more than the best yielding parents, (3) F1 yielding more than the best yielding variety; for the genetic basis of heterosis two hypotheses have received the most attention: dominance hypothesis and overdominance hypothesis *gene* >>> Figure 2 >>> Table 35

heterosomal: a chromosome mutation involving nonhomologous chromosomes *cyto*

heterosome: a chromosome that is deviating from the standard chromosomes in size, shape, or behavior *cyto*

heterospory: the production of spores of two different types of the same plant *bot*

heterostyly: a polymorphism among flowers that ensures cross-fertilization through pollination by visiting insects; flowers have anthers and styles of different length *bot*

heterothallic: describes yeast strains that have a fixed mating type and as a result can only mate when mixed with a strain of the opposite mating type *bot*

heterotroph: an organism that is unable to manufacture its own food from simple chemical compounds and, therefore, consumes other organisms *phys*

heterozygosity: the presence of different alleles of a given gene at a particular gene locus *gene* >>> Tables 9, 10

heterozygote: a diploid or polyploid individual that has different alleles on at least one locus *gene* >>> Tables 9, 10

heterozygotic >>> heterozygous

heterozygous: the condition of having unlike alleles at corresponding loci *gene*

hexaploid: with six sets of chromosomes *cyto*

hexasomic: a cell or individual showing one chromosome six times *cyto*

hexokinase: an enzyme that catalyzes the phosphorylation of hexose sugars *chem phys*

hexose: a monosaccharide sugar that contains six carbon atoms *chem phys*

HI >>> harvest index

hibernaculum: the winter resting body of some plants, generally a bud-like arrangement of potential leaves *bot*

hibernating organ >>> hibernaculum

***Hieracium* type:** apospory where the embryo sac is derived from a somatic cell, usually from the center of the nucellus; three mitoses lead to a mature eight-nucleate, unreduced embryo sac *bot*

hierarchical classification: the grouping of individuals by a series of subdivisions or agglomerations to form characteristic "family trees" or dendrogram of group *stat*

high oleic: in sunflower breeding, seeds that contain a trait for high oleic fatty acid content in its oil; a premium oil used in the snack food industry *agr*

highly repeated DNA sequence: a specific DNA sequence that is present with very high copy number *gene*

high-velocity microprojectile transformation: a procedure for introduction of DNA into plant cells; for example, gold particles are coated with DNA and propelled at high speed through the target cell walls by means of an electrical or gunpowder discharge *biot*

hill plot >>> Figure 33

hilum (hila *pl*): the scar on a seed that marks the point at which it was attached to the plant *bot*

hips: seed pods (fruits), for example, in roses or apples, that are formed after a flower's petals fall if the bloom was pollinated *hort*

histidine (His): a basic, polar amino acid that contains an imidazole group *chem phys*

histogram: a bar graph of a frequency distribution in which the bars are displayed proportionate to the corresponding frequencies *stat*

histone: one of a group of basic, globular, simple proteins that have a high content of the amino acids arginine and lysine; it forms part of the chromosomal material of eukaryotic cells and appear to play an important role in gene regulation *chem phys cyto*

hitchhiking effect: genes favored in a population by close linkage with other genes, which are positively selected *evol gene*

hole seeding >>> dibbling (seed)

holocentric: applied to chromosomes with diffuse centromeres such that the properties of the centromere are distributed over the entire chromosome *cyto*

holokinetic >>> holocentric

homeobox: a characteristic DNA sequence of 180 bp that codes for the 60-amino-acid DNA-binding domain of some developmentally important regulatory genes; mutations in the homeobox can have homeotic effects *gene biot*

homeostatic mechanism: a physiological process that contributes to the maintenance of a relatively stable internal environment in a multicellular organism *phys*

homeotic gene(s): a class of genes that determines the identity of an organ, segment, or other structural unit during development; it controls the identity of, for example, floral organs *gene*

homeotic mutation: a mutation that causes one body structure to be replaced by a different body structure during development *gene*

homoallelic: applied to allelic mutants of a gene that have different mutations at the same site *gene*

homoduplex DNA >>> heteroduplex DNA

homoeologous: partially homologous; chromosomes or genomes that are believed to have originated from ancestral homologous chromosomes *cyto*

homoeologous group: series of two or more chromosomes with similar but not homologous chromosomes (e.g., in hexaploid wheat or oat) *cyto*

homoeostasis: the tendency of a biological system to resist change and to maintain itself in a state of stable equilibrium *bio* >>> genetic equilibrium

homogametic: producing only male or female gametes *bot*

homogamic: of matings between individuals from the same population or species *gene*

homogamous: hermaphroditic flowers showing synchronized function of male and female sex organs *bot*

homogamy: the preference of individuals to mate with others of a similar genotype or phenotype; in botany, the condition in which male and female parts of the flower mature simultaneously *bot*

homogenic: in cytology, sometimes it refers to chromosome pairing between morphologically identical genomes in amphiploids *cyto*

homograft >>> heterograft

homologous: applied to organs and chromosomes; both showing identical structures *cyto bot*

homologous genes: genes with a common ancestor, generally used to describe genes from different species but which are similar and have the same function *gene*

homology: fundamental similarity; in molecular biology, the degree of identity between two sequences from related organisms *biot*

homomeric: genes that control the determination of a trait by joint gene action, and each of them shows a similar contribution to the final product *gene*

homomorphic bivalent: a bivalent that is composed by two homologous chromosomes (i.e., in size and shape) *cyto*

homoplasmic: cells or individuals that carry two or more different types of cytoplasmic components *bot*

homoplasmonic >>> homoplasmic

homopolymer-tailing: attachment of identical nucleotides to the 3' end of a DNA molecule that can be achieved with terminal deoxynucleotide transferase *biot*

homothallic strains: in yeast, strains that are capable of switching mating type and thus of mating with themselves, forming zygotes, asci, and spores under appropriate conditions *bot*

homozygosity: the presence of identical alleles at one or more loci in homologous chromosomal segments *gene* >>> Table 9

homozygote: a cell or organism having the same allele at a given locus on homologous chromosomes *gene* >>> Table 9

homozygous: having identical rather than different alleles in the corresponding loci of a pair of chromosomes and therefore breeding true *gene* >>> Table 9

honeybee: any bee that collects and stores honey (e.g., *Apis mellifera*) and contributes to improved seedsetting in cross-pollinating crops (e.g., rapeseed, fruit trees, etc.) *zoo seed*

honeycomb design: in a honeycomb design, the plant at the center of the hexagon is compared with every other plant within the hexagon; a plant is chosen only if it is superior to every other plant in the hexagon; it was developed for selecting individual plants in a population; seeds or plants are usually spaced equidistantly from one another in a hexagon pattern; plants are spaced far enough apart that they cannot compete with adjacent individuals; homogeneous checks can be included; the size of the hexagon determines the selection intensity; it is used to minimize adverse effects of interplant competition *stat* >>> Figure 32

honeydew: a sticky exudate containing conidia, which is produced during one stage of the life cycle of the fungus *Claviceps purpurea phyt bot* >>> ergot

hook climber: a plant that climbs by the aid of hooks or prickles (e.g., roses) *hort* >>> climbing plants

hordecale: an amphiploid hybrid between cereal species of the genera *Hordeum* and *Secale,* in which *Hordeum* species served as donors of the cytoplasm *bot*

hordein >>> prolamin

***Hordeum bulbosum* procedure:** a method for producing zygotic haploids in barley by crossing *Hordeum bulbosum* with *Hordeum vulgare* genotypes; after formation of zygotes the *H. bulbosum* chromosomes are subsequently eliminated during embryogenesis, the chromosomes of the wild barley are subsequently eliminated, which results in haploid *H. vulgare* plants *biot* >>> Figures 17, 26 >>> Table 7

horizontal resistance: resistance conditioned by polygenes or quantitative genes; it is race nonspecific in nature and does not reveal a gene-for-gene hypothesis; the type of resistance is difficult to identify *phyt*

hormone: a regulatory substance, active at low concentrations, that is produced in specialized cells but that exerts its effect either on distant cells or all cells to which it is conveyed via tissue fluids in the organism *phys*

horticulture: the science or art of cultivating flowers, fruits, vegetables, or ornamental plants (e.g., in a garden, orchard, or nursery) *hort*

hortus siccus: a collection of specimens of plants carefully dried, preserved, and described for botanical purposes and comparisons of mutants, etc. *bot meth* >>> herbarium

host: a living organism harboring a parasite and its cells and metabolism are used for the growth of pathogens; plant host can be classified by: (1) importance: (a) primary or principal hosts, (b) secondary hosts, (c) intermediate hosts, (d) accessory hosts, (e) accidental hosts, and (f) definite or final hosts; (2) season: (a) winter hosts and (b) spring hosts; (3) other functions: (a) alternative, alternate, or differential hosts, (b) transport or transfer hosts *phyt*

host cell: a cell whose metabolism is used for the growth of a pathogen *phyt*

host plant >>> host

host range: the spectrum of genotypes that can infect a specific pathogen or pest; in molecular biology, hosts in which a phage or plasmid can replicate; restriction is one factor that can limit the host range of plasmids or phages *phyt biot*

host species >>> heteroecious

host-mediated restriction: a mechanism by which bacteria prevent infection by phages originating from other bacteria; restriction also acts against unmodified plasmid DNA in transformation experiments; restriction endonucleases cleave the foreign DNA while the host DNA is protected from cleavage by specific methylation *biot*

host-parasite specificity: the ability of a pathogen to pathogenize a specific group of plants *phyt*

hot spot: in genetics, one of sites tending to mutate frequently *gene*

housekeeping enzymes: enzymes present in all cells capable of normal metabolism; they are essential for the synthesis or breakdown of proteins, nucleic acids, and lipids, for glycolysis and respiration and for many standard metabolic pathways *phys*

housekeeping genes: genes whose products are required by all cells at all times *gene* >>> housekeeping enzymes

hull: usually the hard, tightly adhering, outer covering of a seed or caryopsis, which is composed of the pericarp in some species and the lemma and palea in others *seed;* the persistent calyx at the base of some fruits, such as strawberry *bot*

hull: to remove the hull *seed prep* >>> dehull

huller-scarifier: a seed conditioning machine; it removes hulls or pods from seeds by an abrading or rubbing action *seed* >>> Table 11

humic acid: a mixture of dark-brown organic substances that can be extracted from soil with dilute alkali *agr*

humification: the development of humus from dead organic material *agr*

humus: decomposed organic matter of soils *agr*

husk: the leaf sheaths of an ear of maize; the lemma and palea in other grass species or the dry outer cover of a coconut; the dry outer covering of some fruits or seeds (e.g., cereal grain) *bot* >>> dehusk

hyaline: clear, transparent *bot*

hyaline grain >>> hyaline

hybrid: any sort of sexual or somatic combination of genetically more or less differentiated parental cells, individuals, or taxa; specifically, an individual plant from a cross between parents of differing genotypes; any heterozygote represents dissimilar alleles at a given locus; or a hybrid graft *gene* >>> Figures 2, 18, 31

hybrid breakdown >>> hybrid lethality

hybrid breeding (*syn* heterosis breeding): the discovery of heterosis has been recognized as one of the major landmarks of plant breeding; in comparison to inbred lines and homozygous material, the phenotypic superiority of heterozygotes is the basis of hybrid breeding; it is exploited for production of hybrid, synthetic, and composite varieties; hybrid breeding can be performed using traditional breeding techniques, or the process can be hastened using gene marker technology to rapidly identify parents with desired genes for certain attributes; numerous commercial crops are hybrids with increasing tendency; seeds from a hybrid variety, if planted, will not deliver the same benefits as the original seeds and after several offspring will have lost the desired qualities from the original hybridization *meth*

hybrid chlorosis: plant and/or leaf chlorosis due to interacting genes and/or cytoplasm of parental lines in a hybrid *gene*

hybrid complex: masking morphological differences of parental lines in a hybrid plant *gene*

hybrid heterosis >>> heterosis

hybrid inviability: reduced vigor of hybrid plants compared to their crossing parents *gene*

hybrid lethality *syn* **hybrid sterility:** the failure of hybrids to produce viable offspring *gene*

hybrid necrosis >>> hybrid lethality

hybrid plant >>> hybrid

hybrid seed production: the production of hybrid seeds by combination of more or less defined parental forms; usually those hybrid seeds are more productive or more suitable than pure lines; they are used for subsequent growing and commercial production of a crop *seed* >>> Figures 2, 18, 22, 23, 29 >>> Table 5

hybrid selection: the process of choosing plants possessing desired traits among a hybrid population *meth*

hybrid sterility *syn* **hybrid lethality:** the failure of hybrids to produce viable offspring *gene*

hybrid variety: a variety produced from the cross fertilization of inbred lines with favorable combining ability; the progeny is homogeneous and highly heterozygous *seed* >>> Figures 2, 29

hybrid vigor: the increase in vigor of hybrids over their parental inbred types *gene* >>> heterosis >>> Figure 18

hybrid zone: a geographic area where different populations of species meet and hybridize after a period of geographic isolation *eco*

hybridization: a method of breeding new varieties that applies crossing to obtain genetic recombination; in genetics, the fusion of unlike genetic material, such as sexual organs or DNA; in molecular biology, pairing of complementary DNA and/or RNA *meth biot* >>> Figure 31

hybridization probe: labeled nucleic acid molecule used to detect complementary DNA sequences after hybridization *biot*

hydathode: an epidermal structure specialized for the secretion or exudation of water *bot* >>> guttation

hydratation: the status of imbibition of the cytoplasm *phys*

hydrate: any of a class of compounds containing chemically combined water *chem*

hydration: the process whereby a substance takes up water *agr*

hydraulic seeding: a method of planting grass seed by spraying it in a stream of water, which may contain other materials such as nutrients *meth agr*

hydrogen peroxide test: a quick test to determine seed viability; in response to a hydrogen peroxide soak, viable seeds elongate their roots through a cut in the seedcoat; frequently used in conifer seeds *seed*

hydrolase: an enzyme that catalyzes reactions involving the hydrolysis of a substrate *chem phys*

hydrolysat(e): any compound formed by hydrolysis *chem*

hydrolysis: in soil science, the process whereby hydrogen ions from water are exchanged for cations such as sodium, potassium, calcium, and magnesium, and the hydroxyl ions combine with the cations to give hydroxides *chem agr*

hydrophytes >>> hygrophytes

hydroponics: the cultivation of plants by placing the roots in liquid nutrient solutions rather than in soil *hort*

hydroseeding: dissemination of seed hydraulically in a water medium, supplemented by mulch, lime, and fertilizer *meth fore hort*

hydrotaxis: movement of an organism toward or away from water *bot*

hydroxide: a chemical compound containing the hydroxyl group *chem*

hygrometer: any instrument for measuring the water-vapor content of the atmosphere *prep*

hygrophytes: plants that can tolerate excess of water *bot*

hygroscopic(al): water attracting; plants or part of them becoming soft in wet air and hard in dry air *bot*

hygroscopic water: water that is adsorbed onto a surface from the atmosphere *agr*

hypanthium: a cuplike or tubelike enlargement of the floral receptacle or base of the perianth that surrounds the gynoecium and fruits *bot*

hypermorph: a mutant gene which causes an increase in the activity that it influences *gene*

hyperplasia: the enlargement of tissues by an increase in the number of cells by cell division *bot*

hyperploidy: having additional chromosome complements compared to the standard chromosome set *cyto*

hypersensitive resistance >>> hypersensitivity

hypersensitive site: a region of DNA located in a chromatin structure that makes it more sensitive to attack by endonucleases than DNA sites, located elsewhere in the chromatin; the presence of hypersensitive sites is correlated with transcription of adjacent DNA sequences in eukaryotic cells *gene*

hypersensitivity: the response to attack by a pathogen of certain host plants in which the invaded cells die promptly and prevent further spread of infection *phyt;* resistance

hypertonia >>> hypertonicity

hypertonic: a solution whose osmotic potential is less than that of living cells, causing water loss, shrinkage, or plasmolysis of cells *phys*

hypertonicity >>> hypertonic

hypertrophy: the enlargement of tissues by an increase of the size of the cells *bot* >>> hyperplasia

hypha (hyphae *pl*): a tubular, threadlike filament of fungal mycelium *bot*

hypocotyl: part of the embryonic shoot or seedling located below the cotyledon and above the radicle *bot*

hypogeal: living or growing underground *bot*

hypogean >>> hypogeal

hypoplasia: abnormal deficiency of cells or structural elements *bot*

hypoploidy: having missing chromosome complements compared to the standard chromosome set *cyto*

hypostatic epistasis >>> epistasis

hypotonic: of or designating a solution of lower osmotic pressure than another, as opposed to hypertonic *phys*

I

I1, I2, I3, etc.: the first, second, third, etc., generations obtained by inbreeding *gene meth*

IAA >>> indole-acetic acid

IAEA >>> International Atomic Energy Agency, Vienna, Austria

IBPGR: International Board for Plant Genetic Resources, Roma, Italy; cocoordinating international plant conservation, recently renamed IPGRI

ICARDA >>> International Center of Agricultural Research in the Dry Areas

ICBN: International Code of Botanical Nomenclature

ICRISAT >>> International Crops Research Institute for the Semi-Arid Tropics

identical by descent: two genes that are identical in nucleotide sequence because they are both derived from a common ancestor *gene*

identical in structure: two genes that are identical in nucleotide sequence, regardless of whether or not they are both derived from a common ancestor *gene*

identity preservation (IP): a system of crop or raw material management that preserves the identity of the source or nature of the materials *agr*

ideotype: crop plant with model characteristics known to influence photosynthesis, growth, and grain production *phys gene*

ideotype breeding: a method of breeding to enhance genetic yield potential based on modifying individual traits where the breeding goal for each trait is specified *meth*

idioblast: a plant cell committed to develop into a cell type that differs from the surrounding tissue *bot*

idiochromosome: a chromosome that contributes to the determination of sex *cyto*

idiogamy: combination of male and female gametes from the same individual *bot*

idiogram: a diagrammatic representation of the karyotype of a plant *cyto* >>> Figure 12

idioplasm: all hereditary determinants of a plant including genotype and plasmotype *gene* >>> germplasm

idiotype: the sum of the hereditary determinants of a cell or plant consisting of the genotype and plasmotype *gene*

IITA: International Institute of Tropical Agriculture, Ibadan, Nigeria; responsible for groundnut, soybean, sweet potato, cassava, cowpea, and rice research

I-line >>> inbred line

illegitimate crossing-over >>> unequal crossing-over

illegitimate recombination: recombination between DNA fragments that do not share extensive DNA sequence homology; transposons and insertion sequences have special functions that catalyze illegitimate recombination *biot*

Illinois method: the separate sowing of lines or families and selection of best plants from the best families *meth* >>> ear-to-row selection

image processing: various mathematical procedures to improve the signal-to-noise and contrast, and to obtain quantitative intensity data from images *micr*

imbibe >>> imbibition

imbibition: the adsorption of liquid, usually water, into ultramicroscopic spaces or pores found in material such as cellulose, pectin, and cytoplasmic proteins in seeds *bot*

imidazole: a compound whose molecule forms a pentagonal ring of C and H atoms with an N and NH group attached *chem*

imino acid: an acid derived from an imine in which the nitrogen of the imino group and the carboxyl group are attached to the same carbon atom *chem*

immature: not mature or ripe *phys*

immaturity >>> immature

immersed: embedded in a substrate *meth*

immersion lens: a special microscopic lens adjusted to place material between the uppermost surface of a microscopic sample (slide or coverslip) and the objective (e.g., immersion oil) *micr*

immersion medium: material placed between the uppermost surface of a microscopic sample (slide) and the objective (e.g., immersion oil) *micr*

immigration: in genetics, the movement or flow of genes into a population, caused by immigrating individuals, which interbreed with the residents *gene eco*

immobilization: the conversion of a chemical compound from an inorganic to an organic form as a result of biological activity *phys agr*

immune: not affected by pathogens; exempt from infection; the condition of having qualities that do not allow the development of a disease *phyt*

immune reaction: the reaction between a specific antigen and antibody; when plants are inoculated (e.g., with the BUK strain of tomato blacking nepovirus) they are subsequently protected against secondary infection with a similar viral strain, but not against a dissimilar strain of virus *phyt*

immune response >>> immune reaction

immune system: active defense of plants against infections and other invasive aggressions; it detects antigens on the invading entities and creates new antibodies to destroy them *phys phyt*

immunity: a natural or acquired resistance of a plant to a pathogenic microorganism or its products *phyt*

immunity breeding >>> resistance breeding

immunize: to make immune *phyt*

immuno-electrophoresis: a technique for the differentiation of proteins in solution, based on both their electrophoretic and immunological properties;

initially the proteins are separated by gel electrophoresis; they are then reacted with specific antibodies by double diffusion through the gel; the pattern of precipiting arcs thus formed can be used to identify the proteins *meth sero*

immunofluorescence: any of various techniques for detecting an antigen or antibody in a sample by coupling its specifically interactive antibody or antigen to a fluorescent compound, mixing with the sample and observing the reaction under an ultraviolet-light microscope *meth*

immunogenetics: studies using a combination of immunologic and genetic techniques, as in the investigation of genetic characters detectable only by immune reactions *gene*

immunological screening: use of an antibody to detect a polypeptide synthesized from a clone *biot*

immunoprecipitation: precipitation of antigens with the help of antibodies *biot*

immunosuppressants: a substance that results in or affects immunosuppression *meth sero*

impeder: an individual of any value actually impeding the development of another individual of higher grade *gene meth*

imperfect flower: unisexual flowers; flowers lacking either male or female parts *bot*

imperfect state (of fungi): the asexual state of a fungus (i.e., the state in which no sexual reproduction occurs) *phyt bot*

implant: material artificially placed in an organism *biot*

improvement planting: any planting done to improve the value of a stand and/or experiment and not to establish a regular plantation *meth*

in situ: in place; where naturally occurring *meth*

in situ hybridization (ISH): a technique to locate those segments complementary to specific nucleic acid molecules; chromosomes, which are treated to denature the DNA and remove RNA and proteins, are then incubated with radioactive labeled nucleic acids or nonradioactive labeled probes of special properties; the hybridized segments are then visualized by autoradiography or directly through the microscope and photography of fluorescent signals *cyto meth*

in vitro: literally, "in glass;" but applied more generally to studies and propagation of living plant material that are performed under artificial conditions in tubes, glasses, dishes, etc. *prep*

in vitro collection: a collection of germplasm maintained as plant tissue grown in active culture on solid or in liquid medium; it can be maintained as plant tissue ranging from protoplast and cell suspensions to callus cultures, meristems, shoot-tips, and embryos *meth*

in vitro culture: the cell, organ, or tissue culture performed under artificial conditions in tubes, glasses, dishes, etc. *biot*

in vitro fertilization: pollination performed aseptically in vitro by direct application of the pollen to the ovule; it is used to overcome prezygotic incompatibility *meth*

in vitro marker: a mutation that allows identification in vitro of a cell line possessing the marker *biot*

in vitro mutagenesis: methods for altering DNA outside the host cells; mutagenesis can be random or specific for the site and base change depending on the technique used *biot*

in vitro pollination >>> in vitro fertilization

in vitro propagation: propagation of plants under a controlled and artificial environment, usually applying plastic or glass vessels, aseptic techniques, and defined growth media *biot*

in vitro screening: search and selection for particular characters of cells, organs, or tissues performed under artificial conditions in tubes, glasses, dishes, etc., usually in combination with special nutritional media, which allow a differentiated growth of the cells, etc. *biot*

in vivo: literally, "in life;" applied to studies and propagation of whole, living organisms, on intact organ systems therein or on populations of microorganisms *meth phyt*

inarable: a field or land not arable and/or not capable of being ploughed or tilled *agr*

inarching *syn* **side grafting:** a method of grafting; usually a new plant growth onto a stronger root system; it is carried out by establishing young plants (sometimes, one that is in a pot) near an existing tree; at the point where they meet, at the matching areas the bark is removed; the two cut surfaces are then fitted together and bound with soft tying material until they grow together; later they can be gradually separated with the new branches attached to the older rootstock *hort* >>> graft

inbred: a plant resulting from successive self-fertilization of parents throughout several generations *gene*

inbred line: a line produced by continued inbreeding; usually a nearly homozygous line originating by continued self-fertilization, accompanied by selection *gene*

inbred pure lines: involves inbreeding of annual seed-propagated material; homogeneous, homozygous isolated by selection of desired recombinants or segregates in F2 to F7 generations of crosses between parental pure lines (generally monogenotypic, can be blended to form multilines, e.g., tomato, lettuce, soybean, pea, cowpea, snapbean, field bean, Arabian coffee, *Capsicum* pepper, eggplant, okra, lentil, and papaya) *meth*

inbred-variety cross: the F1 cross of an inbred line with a variety *meth*

inbreeding: the crossing of closely related plants; one important purpose of induced inbreeding is the development of genotypes that can be maintained through multiple generations of seed production; self-pollinated cultivars are reproduced for many generations by inbreeding; inbreeding is also used to reduce the frequency of deleterious recessive alleles in genotypes that serve as parents of a synthetic or a vegetatively propagated cultivar; inbreeding increases the genetic and phenotypic variability among individuals in a population; four mating systems are used to increase the homozygosity in a breeding population *meth*

inbreeding coefficient: the probability that the two genes at any locus in a diploid individual are identical by descent (i.e., they originated from the replication of one gene in a previous generation) *gene meth* >>> Table 10

inbreeding depression: the reduction in vigor often observed in progeny from matings between close relatives; it is due to the expression of recessive deleterious alleles; it is usually severe in open-pollinated outcrossing species; an effect opposite to heterosis *gene* >>> heterosis

inbreeding load: the extent to which a population is impaired by inbreeding *bio evol* >>> Table 10

inbreeding population >>> inbreeding

incertae sedis: of uncertain taxonomic position *tax*

inch (in): equals 2.54 cm

incipient species: populations that are too distinct to be considered as subspecies of the same species, but not sufficiently differentiated to be regarded as different species; sometimes called "semispecies" *tax*

inclined draper: a device for separating seeds using an inclined endless belt onto which seeds are metered; seeds are separated on the basis of their different tendencies to roll down the plane or to catch and be carried up and into a separate discharge spout *seed*

incompatibility (homomorphic or heteromorphic): a genetically determined inability to obtain fertilization and seed formation after self-pollination or cross-pollination; there are several types of progamous or postgamous incompatibility; in contrast to heteromorphic incompatibility (e.g., heterostyly in *Primula* spp.), homomorphic incompatibility is not associated with morphological differences *gene* >>> cross-sterility >>> crossbreeding barrier

incompatibility group: plasmids that are incompatible with each other belong to the same incompatibility group *biot*

incomplete diallel: a partial sampling; any individual family or type of family may be omitted *meth* >>> complete diallel

incomplete dominance >>> partial dominance

incomplete resistance: a type of resistance that is not complete and shows slow susceptibility to the pathogen *phyt*

increase (seed): to multiply a quantity of seed by planting it, thereby producing a larger quantity of seeds *seed*

incubation: the act or process of incubating *meth*

incubation period: the period between infection and the appearance of visible disease symptoms *phyt*

incubator: an apparatus in which media inoculated with microorganisms are cultivated at a constant temperature and/or air humidity *prep*

indehiscent: applied to fruits that do not open to release their seeds *bot*

indehiscent fruit >>> indehiscent

indent cylinder separator: a seed separator utilizing a rotating indented cylinder through which seeds are passed for cleaning; it lifts shorter seeds from longer seeds, thus separating them *seed*

indent disk separator: a seed separator utilizing multiple rotating disks inside a cylinder through which seeds are moved; it lifts seeds from longer seeded types, thus separating them *seed*

independent assortment of genes: the random distribution in the gametes of separate genes; if an individual has one pair of alleles *A* and *a*, and another pair *B* and *b* then it should produce equal numbers of four types of gametes: *AB*, *Ab*, *aB*, and *ab;* it is asserted in MENDEL's second law—the law of independent assortment *gene* >>> Figure 6

independent variables: if two random variables "a" and "b" are independent, then the probability of any given value of "a" is unchanged by knowledge of the value of "b" *stat*

indeterminate: descriptive of an inflorescence in which the terminal flower is last to open; the flowers arise from axillary buds and the floral axis may be indefinitely prolonged by a terminal bud *bot*

index selection: a form of intentional simultaneous selection; with the index selection some index value is assigned to each candidate; the index value indicates the aggregate value of each candidate across several traits; the index selection consists of truncation selection with regard to the index values *meth*

indexing: *US* the process used to test vegetatively reproduced plants for freedom from virus diseases before multiplying them *seed*

indicator plant: plants that are indicative of specific site or soil conditions *bot eco*

indigenous: an organism existing in, and having originated naturally in, a particular area or environment *bot eco*

indirect embryogenesis: embryoid formation on callus tissues derived from zygotic or somatic embryos, seedling plants, or other tissues in culture *biot*

indirect fluorescence: fluorescence emitted by fluorophores that are not an endogenous part of the specimen; usually introduced into a specimen as a stain or probe *micr*

indirect organogenesis: organ formation on callus tissues derived from explants *biot*

indirect selection: the direct selection for specific traits may imply unintentional indirect selection with regard to many other traits *meth*

indole-3-acetic acid (IAA): a substance that acts as a growth hormone or auxin in plants, where it controls cell enlargement and, through interaction with other plant hormones, also influences cytokinesis *phys*

indoor culture: growing plants indoors using natural and/or artificial light and additional heating; it is used for subtropical or tropical plants or for plant propagation *hort meth*

indoor plant >>> indoor culture

induced mutation: a change in a gene caused by a treatment *gene*

inducer: an effector molecule responsible for the induction of enzyme synthesis *phys*

induction: transcription of genes can be induced by inactivation of a repressor or by the action of an activator *biot*

induction media: media that can induce organs or other structures to form or media, which will cause variation and/or mutation in the tissue exposed to it *biot*

induction of flowering: the initiation of the production of flowers, possibly stimulated by florigen *phys* >>> florigen

induction of mutation: the process of causing a variation or mutation *gene*

industrial crop: crops that are processed on a large scale by industrial means (e.g., potato, pea, lupin, dwarf bean, or cereals for starch production; sugarbeet, beets, sweet sorghum, chicory, or Jerusalem artichoke for sugar processing; linseed, false flax, common marigold, crambe, caper spurge, *Cuphea* spp., meadowfoam, or jojoba for oil processing; flax, hemp, or nettles for fiber processing; foxglove, poppy, yellow bark, or cocoa for pharmaceutical utilization); industrial uses account for a relatively small but a growing and potentially much larger share of the market for agriculture commodities *agr*

inert: a chromosomal segment that is supposed to be genetically inactive or without coded genetic information *gene*

inert matter: one of the four components of a purity test; it includes non-seed material and seed material that is classified as inert according to the rules for testing seeds *seed*

infect: of a pathogen, to enter and establish pathogenic relationship with an organism; to enter and persist in a carrier; to make an attack on a plant *phyt*

infection: the invasion of the tissue of a plant by a pathogenic microorganism *phyt*

infection court: the site on a host plant at which infection by a parasitic organism is affected *phyt*

infection peg: a thickening of the host cell wall in the vicinity of the penetrating hypha; lignin, callose, cellulose, or suberin may be deposited at this site *phyt*

infection thread: specialized hypha of a pathogenic fungus that invades tissue of the susceptible plant *phyt*

inferior: applied to an ovary when the other organs of the flower are inserted above it *bot*

inferior pelea >>> lemma

infertile: not able to reproduce or not able to produce viable gametes *bot*

infertility: the situation in which a plant is unable to produce viable offspring *bot*

infest: attacked by animals (e.g., insects), or sometimes used of fungi in soil in the sense of contaminated *phyt*

inflected: when the keel of the, for example, wheat glume is bent inward in the upper third *bot*

inflorescence: a flower structure that consists of more than a single flower; the flower head terminates the culm in grasses; it may be determinate (solitary flower, simple cyme, compound cyme, scorpioid cyme, glomerule) or indeterminate (raceme, panicle, spike, catkin, spadix, umbel, head); determinate flowers are those in which the axis terminates as a flower; indeterminate flowers terminate in a bud, which continues to grow and produce flowers throughout the growing season; the latter results in flowers of different maturity within the same inflorescence *bot*

inflorescence meristem: the relatively undifferentiated, dividing plant tissue that gives rise to the inflorescence *bot*

infrared light: the part of the invisible spectrum that is contiguous to the red end of the visible spectrum and that comprises electromagnetic radiation of wavelengths from 800 nm to 1 mm *phy*

infructescence: a fruiting structure that consists of more than a single fruit *bot*

infundibular: funnel-shaped *bot*

infundibuliform >>> infundibular

ingraft: to insert, as a scion of one tree or plant into another, for propagation *hort meth*

inheritance: the transmission of genetic information from parents to progeny *gene*

inhibitor: a chemical substance that retards or prevents a growth process such as germination *phys*

injection: the act of injecting *meth*

inoculant: a preparation containing specific nitrogen-fixing bacteria that is added to legume seed prior to planting to assure that the resulting crop will have nitrogen fixation ability *seed*

inoculate: to place inoculum deliberately where it will reproduce *meth*

inoculation: the act or process of inoculating *meth;* in agriculture, addition of effective *Rhizobia* (bacteria) to legume seed prior to planting for the purpose of promoting nitrogen fixation *agr* >>> inoculate

inoculum: spores of other diseased material that may cause infection *phyt*

inositol: a carbocyclic or sugar alcohol that is widely distributed in plants *chem phys*

inositol triphosphate (InsP3): a chemical compound that contributes to the spatial orientation of a plant; for example, when wheat or maize plants are pressed down to the ground, a change of orientation of starch granules occurs in the cells; after a short time (30-120 minutes) InsP3 is accumulated on the lower side of the leaves; so-called motor cells are activated; they grow longitudinal and stepwise upright the plant *phys*

inprinting >>> genomic inprinting

insecticide: a substance or preparation used for killing insects *phyt*

insect-pollinated plant >>> cross-pollination

insert: a piece of foreign DNA introduced into a phage, plasmid, or other vector DNA *biot*

insertion: a genetic mutation in which one or more nucleotides are added to DNA, or the process and the result of transferring a foreign DNA or chromosome fragment into a recipient *biot*

insertion sequence: DNA sequence, which can excise and integrate into DNA without the need for extensive DNA homology *biot*

insertion vector: cloning vector where the cloned DNA is inserted into a restriction site, as opposed to replacement vectors where a piece of DNA is replaced in the process of cloning *biot*

insertional duplication: insertion of extra homologous base pairs into a recipient genome, which results in a mutation (by duplicated segments) *cyto gene*

insertional inactivation of a gene: insertion of a DNA fragment into the coding sequence of a gene usually leads to the inactivation of this gene *biot*

instability: variation that appears to be random and occurs constantly *gene*

intake auger (at a harvester): the auger tines guide the crop to the chain conveyor, which delivers it to the threshing section; any foreign bodies that may have been ingested fall into the stone trap, which is located between the conveyor and the concave *agr*

integrated control >>> integrated plant protection

integrated plant protection: disease and pest control by combining all available techniques, such as agronomic control, biological control, chemical control and sanitary procedures *phyt*

integument: the coats of the ovule (mostly two), which develop into the seed coat (testa) after fertilization *bot*

intellectual property rights: a system of patents that allows ownership over the applications of research *biot agr seed*

interbreeding: intercrossing of individuals within a population *meth*

intercalary: chromosomal segments located beside terminal regions *cyto*

intercalary meristem: an internodal meristem, situated between differentiated tissues; it produces cells perpendicular to the growth axis and causing internode elongation *bot biot*

intercalary segment >>> interstitial segment

intercalating agent: a chemical that can insert itself between the stacked bases at the center of the DNA double helix, possibly causing a frameshift mutation *chem meth*

interchange: an exchange of segments between nonhomologous chromosomes resulting in translocations *cyto* >>> translocation

interchange trisomic: an additional chromosome to the diploid set, which is composed by two different chromosomes via translocation *cyto* >>> balanced tertiary trisomic

interchromosomal: effects and processes between chromosomes *cyto*

interclass variance >>> variance

intercropping: two or more crops produced on the same field at the same time *agr*

intercrossing: mating of heterozygotes *meth*

interference: the effect of recombination in one interval on the probability of recombination in an adjacent interval *gene*

interference distance: the distance within which further crossing-overs may be formed after the previous has been produced away from the centromere *gene*

interference microscopy: like the phase microscope, the interference microscope is used for observing transparent structures *micr*

interference range: the distance large enough for forming two crossing-overs without mutual interference *cyto*

intergeneric cross: spontaneous or experimental crosses of individuals of different genera, for example, wheat *(Triticum aestivum)* and rye *(Secale cereale);* this cross even resulted in a human-made new crop plant "triticale" *meth* >>> Figure 3 >>> *cf* Important List of Crop Plants

intergeneric hybrid >>> intergeneric cross

intergenic: effects and phenomenons between genes *gene*

intergenic suppressor: a mutation that suppresses the phenotype of another mutation in a gene other than that in which the suppressor mutation resides *gene*

intergenotypic competition >>> allocompetition

interkinesis: a resting stage that may occur between the first and second meiotic division *cyto* >>> interphase

interlocking: during meiotic pairing, the intertwisting of nonhomologous chromosomes and/or chromosome configurations *cyto* >>> Figure 13

intermated recombinant inbreds: another structure of mapping population alike F2, backcross, or near-isogenic lines; the intermating of F2 individuals result in new recombination events; therefore, intermated-recombinant-inbreds populations have improved genetic resolution *biot*

intermediary: a plant trait controlled by a heterozygous pair of alleles, which result in an intermediate phenotype as compared to the corresponding homozygous genotypes *gene*

intermediate host: a host essential to the completion of the life cycle of a parasite, but in which it does not become sexually mature *phyt* >>> host

internal hairs (of the glume): the hairs situated across the upper part of the internal surface of the broad wing in the glumes of, for example, wheat *bot*

internal inprint: the mark on the inner surface of, for example, wheat glume caused by the pressure of the enclosed lemma and grain *bot*

International Center of Agricultural Research in the Dry Areas (ICARDA): Aleppo, Syria; responsible for wheat, durum wheat, barley, faba beans, lentil, chickpeas, alfalfa >>> lentil

International Crops Research Institute for the Semi-Arid Tropics (ICRISAT): this institute has a global mandate for the improvement of its mandate crops, such as sorghum, pearl millet, chickpea, pigeonpea, and groundnut; these crops are grown on large area worldwide, but generally grown on marginal land by resource-poor farmers *agr*

International Rice Research Institute (IRRI): the institute is located at Los Banos, Philippines; IRRI is a nonprofit agricultural research and training center established to improve the well-being of present and future generations of rice farmers and consumers, particularly those with low incomes; it is dedicated to helping farmers in developing countries produce more food on limited land using less water, less labor, and fewer chemical inputs, without harming the environment *agr*

International Seed Testing Association (ISTA): Bassersdorf (Switzerland); the primary purpose of ISTA is to develop, adopt, and publish standard procedures for sampling and testing seeds and to promote uniform application of these procedures for evaluation of seeds moving in international trade; the secondary purpose of ISTA is to actively promote research in all areas of seed science and technology (sampling, testing, storing, processing and distributing seeds), to encourage variety (cultivar) certification, to participate in conferences and training courses aimed at furthering these objectives and to establish and maintain liaison with other organizations having common or related interests in seed *seed*

internodal cell >>> internode

internode: the part of a stem between two consecutive nodes *bot*

interphase: a stage in the cell cycle in which there is no visible evidence of nuclear division; therefore sometimes it is called "resting phase" but in which there is intense activity, including replication of chromosomes *cyto*

interphase nucleus: a nucleus during the stage of interphase in which there is no visible dividing activity, but in which metabolic and synthetic activities are going on *cyto*

interplot competition: it can be avoided and/or decreased by use of plots with multiple rows in which only plants in the center rows are evaluated; in plots with three or more rows, the outermost rows are designated as the border or guard rows; they may prevent plants in adjacent plots from influencing the performance of plants in the center of the plot *stat*

interseeding: seeding between sod plugs, sod strips, rows, or sprigs *agr*

intersex: a class of individuals of a bisexual species that have sexual characteristics intermediate between the male and the female *gene*

interspecific: effects and phenomenons between species

interspecific cross: a crossing between two species *meth* >>> Figure 3

interspecific hybrid: a hybrid between two or more species *meth* >>> Figures 2, 3

interspecific hybridization: crossing between species *meth* >>> species hybridization >>> Figures 2, 3

interstitial segment: a chromosome region between the centromere and a site of rearrangement *cyto*

intervarietal: effects and phenomenons between varieties (cultivars)

intervening sequence: a noncoding nucleotide sequence in eukaryotic DNA, separating two portions of nucleotide sequence found to be contiguous in cytoplasmic mRNA *gene*

intrabreeding: a mating type in which only individuals of the same populations are combined *meth*

intrachromosomal: within a chromosome *cyto*

intragenic: effects and phenomenons within a gene or its physical unit *gene*

intragenic suppressor: a mutation that suppresses the phenotype of another mutation in the same gene as that in which the suppressor mutation resides *gene*

intragenotype competition >>> isocompetition

intraspecific: effects and phenomenons within a species

intravarietal: effects and phenomenons within a variety (cultivar)

introduced species: species not part of the original flora of a given area, rather, brought by human activity from another geographical region *eco agr*

introgression: the incorporation of genes of one species into the gene pool of another; if the ranges of two species overlap and fertile hybrids are produced, the hybrids tend to backcross with the more abundant species; it results in a population in which most individuals resemble the more abundant parents but also possess some of the characters of the other parent species *meth*

introgressive hybridization: crossbreeding of plants from different species that results in introgression *meth*

intron: a segment of DNA of unknown function within a gene; it may be transcribed in precursor RNA, but cannot be found in functional mRNA *gene*

inulin: a polysaccharide in which about 32 b-fructose units are joined in a chain by glyosidic linkages between the first and second carbon atoms on neighboring sugar units; it is found as a storage compound in roots, rhizomes, and tubers of many species of Compositae *chem phys*

invasion: the spreading of a pathogen through tissues of a diseased plant *phyt*

invasiveness: ability of a plant to spread beyond its introduction site and become established in new locations where it may provide a deleterious effect on organisms already existing there *eco*

inversion: a change in the arrangement of genetic material involving the excision of a chromosomal segment that is then turned through 180° and reinserted at the same position in the chromosome *cyto*

inversion polymorphism: the presence of two or more chromosome sequences, differing by inversions, in the homologous chromosomes of a population *gene*

inviability: the inability to survive *bot*

involucre: a whorl of bracts below an inflorescence *bot*

involute: having edges that roll under or inwards *bot*

iodine: a nonmetallic halogen element occurring as a grayish-black crystalline solid that sublimes to a dense violet vapor when heated; used in radiolabeling *chem cyto meth*

iojap: an idiomatic description of a mutant locus in maize that produces variegation *gene*

ion: an atom that has acquired an electric charge by the loss or gain of one or more electrons *chem*

IPGRI >>> International Plant Genetic Resources Institute, formerly IBPGR >>> IBPGR

iris diaphragm: a composite diaphragm with a central aperture readily adjustable for size in order to regulate the amount of light admitted to a lens or optical system *micr*

iron (Fe): an element required by plants; it is used in reactions in which rapid reductions occur by the transfer of electrons as in photophosphorylation and oxidative phosphorylation; iron-deficient plants have chlorotic young leaves; at first the veins remain green but later they too become chlorotic; fertilizers containing iron chelates can be added to the soil or sprayed on the leaves to make iron available to the roots and foliage *chem phys*

irradiate: expose to radiation that may increase the mutation rate of some genes and hence may increase genetic variation *meth*

irradiated callus: callus that have been exposed to radiation *biot*

IRRI >>> International Rice Research Institute

irrigation: to supply land with water by artificial means, as by diverting streams, flooding, or spraying; main types of irrigation are (1) sprinkler irrigation, (2) surface irrigation, (3) subsurface irrigation *agr*

isoallele: an allele whose effect can only be distinguished from that of the normal allele by special tests *gene*

isobrachial: a chromosome with a metacentric centromere position resulting in two chromosome arms with equal length *cyto* >>> metacentric >>> Figure 11

isochromocentric: nuclei showing as many chromocenters as chromosomes *cyto*

isochromosome: a chromosome with two identical arms; it usually derives from telocentric chromosomes *cyto* >>> Figure 37

isocompetition: cultivation at high plant density implies the presence of strong interplant competition; when there is no genetic variation, the competition is called isocompetition *stat*

isodicentric chromosome: a structurally abnormal chromosome containing a duplication of part of the chromosome including the centromere; the resulting structure contains two centromeres and a point of symmetry that depends on the position of the breakpoint *cyto* >>> Figure 37

isoelectric focu(s)sing (IEF): a technique for the electrophoretic separation of amphoteric molecules in a gradient of pH, usually formed from a combination of buffers held on a polyacrylamide gel support medium; the molecules will move in the gradient, under the influence of an electric field, until they reach their isoelectric pH, where they form a sharp band; separation is achieved because the various molecular species will have different isoelectric values of pH *meth*

isoenzyme: a species of enzyme that exists in two or more structural forms, which are easily identified by electrophoretic methods *phys* >>> Table 29

isogamete: male and female gametes that are similar to each other *bot*

isogamy: the fusion of gametes that are morphologically alike *bot*

isogeneic: applied to a graft that involves a scion and stock that are genetically identical *hort*

isogenic: a group of individuals showing the same genotype *gene*

isogenic lines (vs. random lines): two or more lines differing from each other genetically at one locus only *gene*

isogeny: the situation that a group of individuals shows the same genotype *gene*

isograft: a graft or transplant among isogenic (i.e., genetically identical) individuals on the same organism *hort*

isolate: in general, to make an isolation; in genetics, a segment of a population within which assortative mating occurs *gene;* in plant pathology, to remove an organism (e.g., a fungus) from the plant in pure form *phyt*

isolation: the separation of one group from another so that crossing between groups is prevented *meth* >>> Table 35

isolation requirement: the spatial separation required between a seed field and other sources of mechanical and genetic contamination, especially between cross-pollinated varieties *seed* >>> Tables 29, 35

isoleucine (Ile): a crystalline amino acid, $C_6H_{13}O_2$, present in most proteins *chem phys*

isolines >>> isogenic lines

isomer: a chemical compound or nuclide that displays isomerism *chem*

isomerase: an enzyme that catalyzes a reaction involving the interconversion of isomers *chem phys*

isomeric: genes that can each produce the same or similar phenotype *gene*

isoprene (2-methyl butadiene): a 5-carbon compound that forms the structural basis of many biologically important compounds, such as terpenes, etc. *chem phys*

isoschizomer: restriction endonucleases with identical recognition sequence and cleavage sites, isolated from different bacterial species; it can differ in the amino acid sequence, temperature stability, may require different reaction conditions, and may differ in the sensitivity to DNA methylation *biot*

isosome: a chromosome showing morphologically and genetically identical arms *cyto* >>> Figure 37

isosomic: cells or individuals showing isosomes (i.e., chromosomes with genetically and morphologically identical chromosome arms), usually derived from telocentric chromosomes *cyto* >>> Figure 37

isotelocompensating trisomic: a compensating trisomic; a missing chromosome is compensated by one telocentric and one tertiary chromosome *cyto* >>> Figure 14

isotertiary compensating trisomic: a compensating trisomic; a missing chromosome is compensated by one isochromosome and one tertiary chromosome *cyto* >>> Figure 14

isotope: one of two or more varieties of a chemical element whose atoms have the same numbers of protons and electrons but different numbers of neutrons *phy*

isotope labeling >>> isotopic tracer

isotopic dating: an approach of determining the age of certain materials by reference to the relative abundances of the parent isotope and the daughter isotope; if the decay constant and the concentration of the daughter isotope are known, it is possible to calculate an age *meth* >>> ^{14}C dating

isotopic tracer: isotopically labeled precursors of nucleic acids; the labeled compounds are injected or fed to plants; subsequently, the excretions (solid, liquid or gas) or tissues are analyzed to determine by detection of radioactive tracer how the original compound has been changed; many tracers have been used but most common are ^{3}H, ^{14}C, ^{32}P, and ^{35}S *prep meth*

isotrisomic: when the extra chromosome shows identical arms *cyto* >>> Figure 14

isozyme >>> isoenzyme

ISTA >>> International Seed Testing Association

IUCN: International Union for Conservation of Nature and Natural Resources

***Ixeris* type:** diplospory where a syndetic prophase leads to a restitution nucleus, which divides; it is not followed by a cell division; the resulting megaspore contains two unreduced nuclei; two further mitotic divisions lead to an eight-nucleate embryo sac *bot*

J

jarovization >>> vernalization

jasmonate(s): a group of cyclopentanone derivatives, originate biosynthetically from linolenic acid via and inducible octadecanoid pathway consisting of at least seven enzymatic steps; the end product is (+)-7-iso-jasmonic acid, a physiologically active substance that is rapidly converted to its stereoisomer, stable (–)-jasmonic acid *chem phys*

jasmonic acid: it is distributed throughout higher plants, synthesized from linolenic acid via the octaadecanoic pathway; an important role seems to be its operation as a "master switch," responsible for the activation of signal transduction pathways in response to predation and pathogen attack; proteins encoded by jasmonate-induced genes include enzymes of alkaloid and phytoalexin synthesis, storage proteins, cell wall constituents, and stress protectants; the wound-induced formation of proteinase inhibitors is an example, in which jasmonic acid combines with abscisic acid and ethylene to protect the plant from predation *chem phys*

joining segment: a small DNA segment that links genes to yield a functional gene encoding an immunogobulin *gene*

joint >>> node

jointing stage: in cereals, the growth stage at which the first stem node is visible above ground *bot*

jumping gene >>> transposon

juvenile stage: the immature, reproductively incompetent, and, sometimes, phenotypically distinct phase of plant growth *bot*

juvenillody: a condition in which tissues and organs remain immature *bot*

K

kafirin: a storage protein of sorghum *phys* >>> Table 15

kanamycin: an aminoglycoside antibiotic; a kanamycin resistance is used as a selection marker in genetic experiments *phys biot* >>> kanamycin-resistant tissue

kanamycin-resistant tissue: tissue that is resistant to the lethal effects of the aminoglycoside antibiotic, kanamycin; some cloning vectors have a kanamycin-resistant gene as a selectable marker *biot*

karnal bunt: a fungus disease of, for example, wheat that reduces yields and causes an unpalatable but harmless flavor in flour milled from infected grains *phyt*

karyogamy: the fusion in a cell of haploid (n) nuclei to form a diploid (2n) *cyto*

karyogenesis: formation of the nucleus (the central structure) of a cell—the smallest, most basic unit of life that is capable of existing by itself; karyogenesis comes from the Greek word *karyon* meaning "nucleus," and the Greek word *genesis* meaning "production"; in a narrow sense, the division of the cell nucleus is distinguished from cytoplasmic division or cytokinesis; it represents a system by which the genetic information contained in the chromosomes of eukaryotes is distributed to the daughter nuclei, which are generally identical to the mother cell nucleus *cyto*

karyogram >>> idiogram

karyology: the study of the nucleus and its components *cyto*

karyolysis: the disappearance of the interphase nucleus during karyogenesis *cyto*

karyosome: any of several masses of chromatin in the reticulum of a cell nucleus *cyto*

karyostasis: the stage of cell cycle in which there is no visible dividing activity of the nucleus, but a metabolic and synthetic activity *cyto phys*

karyotype: the entire chromosomal complement of an individual cell or individual, which may be observed during mitotic metaphase *cyto*

keel: the main nerve of, for example, the wheat glume, shaped somewhat like a keel of a boat; in legumes also a boatlike formation of the flower *bot*

keel flower: boatlike shape of a flower (e.g., in legumes such as pea) *bot*

keiki: a vegetative offshoot formed at a node (e.g., in some orchids) *bot*

kernel: a whole grain or seed of a cereal plant or the part of the seed inside the pericarp *bot*

ketone: any of a class of organic compounds containing a carbonyl group, CO, attached to two alkyl groups, as CH_3COCH_3 *chem*

key gene >>> oligogene

killing frost: a sharp fall in temperature that damages a plant so severely as to cause its death *env phys*

kilning: the heating and/or drying process used in the production of malt to stop germination and kill the grain *prep*

kilobases (kb): 1,000 base pairs/bases in a single- or double-stranded nucleic acid, which is used as a common unit of length in molecular genetics *gene*

kilogram: equals 1,000 grams

kilometer: equals 1,000 meters

kinase: an enzyme that catalyzes reactions involving the transfer of phosphates from a nucleoside triphosphate (e.g., ATP) to another substrate *phys*

kinetin (6-fururylaminopurine): a degradation product of animal DNA, which does not occur naturally and which has properties similar to those of cytokinins; applied to certain leaves, kinetin delays senescence in its vicinity and attracts nutrients *chem phys*

kinetochore: a dense, plaquelike area of the centromere region of a chromatid, to which the microtubules of the spindle attached during cell division *cyto* >>> centromere >>> Figure 11

kinin >>> cytokinin

KJELDAHL method: a technique often used for the quantitative estimation of.the nitrogen content of plant material (e.g., of cereal grains) *meth*

KLENOW fragment: large fragment of DNA polymerase I after proteolytic digestion; it lacks 5' to 3' exonuclease activity and can therefore not be used for nick translation but is very useful for filling-in reactions and DNA sequencing by the SANGER method *biot*

klon >>> clone

kneading: to work dough into a uniform mixture by pressing, folding, and stretching *meth*

knob: a heavily stainable and quite a big chromomere observed along a chromosome of some plants (e.g., in maize it is used as a marker in pachytene analysis) *cyto* >>> chromomere

knot: a lump or swelling in or on a part of a plant (e.g., the node of grass) *bot*

KOEHLER illumination: illumination optics resulting in the image of the light source being out of focus at the specimen plane; it provides homogeneous illumination of the specimen *micr*

KORNBERG enzyme >>> DNA polymerase I

KOSAMBI formula: recombination fractions and map distances correspond only over relatively short recombinational distance; as genetic distance increases, the probability of a second (and correcting) recombination also increases, hence the measured recombination for two loci is less than would be apparent if a third intervening locus were present; various mapping functions have been suggested to permit single recombination fractions to be converted to map distances; KOSAMBI presented the simplest and probably most general functions, which is given as: $x = 25 \log_n [(1 + 2y)/(1 - 2y)]$; x is the map distance (cM) corresponding to the recombination fraction, y; for example, if the recombination value is 0.05 then the distance amounts 5 cM, is the value 0,1 then the distance is 10.1 cM, etc. *gene*

L

label >>> plant label

labeling: incorporation of an easily detectable signal into a DNA molecule; radioactive labeling is increasingly replaced by nonradioactive methods *biot;* in seed science, attaching labels to seed lots with information on variety identity, purity, and seed quality *seed*

labellum: in Orchidaceae, the lowest of the three flower petals, which differs from the other two; in lipped flowers, the platform formed by the lowest petal or fused petals *bot*

***Lac* operon:** a cluster of structural genes specifying the enzymes acetylase, permease, and beta-galactosidase *gene*

lacinate: deeply cut, into irregular, narrow segments or lobes *bot*

laggard: a chromosome, which is not included in the daughter nuclei after anaphase *cyto* >>> lagging

lagging (of chromosome): delayed movement from the equator to the poles at anaphase of a chromosome so that it becomes excluded from the daughter nuclei *cyto*

lagging strand: DNA strand growing in the 3' to 5' direction, synthesized discontinuously *biot*

lamda phage: lamda temperate bacteriophage, size: 48.5 kb; it infects *Escherichia coli biot*

lamina: a flat, sheetlike structure (e.g., the blade of a leaf) *bot*

lampbrush chromosome: a particular type of chromosome shape, usually found in diplotene stage of animals including flies; it is a type of puffed chromosome; the loops of DNA strands form a lampbrush-like shape;

puffed chromosome regions may also occur in plants (e.g., in *Phaseolus* beans) *cyto* >>> polyteny >>> polytene chromosome >>> puff

land classification: soil that is grouped into special units, subclasses, and/or classes according to their capability for use and treatments that are required for sustained agriculture, horticulture, and forestry *agr*

landmark: a labeled stick of different length and manufacturing used for marking fields, experimental plots, paths, or margins, usually after seed bed preparation *prep meth*

landrace: a set of populations or clones of a crop species produced and maintained by farmers; in breeding, a mixture of a great number of different genotypes, which are well adapted to the environmental conditions of its habitat; it shows only average, but reliable yield; in countries of highly developed agriculture, landraces have been superseded by highly advanced varieties; however, for selection landraces are a suitable material in which a great diversity of useful genotypes may be found *tax* >>> Figure 5

larva: the wormlike immature form of certain insects; some are called caterpillars, grubs, or maggots *zoo phyt*

LASER >>> light amplification by stimulated emission of radiation

late blight, **foliage blight**, **tuber blight** (of potato): a widespread and serious disease *(Phytophthora infestans)* affecting the potato and related species; symptoms include the appearance of brown patches on the leaves, often with white mold on the underside; under damp conditions the entire foliage may collapse; brown lesions also develop on tubers, spreading to involve the entire tuber in a dry brown rot; it reduces yields and marketability and may cause losses in store by encouraging soft rotting *phyt*

late crop: a crop or plant showing late maturation within a given season *agr hort*

late replicating: (often) heterochromatic regions of chromosomes, which show a later replication than the euchromatic once *biot cyto*

late replication: in microbiology, the rolling circle of replication of phage lamda, producing concatemers suitable for packaging in lamda heads *bio;* in cytology, heterochromatic regions of chromosomes, which show a later replication of DNA than the euchromatic ones *cyto* >>> lamda phage >>> heterochromatin

late wood: a result of secondary plant growth; very young branches have hardly any fascicular cambium; interfascicular cambium develops very early during the year, even before the beginning of secondary growth; the activity of the cambium increases branch diameter and the vascular bundles become elongated in cross section; far more xylem than phloem elements are produced; annual rings become clearly visible because at the beginning of each vegetation period (in spring) vessels (conducting function) and fibers (supporting function) with a wide lumen are assembled first, the so-called early wood; in the following season, elements with steadily narrowing volumes are produced; in autumn, only a few vascular elements with narrow lumina (late wood) form *bot hort fore*

latency: the state of being latent; the interval between exposure to a toxin or disease-causing organism and development of a consequent *phyt*

latency period >>> latency

latent >>> latency

latent infection: a chronic infection in which a host-pathogen equilibrium is established without any visible symptoms of disease *phyt*

latent period: the period between infection and the sporulation of the pathogen on the host *phyt* >>> latency

latent virus: a virus that does not induce symptom development in its host *phyt* >>> latent infection

lateral: belonging to or borne on the sides *bot*

lateral meristem: a meristem giving rise to secondary plant tissues, such as the vascular and cork cambia *bot biot*

lateral nerve(s): for example, in wheat, the nerves, which run along the length of the broad and narrow wings of the glume; in barley, the two pairs of nerves (inner and outer) lying toward the margins of the lemma and on either side of the median nerve *bot*

lateral root: roots arising from the main root axis *bot*

lateral shoot: shoots originating from vegetative buds in the axils of leaves or from the nodes of stems, rhizomes, or stolons *bot*

laterite: a weathering product of rock, composed mainly of hydrated iron and aluminum oxides, hydroxides, and clay minerals but also containing some silica *agr*

late-sown: sowing date later than the optimal time for a given crop or variety *agr*

latest safe sowing date: the date till which seeds can be sown without severe yield lost during the following year, usually in cereal crops *agr*

latex: a white, commonly sticky substance produced in specialized tissues within a plant *bot*

latifoliate: broad-leafed *bot*

Latin rectangle: a field design that is similar to the Latin square, just differentiated by the number of replication, which is not equal to the number of variants; the number of replications may be a third, a quarter, or a fifth of the number of variants; thus the number of replications is reduced *stat* >>> Figure 9

Latin square: in general, a set of symbols arranged in a checkerboard in such a fashion that no symbol appears twice in any row or column; it is used for subdividing plots of land for agricultural and breeding experiments, so that treatments can be tested even though the field has soil conditions that might vary in an unknown fashion in different areas; it requires that the field be subdivided by a grid into subplots and the differing treatments be performed at consecutive intervals to plants from different subplots *stat* >>> Figure 9 >>> Table 26

lattice design: an experimental (field) design in which the number of treatments forms a square >>> Latin square

lattice square >>> Latin square

lattice square design >>> Latin square

lawn: a stretch of open, grass-covered land (e.g., one closely mowed, as near a house, on an estate, or in a park) *agr*

laws of inheritance >>> MENDEL's laws of inheritance

layering: covering stems, runners, or stolons with soil causing adventitious roots to form at the nodes, enabling propagation by rooted cuttings; this procedure is used commercially to propagate many plants *hort;* in vitro layering, the horizontal placement of cultured shoots or nodal segments on agar growth medium in order to produce axillary bud formation *biot*

leaching: the washing out of material from the soil, both in solution or suspension *agr*

leader sequence: a nucleotide sequence of the mRNA on which the ribosomes bind; inother words, nontranslated sequence at 5' end of mRNA, or N-terminal sequence of a protein constituting a signal for transport through a membrane, which is later removed *biot*

leading strand: DNA strand synthesized in the 5' to 3' direction *biot*

leaf: a thin, usually green, expanded organ born at a node on the stem of a plant, typically comprising a petiole (stalk) and blade (lamina), and subtending a bud in the axil of the petiole; it is the main site of photosynthesis *bot*

leaf area index (LAI): the total leaf surface area exposed to incoming light energy, expressed in relation to the ground surface area beneath the plant (e.g., LAI = 3, the leaf area exposed to light is three times of the ground surface area) *phys*

leaf axil: the angle between a petiole and the stem *bot*

leaf blight: various diseases that lead to the browning and dropping of leaves *phyt* >>> late blight

leaf bud: a bud producing a stem and leaf, unlike a flower bud, which contains a blossom *bot*

leaf cutting: a cutting made from a single leaf; a method for propagation (e.g., of succulents); a leaf can be knocked off or cut off the plant; either it spontaneously roots on the ground or it is placed in certain media for rooting *hort meth*

leaf fall: leaf abscission

leaf miner: various insects, which, in the larval stage, produce a tunnel through leaves, feeding on the tissue, and leaving conspicuous traces of their paths *phyt*

leaf posture: the characteristic position of the foliage leaves on the stem axis, which imprints the plant habit of the species; it may contribute to optimal utilization of light and thus photosynthesis; there were several approaches to breed for specific leaf posture in order to improve photosynthetic capacity of, for example, cereals *bot*

leaf primordium: a lateral outgrowth from the apical meristem that develops into a leaf *bot biot*

leaf senescence: a type of programmed cell death, during which leaf cells undergo cocoordinated changes in cell structure, metabolism, and gene expression, resulting in a sharp decline in photosynthetic capacity; a cytokinin class of plant hormones plays a role in controlling leaf senescence because a decline in the cytokinin level occurs in senescing leaves; external application of cytokinin often delays senescence *phys*

leaf sheath: a tubular envelope, as the lower part of the leaf in grasses *bot*

leaf spot: it refers to various plant diseases that cause well-defined areas of tissue to die, creating noticeable spots *phyt*

leaf vein: vascular bundles in the leaves; in the petiole and the midvein of the leaf the veins are very large; farther out into the mesophyll the veins may consist of only one xylem or phloem element; in these regions, these very small veins are called veinlets *bot*

leafage >>> foliage

leafstalk: the footstalk or supporting stalk of a leaf *bot* >>> petiole

leaky mutation: mutation that is very prone to reversion *gene*

least squares method: a method of estimation based on the minimization of sums of squares *stat*

lectin: a generic term for proteins extracted from plants (e.g., legumes) that exhibit antibody activity in animals *chem phys*

leghaemoglobin: an iron-containing, red pigment produced in root nodules during the symbiotic association between rhizobia and leguminous plants *phys phyt agr* >>> legume(s)

legume(s): plants showing a simple or single pistil and characterized by a dry fruit pod that splits open by two longitudinal sutures and has a row of seeds on the inner side of the ventral suture (e.g., bean, pea, soybean, locust); there are many valuable food, forage, and cover species, such as peas, beans, soybeans, peanuts, clovers, alfalfas, sweet clovers, lespedezas, vetches, and kudzu; sometimes referred to as nitrogen-fixing plants; legumes are an important rotation crop because of their nitrogen-fixing property *bot agr*

legumin >>> globulin

leguminous plant >>> legume(s)

lemma: flowering glume; the lower or outer of the two bracts of the floret *bot*

lenticular: shaped like a biconvex lens, lentil-shaped *bot*

lentiform >>> lenticular

leptodermous: thin-walled or thin-skinned *bot*

leptokurtic (distribution): a flat-topped, bell-shaped curve of frequency distribution of a given character in a population *stat*

leptonema >>> leptotene

leptotene: during the first meiotic division, the first stage, in which the chromosomes appear as long, widely uncoiled, and single strands; the DNA of each of the chromosomes has replicated; each chromosome consists of two identical members (chromatids) *cyto*

lesion: a visible area of diseased tissue on an infected plant *phyt*

lethal: a gene or genotype that is fatal for the individual *gene*

lethal doses (LD): the concentration of a poison that kills a certain amount of cells, individuals, etc., for example, LD50 = 50 percent of the cells or individuals are killed *meth*

lethal gene: a gene whose expression results in the premature death of the organism carrying it; dominant alleles kill heterozygotes, whereas recessive alleles kill homozygotes only *gene*

lethal mutation: a gene mutation whose expression results in the premature death of the organism carrying it *gene* >>> lethal gene

leucine (Leu): an aliphatic, nonpolar, neutral amino acid that, unlike most amino acids, is sparingly soluble in water *chem phys*

leucoplast: a colorless plastid that is involved in the metabolism and storage of starches and oils *bot*

levulose >>> fructose

liber >>> bast

library: in biotechnology and molecular genetics, a collection of cells, usually bacteria or yeast, that have been transformed with recombinant vectors carrying DNA inserts from a single species (e.g., cDNA), expression, or genomic library *biot*

lid: the cap of a boxlike seed capsule *bot*

life cycle, life history: in fungi, the stage or series of stages between one spore form and the development of the same spore again; there are commonly two stages in the life cycle (the imperfect, which may have more than one kind of spore, and the perfect), but there may be no development of one or the other *phyt*

life span: the longest period over which the life of any organism or species may extend *phys*

ligand: an atom, ion, or molecule that acts as the electron donor partner in one or more coordination bonds or a molecule (e.g., antibody), which can bind to specific sites on cell membranes *chem phys*

ligase: an enzyme that catalyzes a reaction that joins two substrates using energy derived from the simultaneous hydrolysis of a nucleotide triphosphate; in general, a joining enzyme, which closes single-strand breaks in DNA *phys gene*

light amplification by stimulated emission of radiation (LASER): a device that produces a nearly parallel, nearly monochromatic, and coherent beam of light by exciting atoms and causing them to radiate their energy in phase; it can be used for the elimination or manipulation of cell particles *via* a special microscope device *micr*

light leaf spot (of rape, *Pyrenopeziza brassica,* asexual stage *Cylindrosporium concentricum*): it appears as light green or bleached areas on the leaves; small white spore masses bordering the lesions *phyt*

light reaction: during photosynthesis, those reactions that require the presence of light *phys*

light soil: a soil that has a coarse texture and is easily cultivated *agr*

ligneous: woody *bot*

lignification: converting into wood; cause to become woody *bot*

lignin: a complex, cross-linked polymer, comprising phenyl propene units, that is found in many cell walls; its function is to cement together and anchor cellulose fibers and to stiffen the cell wall; it reduces infection, rot, and decay *chem bot*

ligula: a scalelike membrane that covers the surface of a leaf; in some Compositae, a strap-shaped corolla; sometimes, a fringe of epidermal tissue found at the boundary between the sheath and the blade of a maize leaf *bot* >>> Table 30

ligulate: straplike, tongue-shaped *bot*

ligulate flower >>> ligulate >>> ligula

ligule >>> ligula

likelihood: the state of being likely or probable; probability *stat*

lime: compounds of calcium used to correct the acidity in soils *agr*

liming >>> lime

limited backcrossing: instead of complete, at least six cycles of backcrossing, only two or three cycles are coupled with rigorous selection to gain the advantage of transgressive segregation *meth*

line: a group of individuals of a common ancestry and more narrowly defined than a strain of variety; in breeding, it refers to any group of genetically uniform individuals formed from the selfing of a common homozygous parent *gene*

line breeding: a system of breeding in which a number of genotypes, which have been progeny tested in respect of some characters, are composited to form a variety *meth* >>> line

line of breeding >>> line

line of descent >>> line

line of inbreeding >>> line

line variety >>> line breeding

lineage: a chart that traces the flow of genetic information from generation to generation *meth gene*

linear: long and narrow, with parallel margins *bot*

lining out: transplanting seedlings or rooted cuttings in rows in a nursery bed *hort fore meth*

linkage (of genes): the association of genes that results from their being on the same chromosome; linkage is detected by the greater association in inheritance of two or more nonallelic genes than would be expected from independent assortment; the nearer such genes are to each other on a chromosome, the more closely linked they are, and the less often they are likely to be separated in future generations by crossing over; all genes in one chromosome form one linkage group *gene*

linkage desequilibrium: the nonrandom association of alleles at different gene loci in a population (e.g., when two loci occur close together on the same chromosome and selection operates to keep the allele combinations together) *gene*

linkage disequilibrium >>> linkage desequilibrium

linkage group: all genes in one chromosome form one linkage group *gene* >>> linkage

linkage map: an abstract map of chromosomal loci, based on experimentally determined recombinant frequencies, that shows the relative positions of the known genes on the chromosomes of a particular species; the more frequently two given characters recombine, the further apart are the genes that determine them *gene* >>> linkage

linkage value: recombination fraction expressing the proportion of nonparental or recombinant versus parental types in a progeny; in diploids, the recombination fraction can vary between zero to one half *gene* >>> linkage

linked (genes): genes or alleles showing less than 50 percent recombination, which is typical for unlinked (independent) genes; depending on the strength of linkage, the linked genes tend to be transmitted together *gene* >>> linkage >>> linkage value

linker: a synthetic, short, and double-stranded oligodeoxyribonucleotide containing a restriction site, which is ligated to the ends of a DNA fragment *gene*

Linola: a new form of linseed known by the generic crop name "solin," which produces a high-quality edible polyunsaturated oil similar in composition to sunflower oil *agr*

lint (linters): the long fibers of cotton seed; the short fibers generally remain attached to the seed in ginning; sometimes called "fuzz," they are used mainly for batting, mattress stuffing, and as a source of cellulose *agr*

lipase: an enzyme that degrades fats to glycerol and fatty acids *chem phys*

lipid(e): a member of a heterogeneous group of small organic molecules that are sparingly soluble in water but soluble in organic solvents; included in this classification are fats, oils, waxes, terpenes, and steroids; the functions are equally diverse and include roles as energy-storage compounds, as hormones, as vitamins, and as structural components of cells, such as membranes *chem phys*

lipid body >>> lipid(e)

liposome(s): membrane-bound vesicles experimentally constructed to transport biological molecules *biot*

liquid culture: the culturing of cells on or in a liquid medium on supports or in suspension; the culture can be stationary or agitated *prep*

liquid nitrogen: nitrogen gas that has been condensated to a liquid and has a boiling point of –195.79°C; it is used for storage of tissue, organs, cells, or suspensions and for several cytological and molecular preparations *chem* >>> cryopreservation

L-notch planting: a form of slit planting involving two slits at right angles with the seedling placed at the apex of the "L" *meth fore hort*

loam soil: a soil containing sand, silt, and clay *agr*

local >>> indigenous

local infection: an infection just affecting a limited number of a plants *phyt*

local population: a group of individuals of the same species growing near enough to each other to interbreed and exchange genes *tax*

locule: a cavity of the ovary *bot*

locus (loci *pl*): a specific place on a chromosome where a gene is located; in diploids, loci pair during meiosis and, unless there have been translocations, inversions, etc., the homologous chromosomes contain identical sets of loci in the same linear order; at each locus is one gene; if that gene can take several forms (alleles), only one of these will be present at a given locus *gene*

locus-specific: plant characters or genetic activity that is exclusively correlated with a particular chromosomal locus *gene*

lodging: a state of permanent displacement of a stem crop from its upright position; it can cause considerable reduction in yield by storm damage, rots, insects, or excess of nitrogen *agr* >>> Table 34

lodging resistance: plants that can resist lodging by optimal root system, stiffer straw, or other characteristics (e.g., in cereal breeding, the introduction of "semi-dwarf genes" contributed to shorter plants and thus higher lodging resistance even when nitrogen fertilization is increased) >>> lodging >>> near-isogenic lines >>> semidwarf >>> *Rht* gene

lodiculae: two small, translucent, scale-like structures situated at the base of the floret *bot*

lodicule >>> lodiculae

loess (soil): unconsolidated, wind-deposited sediment composed largely of silt-sized quartz particles (0.015-0.05 mm diameter) and showing little or no stratification *agr*

loment(um): a dry schizocarpic fruit in the form of a legume or siliqua with constrictions formed between the seeds as it matures, so that the final fruit is composed of one-seeded, indehiscent loment segments *bot*

long-day plant: a plant in which flowering is favored by long days (>14 h daylight) and corresponding short dark periods; there are two types: species in which there is an absolute requirement for these conditions and others in which flowering is merely hastened by them *bot*

longevity: the persistence of an individual for longer than most members of its species, or of a genus and/or species over a prolonged period of geological time *bot phys evol*

long-plot design: a specific type of field experiment using preferentially long plots; a plot, the area to which an individual treatment is applied, can be any size, including a single plant growing in a pot, a five-acre field, or more; however, there are some considerations, including the equipment to be used in planting, harvesting, and treatment application, that determine size and shape of plots (e.g., space for the experiment, number of treatments, or specificity of character to be tested); if there is equipment to plant, harvest, and apply treatments to four rows at a time, then the logical plot width would be some multiple of four rows; the lengths of plots are more flexible than their widths (e.g., if the harvest from each plot has to be weighed, the scales may influence the length of plots; if the scales are designed to weigh hundreds of pounds, the plots must be large enough to provide a harvest weight that can be accurately determined by the equipment; increasing the length of plots is an easy way to do that); in general, once the plots are large enough to be representative of a much larger area, further increasing plot size will not significantly improve the accuracy of the results; plots that are larger than necessary take more field space and may increase the amount of work required for an experiment, but they usually will not adversely affect the test results unless the plots get so large that the plots within a block are no longer uniform; plots that are too small may prevent the accurate assessment of treatment effects; if the space available for an experiment is limited, more replications are usually more beneficial than having larger plots as long as plot size allows accurate assessment of treatment effects *meth agr*

long-term gene pool: a population with wide genetic variability established for long-term breeding objectives; lower selection pressure is applied when it is improved through recurrent selection; it can also provide genetic variability to other gene pools *meth*

long-term storage: storage of seeds in a gene bank longer than ten years *seed*

loose smut: a disease of plants caused by a fungus of the *Ustilaginales* in which the masses of spores are exposed at maturity and can be dispersed freely by wind *phyt*

lopping: a procedure by which all the branches of a tree are cut off, except the leading shoot, as opposed to pruning, in which only some of the branches are cut *hort meth*

lower palea >>> lemma

low-input variety: a crop variety with low claims at macro- and micronutrient fertilizers, pest control, and agronomic measures *agr*

luciferin: a pigment of bioluminescent organisms that emits light while being oxidized *chem phys*

lumen: the central cavity of a cell or other structure *bot*

luminescence: the emission of light without accompanying heat *phy*

lunate: shaped like a half-moon *bot*

Lux: a unit of light measurement (= 0.0929 foot candles) once widely employed but now largely supplanted by photosynthetically active radiation units, such as $\mu molm^{-2s-1}$ ($\mu molEm^{-2s-1}$) and Wm^{-2} *phy*

luxuriance: hybrids that are larger, faster growing, or otherwise exceed the parental forms in some traits; it is usually brought by complementary gene action present in the parents and combined in the hybrid *gene* >>> heterosis

lyase: an enzyme that catalyzes nonhydrolytic reactions in which groups are either removed or added to a substrate, thereby creating or eliminating a double bond, especially between carbon atoms or between carbon and oxygen *phys*

lyse: to destroy or disorganize cells by enzymes, viruses, or other means *meth*

lysimeter: an apparatus for electronically measuring water balance *meth*

lysis: cell rupture and death; it is applied if a bacterial cell is killed upon the release of phage progeny *phys*

lysosome: a membrane-bound vesicle in a cell that contains numerous acid hydrolases capable of digesting a wide variety of extra- and intracellular materials *phys bot*

lysozyme: an enzyme that is destructive of bacteria and functions as an antiseptic, found in certain plants *phys*

M

M1, **M2**, **M3**, etc.: symbols used to designate first, second, third, etc. generations after treatment with mutagenic agents *meth* >>> Figure 1

M2 population: the progeny derived from selfing M1 plants, which themselves are progeny that arise by selfing plants grown from mutagenized seed; recessive mutations, resulting from the seed mutagenesis, are detected in M2 plants, which are homozygous for the mutation *gene* >>> Figure 1

macerate >>> maceration

maceration: softening of plant tissue by using of enzymes, hydrolic acid, or other means; usually, the middle lamella of the cell walls is degraded without modification of the cell content *meth cyto*

macerozyme: an enzyme or a mixture of enzymes able to soften plant tissue *phys* >>> maceration

machinability: in quality testing of cereal flour, a test that measures the stickiness of the dough *meth*

macrocarpous: carrying or forming big fruits *bot*

macroclimate: the general climate of a large area, as that of a continent or country, as opposed to microclimate *env eco*

macroelement: chemical elements, such as nitrogen or phosphorus, that are needed in large amounts as nutrients for plant growth *phys agr*

macroevolution: evolution above the species level (i.e., the development of new species, genera, families, orders, etc.) *evol*

macromolecule: a molecule that has a high molecular weight, often a polymer *chem*

macromutant >>> macromutation

macromutation: a mutation that results in a profound change in an organism, as a change in a regulatory gene that controls the expression of many structural genes, as opposed to micromutation *gene*

macronutrient: an inorganic element or compound that is needed in relatively large amounts by plants *phys*

macroscopic: visible to the naked eye *micr*

macrospore >>> megaspore

macrostylous: showing long stamen *bot*

maculate: spotted or blotched *bot*

magnesium: an element that is found in high concentrations in plants; it plays an important role in the chemical structure of chlorophyll and of membranes and is involved in many enzyme reactions, especially those catalyzing the transfer of phosphate compounds; deficiency can produce various symptoms, including chlorosis and the development of other pigments in leaves *chem phys*

magnesium chlorate >>> chemical desiccation

magnification: the ratio of the distance between two points in the image to the distance between the two corresponding points in the specimen; the apparent size of the specimen at 25 cm from the eye is considered to be at 1 × *micr*

maintainer: it is used for maintaining and multiplication of a cytoplasmic male sterile line; usually genotypes containing the normal cytoplasm and recessive at the restorer locus *seed* >>> Figure 23 >>> hybrid breeding >>> heterosis >>> cytoplasmic male sterility (CMS)

maize gluten: a byproduct of wet milling; it is used as a medium-protein (20-24 percent) and medium-fiber (10 percent) foodstuff *agr* >>> Table 15

maize picker: a device to harvest maize *agr*

maize streak virus (MSV): a geminivirus, which has a single-stranded DNA genome that replicated in the nucleus of the host cells to provide a double-stranded replication intermediate and transcription template *phyt*

major gene: a gene with pronounced phenotypic effects, in contrast to modifier gene, which modifies the phenotypic expression of another gene *gene* >>> oligogene

male parent >>> father plant

male sterility: producing no functional pollen *bot* >>> Figure 23

malformation: faulty or anomalous formation or structure *bot gene*

Malpighian layer: a protective layer or layers of cells present in the coats of many seeds; it is characteristically made up close-packed, radially placed, heavy-walled in columnar cells without intercellular spaces; the cells often are heavily cutinized or lignified and are relatively impervious to moisture and gases *bot*

malt: germinated grain used in brewing and distilling *meth*

maltose: a disaccharide that consists of two alpha-glucose units linked by an alpha-1,4-glycosidic bond *chem phys*

manganese: an element that is required in small amounts by plants; it is involved in the light reaction of photosynthesis and also binds proteins; deficiency causes interveinal chlorosis and malformation *chem phys*

mannitol: a polyhydroxy alcohol that can be synthesized chemically by the reduction of mannose and is present in many plants *chem phys*

mannose: a hexose, $C_6H_{12}O_6$, obtained from the hydrolysis of the ivory nut and yielding mannitol upon reduction *chem phys*

MANN-WHITNEY test: a statistical test of differences in location for an experimental design involving two samples with data measured on an ordinal scale or better *stat*

manure: animal excreta with or without a mixture of bedding or litter *agr*

map (chromosomes, genes): as a verb, to determine the relative or physical position of a gene, DNA molecule, or chromosome segment *gene*

map distance: the distance between any two markers on a genetic map, based on the percentage of crossing-over; the minimum distance between linked genes is 1 percent and maximum 50 percent *gene*

map length >>> map distance >>> MORGAN unit

map-based cloning: the isolation of important genes by cloning the gene in question on the basis of molecular maps *biot gene*

mapmaker: an idiomatic description of computer software developed for detection and estimation of linkages; this calculation is based on the maximum likelihood method and often applied for construction of molecular marker maps *meth stat biot* >>> KOSAMBI formula

mapping: the process and the result of determination of map distances within or between linkage groups; sometimes it refers to the localization of genes or chromosome segments *gene* >>> physical map

marginal farmland: land repeatedly farmed without benefit of humus or chemical replacements *agr*

marker: a gene of known function and location, or a mutation within a gene that allows studying the inheritance of that gene *gene* >>> Table 29

marker gene >>> marker

marker-aided selection, marker-assisted selection (MAS): indirect selection exploiting the association between the qualitative variation in a trait (isoenzymes, DNA marker) and the quantitative variation in another trait; it is a strategy permitting plant selection at the juvenile stage from early generations; the essential requirements for MAS in plant breeding are: (1) marker(s) should cosegregate or be closely linked (1 cM or less) with the desired trait, (2) an efficient means of screening large populations for molecular markers should be available, (3) the screening technique should have high reproducibility across laboratories, be economical to use, and should be user friendly *meth* >>> Table 29

marsh >>> fen

MAS >>> marker-aided selection

masked symptoms: plant symptoms (e.g., caused by a virus) that are absent under some environmental conditions but appear when the host is exposed to certain conditions of light and temperature *phyt*

mass emasculation: in hybrid breeding (e.g., in maize), the emasculation of the male flowers by mechanical means *meth* >>> emasculate >>> detasseling >>> Table 35

mass pedigree selection: a system of breeding in which a population is propagated in bulk until conditions favorable for selection occur; usually, after mass pedigree selection pedigree selection is followed *meth* >>> Tables 5, 35

mass selection (positive or negative): a form of breeding in which individual plants are selected on their individual advantages and the next generation propagated from the aggregate of their seeds; the easiest method is to select and multiply together those individuals from a mixture of phenotypes, which correspond to the breeding aim (positive mass selection), it is still applied in cross-pollinating of vegetable species, such as carrots, radishes, or beetroots, in order to improve the uniformity; when all undesired off-types are rouged in grown crop population and the remaining individuals are propagated further, the method is termed negative mass selection; negative mass selection is no longer an adequate breeding method for highly advanced varieties; it is usually applied in multiplication of established varieties (i.e., for seed production in order to remove diseased plants, casual hybrids, or other defects) *meth* >>> Figures 39, 40 >>> Tables 5, 35

mass spectrometry: a technique that allows the measurement of atomic and molecular masses; material is vaporized in a vacuum (ionized) and then passed first through a strongly accelerating electric potential, and then through a powerful magnetic field; it serves to separate the ions in order of their charge (i.e., mass ratio); detection is made using an electrometer, which measures the force between charges and hence the electrical potential *meth phy* >>> gas chromatograph

mate: a unisexual individual that is involved in sexual reproduction *gene* >>> Table 35

maternal effect: any nonlasting environmental effect or influence of the maternal genotype or phenotype on the immediate offspring *gene*

maternal inheritance: phenotypic differences found between individuals of identical genotype due to an effect of maternal inheritance *gene*

mating: the combination of unisexual individuals with the aim of sexual reproduction *gene* >>> Table 35

mating group: a group of individuals that gives the chance for mating among one another on the basis of genetic prerequisites *gene*

mating system: the pattern of mating in sexually reproducing organisms; two types of mating systems are: (1) random mating and (2) assortative mating (genetic assortative mating, genetic disassortative mating, phenotypic assortative mating, phenotypic disassortative mating) *gene* >>> Table 35

mating type: the genetic properties of an individual for a particular type of mating *gene* >>> Table 35

matroclinal: with hereditary characteristics more maternal than paternal (e.g., in certain banana hybrids) *bot*

matroclinous >>> matroclinal

matromorphy: resembling the female parent in morphology *bot*

maturation: the completion of development and the process of ripening *phys*

maturation division >>> meiosis

mature: fully differentiated and functionally competent cells, tissues, or organisms *phys*

mature resistance >>> adult resistance

maturity >>> mature

maximum likelihood method: a statistic method of linkage estimation depending on maximizing of the log likelihood; the method leads to an efficient and sufficient statistic if one exists *stat*

mDNA >>> messenger DNA

ME >>> mega-enviroment

meal: the byproduct of oilseeds; used as a high-protein animal feed *agr*

mean: the sum of an array of quantities divided by the number of quantities in the group *stat*

mean square: the square of the mean variation of a set of observations around the sample mean *stat* >>> variance

mechanical inoculation: a method of transmitting the pathogen from plant to plant; for example, sap from diseased plants or a defined inoculum are rubbed on test-plant leaves that usually have been dusted with carborundum or other abrasive materials; it is applied in experimental testing of plant resistance *phyt*

media composition: different supplements to the nutritive substance provided for the growth of a given plant in the laboratory *biot*

median centromere: a centromere that is located midway of chromosomes resulting in two equal-long arms *cyto*

medical plants: plants that are or have been used medicinally (e.g., chamomile) *hort* >>> chamomile

medium (media *pl*): any material in or on which cultures are grown *prep*

medium-term seed storage: with a storage time of about ten years *meth* >>> long-term storage

megabase cloning: the molecular cloning of very large DNA fragments (>500 bp) *meth biot*

mega-environment (ME): a broad, not necessarily contiguous area, occurring in more than one country and frequently transcontinental, defined by similar biotic and abiotic stresses, cropping system requirements, consumer preferences, and by a volume of production; the concept was introduced by CIMMYT in 1988 to address the needs of diverse wheat-growing areas of the world *eco agr* >>> CIMMYT

megagametogenesis: the development of the female gametophyte from a functional megaspore *bot*

megagametophyte >>> embryo sac

megaspore: one of the four cells formed in the ovule of higher plants as a result of meiosis or sexual cell reduction division; one of these later undergoes mitosis to give rise to the female gamete *bot*

megaspore mother cell: a diploid cell in the ovary that gives rise, through meiosis, to four haploid megaspores *bot*

megasporocyte >>> embryo sac

megasporogenesis: the development of the megaspore from the archesporial cell *bot*

mega-yeast artificial chromosomes (mega YAC): a large (>500 bp) piece of DNA that has been cloned inside a living yeast cell; while most bacterial vectors cannot carry DNA inserts that are larger than 50 bp, and standard YACs typically cannot carry DNA pieces that are larger than 500 bp, mega YACs can carry DNA pieces (chromosomes) as large as one million bp *biot*

meiocyte: the sporocyte giving rise to the embryo sac and to pollen grains *bot*

meiosis: a type of nuclear division that occurs at some stage in the life cycle of sexually reproducing organisms; by a specific mechanism the number of chromosomes is halved to prevent doubling in each generation; genetic material can be exchanged between homologous chromosomes; there are different stages: leptotene, zygotene, pachytene, diplotene, diakinesis, metaphase I, anaphase I, telophase I, interphase, metaphase II, anaphase II, telophase II, microspore formation *bot cyto* >>> Figures 15, 28

meiotic crossing-over >>> chiasma

meiotic cycle >>> meiosis

meiotic duration: the time needed for completing the cell cycle from prophase to telophase under certain conditions *cyto*

membrane: a sheetlike structure, 7-10 nm wide, that forms the boundary between a cell and its environment and also between various compartments within the cell; it is composed of lipids, proteins, and some carbohydrates; it functions as a selective barrier and also as a structural base for enzymes *bot*

Mendelian character: a character that follows the laws of inheritance formulated by G. MENDEL *gene*

Mendelian inheritance >>> MENDEL's laws of inheritance

Mendelian population: an interbreeding group of organisms that share a common gene pool *gene*

Mendelian ratio: the segregation rations according to MENDEL's laws of inheritance *gene* >>> Figure 6 >>> Table 2

Mendelism >>> MENDEL's laws of inheritance

mendelize: to segregate according to MENDEL's laws of inheritance *gene*

MENDEL's laws of inheritance, Mendelian laws of inheritance: the inheritance of chromosomal genes on the basis of chromosome theory of heredity; three laws are considered: (1) law of dominance or of uniformity of hybrids (2) law of segregation (3) law of independent assortment *gene* >>> Figure 6 >>> Table 2

mercaptan >>> thiol

mericlinal: it refers to a chimera in which the inner tissue has a different genetic constitution than the surrounding outer tissue *bot*

meristem: a group of plant cells that is capable of dividing indefinitely and whose main function is the production of new growth; meristematic cells are found at the growing tip of a root or a stem (apical meristem), in cambium (lateral meristem) and also within the stem and leaf sheaths (intercalary meristem of grasses) *bot phys*

meristem (tip) culture: the culture of an explant consisting only of a meristematic part *biot*

meristematic: pertaining to the meristem *bot*

meristematic tip: the meristematic dome and one pair of leaf primordia; it is commonly used as explants, particularly to produce virus-free plant material *bot hort* >>> meristem

merogony: an individual with the egg cytoplasm from one parent and the egg nucleus from the other parent *gene*

mesenchyme: an embryonic type of connective tissue *bot*

mesocarp: middle layer of the fruit wall *bot* >>> pericarp

mesocotyl: an elongated portion of the seedling axis between the point of attachment of the scutellum and the shoot apex of, for example, a grass seedling; it is recognized as a compound structure that is formed by the growing together of the cotyledon and the hypocotyl *bot*

mesoderm: the middle layer of embryonic cells between the ectoderm and the endoderm *bot*

mesophyll: internal parenchyma tissue of a plant leaf that lies between epidermal layers; it functions in photosynthesis and in storage of starch *bot*

mesophyll explant: an explant prepared from internal parenchyma tissue *biot* >>> mesophyll

mesophyte: a plant with an intermediate water requirement *bot*

messenger DNA (mDNA): a single-stranded DNA that acts as a messenger of the protein biosynthesis *gene*

messenger RNA (mRNA): a single-stranded RNA molecule responsible for the transmission to the ribosomes of the genetic information contained in the nuclear DNA; it is synthesized during transcription and its base sequencing exactly matches that of one of the strands of the double-stranded DNA molecule *gene*

metabolic pathway: a sequential series of enzymatic reactions involving the synthesis, degradation, or transformation of a metabolite; the pathway can be linear, branched, or cyclic and directly or indirectly reversible *gene*

metabolism: the chemical changes within the living cell; it is sum of all the physical and chemical processes by which the living protoplasm is produced and maintained and by which energy is made available for the use of the organism *phys*

metabolite: a substance taking part in metabolism *phys*

metacentric: applied to a chromosome that has its centromere in the middle *cyto* >>> Figure 11

metaphase: a stage of mitosis or meiosis at which the chromosomes move about within the spindle until they eventually arrange themselves in its equatorial region; in metaphase I (MI) of meiosis, the chromosomes of a genome line up within the cell at a position referred to as the equatorial plate; spindle fibers form, which link each chromosome of a homologous pair to a different pole of the cell; the orientation of the chromosomes relative to the two poles seems to be random; the number of different combinations of chromosomes that can occur due to their orientation at MI is defined by the formula $2n{-}1$, where n is the number of chromosomes in the genome; for example, there are two combinations possible with two chromosome pairs, four combinations with three chromosome pairs, and eight combinations with four chromosome pairs *cyto* >>> Figure 15

metaphase arrest: the stopping of cell division at mitotic or meiotic metaphases, usually by application of specific agents *cyto meth*

metaphase plate: the grouping of the chromosomes in a plane at the equator of the spindle during the metaphase stage of mitosis and meiosis *cyto* >>> Figure 15

metaxenia: the influence of pollen on maternal tissue of the fruit *bot*

methionine (M): sulfur containing nonpolar amino acid *chem phys*

method of overstored seeds: in pedigree breeding of allogamous crop plants (e.g., in rye), an effective method of regulating cross-fertilization; usually, a greater number of individual plants is harvested from a genotypic mixture of a certain population; their progenies are sown as A families in smaller plots while a half of the seed of all elite plants is retained in reserve; those A families meeting all the requirements are not directly multiplied but the remaining seed of the corresponding elite plants is sown in the following year; it enters a so-called A' family trial; in this way the economically valuable traits of the A family can be definitely evaluated after maturity; the best A families determined for further breeding have already been pollinated by a pollen mixture that also contains pollen of less valuable plants *meth* >>> Figure 4 >>> Table 35

methotrexate: a toxic folic acid analogue, $C_{20}H_{22}N_8O_5$, that inhibits cellular reproduction *chem phys biot*

methylation: the introduction of a methyl group into an organic compound; methylation of specific nucleotides within a target site of a restriction enzyme can protect the DNA against attack by that enzyme *chem phys*

2-methyl-butadiene >>> isoprene

methylene blue: a vital staining agent for chromosomes *micr*

methyl-methanesulfonate: a frequently used, very efficient chemical mutagen; it acts by adding methyl to guanine; thus, it causes base pairing errors as it binds to adenine *prep gene*

metric character: a trait that varies more or less continuously among individuals, which are therefore placed into classes according to measured values of the trait; it is also called "quantitative character" *gene*

microbe: a microorganism, especially a disease-causing bacterium *bot phyt* >>> microorganism

microbial >>> microbe

microclimate: the atmospheric characteristics prevailing within a small space *eco env* >>> macroclimate

microdissection >>> micromanipulation

microelement >>> trace element >>> macroelement >>> micronutrient

microevolution: evolutionary change within species that results from the differential survival of the constituent individuals in response to natural selection; the genetic variability on which the selection operates arises from mutation and sexual recombination in each generation *evol*

microgametogenesis: the development of the microgametophyte (pollen grain) from a microspore *bot*

microinjection: injection performed under a microscope using a fine microcapillary pipette (e.g., into a single cell or cell part) *biot*

micromanipulation: manipulation or surgery done while viewing the object through a microscope and often carried out with the aid of an injection or dissection of substances or particles *biot* >>> microinjection

micromanipulator: the facility usually attached to a miçroscope in order to carried out micromanipulation *micr* >>> micromanipulation

micrometer: a unit of measurement frequently used in microscopy (1 µm = 10^{-6} meter); in older usage also known as micron (1µ) *meth*

micromutant >>> macromutant

micron >>> micrometer

micronucleolus: a small nucleolus that may arise by nucleolar budding in the course of nucleolar degeneration *cyto*

micronucleus: the smaller nucleus as distinguished from the larger nucleus, produced during telophase of mitosis or meiosis by lagging chromosomes or chromosome fragments derived from spontaneous or induced chromosome aberrations *cyto*

micronutrient: an inorganic element or compound that is needed in relatively small amounts by plants *phys* >>> microelement >>> trace element

microorganism: literally, a "microscopic organism"; the term is usually taken to include only those organisms studied in microbiology (bacteria, fungi, microscopic algae, protozoa, viruses) *bio*

microphotography: photography requiring optical enlargement *meth micr*

microplot: microplots are used to minimize the amount of seed or space required to evaluate a group of individuals; the number of plants in a microplot differs among crops; when short rows are used as microplots, the plant density is comparable to that of larger row plots; in unbordered microplots, the effect of interplot competition has to be considered when determining an appropriate distance among the plots *meth*

microprojectile >>> high-velocity microprojectile transformation

microprojectile bombardment >>> high-velocity microprojectile transformation

micropropagation: the in vitro culture or vegetative propagation of a plant *seed biot* >>> in vitro propagation

micropylar >>> micropyle

micropyle: a canal in the coverings of the nucellus through which the pollen tube usually passes during fertilization; later, when the seed matures and starts to germinate, the micropyle serves as a minute pore through which water enters *bot*

microsatellite DNA: pieces of small DNA sequences that are repeated (appear repeatedly in sequence within the DNA molecule) adjacent to a specific gene within the DNA molecule; thus, microsatellites are linked to that specific gene *biot* >>> microsatellite marker

microsatellite marker: microsatellites or simple sequence repeats are a type of molecular markers; microsatellites consist of tandem repeats of 1-6 nucleotide motifs; the repeats usually are in units of ten or more, although repeats as small as six units have been found; the repeats can be (1) perfect tandem repeats, (2) imperfect (interrupted by several non-repeat nucleotides), or (3) compound repeats; they are well-distributed throughout a genome; microsatellites can be amplified by the polymerase chain reaction (PCR) using a pair of primers flanking the repeat sequence; the polymorphism between different individuals is due to the variation in the number of repeat units; each locus can have many alleles; one advantage of microsatellites is that they are mostly codominant, which make them easily transferable between genetic maps of different crosses in the same or closely related species, in contrast with RAPDs, which are dominant and therefore new maps have to be generated for every cross; several microsatellite-primer pairs may be used simultaneously, thus reducing time and costs; the relatively simple interpretation and genetic analysis of single-locus markers make them superior to multi-locus DNA marker types such as RAPDs; microsatellites are also called "simple sequence repeats" (SSRs), "simple tandem repeats" (STRs) or "simple sequences" (SSs) *gene* >>> Table 29

microsatellite-primed PCR (MP-PCR): a technique resulting in RAPD-like patterns after agarose gel electrophoresis and ethidium bromide staining; the MP-PCR technique is more reproducible than RAPD analysis because of higher stringency *biot* >>> random amplified polymorphic (RAPD) technique

microscope: an optical instrument having a magnifying lens or a combination of lenses for inspecting objects too small to be seen distinctly by the unaided eye *micr*

microscopic slide >>> glass slide

microsome >>> minichromosome

microsporangium: a sporangium (e.g., pollen sac) that produces the microspores (pollen) *bot*

microspore: the first cell of the male gametophyte generation of Angiospermae and Gymnospermae, later to form the pollen grain *bot*

microspore culture: it refers to the in vitro culture of pollen grains to obtain haploid callus or haploid plantlets directly from the pollen grains; microspore cultures differ from pollen cultures by the stage of development in gametogenesis *biot* >>> Figures 17, 26 >>> Table 7

microspore mother cell: one of the many cells in the microsporangium (anther) that undergo microsporogenesis to yield four microspores, as opposed to megaspore mother cell *bot*

microsporocyte >>> pollen mother cell (PMC)

microsporogenesis: the development of microspores from the microspore mother cell *bot* >>> meiosis

microstylous: showing a short stamen *bot*

microsurgery >>> micromanipulation

microsynteny: genomic relationships at gene level >>> synteny

microtome: a machine for cutting thin slices of embedded tissue; these sections may be stained and examined with the light or electron microscope *prep*

microtubule: a tubular structure, 15-25 nm in diameter, of indefinite length and composed of subunits of the protein tubulin; it occurs in large numbers in all eukaryotic cells, either freely in the cytoplasm or as a structural component of organelles; they form part of the structure of the mitotic spindle, which is responsible for the movement of chromosomes during cell division *cyto*

micrurgy >>> micromanipulation

mictic >>> amphimictic

mid rip: the central, thick, linear structure that runs along the length of a plant lamina; it occurs in true leaves as a vein running from the leaf base to the apex; it provides support and is a translocative vessel *bot*

migration: the movement of individuals or their propagules from one area to another *eco*

mildew: a plant disease in which the pathogen is seen as a growth on the surface of the host; a powdery (true) mildew is caused by one of the Erysiphaceae;

a downy (false) mildew by one of the Peronosphaceae; the first may be controlled by sulfur, the second by copper *phyt*

milky stage: in cereals, the stage of "milk" ripening of caryopses at which the endosperm shows a milky consistency *phys* >>> Table 13

milling: the processes in which cereal grains are subjected to grinding followed by sifting, sizing, or other separation techniques, for example, in wheat, the grain is tempered to approximately 15-17 percent moisture, which facilitates the separation of the endosperm, the pericarp, and embryo *meth agr*

Millipore filter: a disc-shaped synthetic filter having holes of specified diameter (0.005-8 μ) through its surface *prep*

mineral soil: a soil containing <20 percent organic matter or having a surface organic layer <30 cm thick *agr*

mineralization: the conversion of organic tissue to an inorganic state as a result of decomposition by the organic content *agr*

minichromosome: a very small chromosome, usually as a result of chromosome aberrations *cyto*

miniprep: an abbreviation for minipreparation; it refers to a small-scale preparation of plasmid or phage DNA commonly used after cloning to analyze the DNA sequence inserted into a cloning vector *biot*

minor crops: crops that may be high in value but that are not widely grown (e.g., many fruits, vegetables, and trees) *agr*

minor element >>> trace element

minor gene: a gene that individually exerts a slight effect on the phenotype *gene* >>> modifying (modifier) gene

minor oilseeds: oilseed crops other than soybeans and peanuts (e.g., in some countries, sunflower seed, canola, rapeseed, safflower, mustard seed, and flax seed) *agr*

minute fragment (of a chromosome): usually very tiny chromosome segments as a result of chromosome aberrations *cyto*

misdivision: aberrant chromosome division in which no longitudinal but transversal separation of the centromere occurs; the consequence may be telocentric chromosomes *cyto* >>> Figure 37

mismatch repair: any of the several cellular mechanisms for correction of mispaired nucleotides in double-stranded DNA *gene* >>> mismatching

mismatching: a region of DNA in a heteroduplex where bases cannot pair *gene* >>> mismatch repair

mispairing: the presence in one chain of a DNA double helix of a nucleotide that is not complementary to the nucleotide occupying the corresponding position in the other chain *gene*

missense mutation: a mutant in which a codon has been altered by mutation so that it encodes a different amino acid; the result is almost always the production of an inactive or possibly unstable protein and/or enzyme *gene*

mites: very small insects of the genus *Arachnida,* which includes spiders; they occur in large numbers in many organic surface soils or as pest on plants; they are not an important problem, possibly with the exception of the wheat curl mite, which is a vector of wheat streak mosaic virus (WSMV) zoo phyt >>> acarides

mitochondrion (mitochondria *pl*): an oval, round, or thread-shaped organelle, whose length averages 2 μm and that occurs in large numbers in the cytoplasm of eukaryotic cells; it is a double-membrane-bound structure in which the inner membrane is thrown into folds (cristae) that penetrate the inner matrix to varying depths; it is a semi-autonomous organelle containing its own DNA and ribosomes and reproducing by binary fission; it is the major site of ATP production and thus of oxygen consumption in cells *bot*

mitomycin C: a form of a family of antibiotics produced by *Streptomyces caespitosus;* it prevents DNA replication by crosslinking the complementary strands of the DNA double helix *chem phys*

mitosis: the process of nuclear division by which two daughter nuclei are produced, each identical to the parent nucleus; before mitosis begins each chromosome replicates to two sister chromatids; these then separate during mitosis so that one duplicate goes into each daughter nucleus; in contrast to the prophase of meiosis I, the prophase of mitosis does not involve pairing of chromosomes or crossing-over between the homologous chromosomes; during metaphase, the individual chromosomes line up at the equatorial plate of the cell and a spindle fiber develops that links each of their chromatids to one of the two poles in the cell; the chromatids of each chromosome separate and move to opposite poles at anaphase; at telophase, a nuclear membrane develops around the chromosomes to define the nucleus of the cell; a cell wall is formed *cyto*

mitotic apparatus: an organelle consisting of three components: (1) the asters, which form the centrosome, (2) the gelatinous spindle, and (3) the traction fibers, which connect the centromeres of the various chromosomes to either centrosome *bot*

mitotic crossing-over >>> somatic crossing-over

mitotic cycle: the sequence of steps by which the genetic material is equally divided before the cell division into two daughter cells happens *cyto*

mitotic index (MI): the fraction of cells undergoing mitosis in a given sample; usually the fraction of a total of 1,000 cells that are undergoing division at one time *cyto meth*

mitotic inhibition: induced or spontaneous inhibition of the mitotic division *phys cyto*

mitotic poison: any substance that hampers the proper mitosis *cyto phys*

mitotic recombination: the recombination of genetic material during mitosis and the process of asexual reproduction; the mechanism for the production of variation in heterokaryons *cyto*

mitotic spindle: the spindle-shaped system of microtubules that, during cell division, traverses the nuclear region of eukaryotic cells; the chromosomes become attached to it and it separates them into two sets, each of which can be enclosed in the envelope of a separate daughter nucleus *cyto*

mixoploid: cell populations in which different cells show different chromosome numbers *cyto* >>> mosaicism

mobile element >>> transposable element

modal value >>> mode

mode: the value of the variate at which a relative or absolute maximum occurs in the frequency distribution of the variate *stat*

mode of reproduction: two general modes of reproduction are distinguished: (1) sexual and (2) asexual; sexual reproduction involves the union of male and female gametes derived from the same or different parents; asexual reproduction occurs by multiplication of plant parts or by seed production that does not involve the union of sexual gametes; the breeding procedures are dependent on the mode of reproduction *bio gene*

modern varieties: varieties developed by breeders in the formal system *seed*

modificability: the ability of phenotypic variation of a particular genotype in response to varying environmental conditions *gene*

modification: nonheritable morphological or physiological changes induced by varying abiotic or biotic influences *gene*; in molecular biology, modification of DNA by DNA methylases occurs after replication; site-specific methylation protects the DNA (e.g., of bacteria), which synthesize restriction endonucleases *biot*

modifying (modifier) gene: a gene that modifies the phenotypic expression of another gene *gene*

moisture meter >>> hygrometer

moisture tension: the force at which water is held by soil: it is expressed as the equivalent of an unit column of water in centimeters *agr*

mol(e) (Mmol): gram molecular weight *chem* >>> molecular weight (MW)

molasses: residue of the beet sugar or sugarcane production *agr*

mold: any profuse or woolly fungus growth on damp or decaying matter or on surfaces of plant tissue *phyt;* any fungus of "moldy" appearance (i.e., with abundant, visible, woolly mycelium upon which dusty or powdery conidia can be seen); for example, anther mold of clover *(Botrytis anthophila),* black mould *(Aspergillus niger),* blue mold (*Penicillium* spp.) of apple *(P. expansum),* of citrus *(P. italicum),* of tobacco *(Peronospora tabacina),* bread mold *(Monilia sitophilo),* green mold of citrus *(Penicillium digitatum),* gray mold *(Botrytis cinera),* snow mold *(Calonectria nivalis),* tomato leaf mold *(Cladosporium fulvum),* or white mold of sweet pea *(Hyalodendron album) phyt*

moldboard: the curved metal plate in a plough that turns over the earth from the furrow *agr*

moldboard plough: a plough with a point and a heavy curved blade for breaking the soil *agr*

molecular cloning: DNA segments of different sizes of prokaryotic or eukaryotic origin are identically multiplied as a part of bacterial plasmids or phages: the alien DNA is incorporated into the host DNA ring molecule; by the rapid division of the host cells the alien DNA segments are simultaneously cloned; subsequently, those cloned DNA segments can be excised, separated and purified for further utilization *biot*

molecular genetics: a branch of genetics that deals with molecular aspects of genetic mechanisms *gene*

molecular hybridization: the annealing of previously purified and denatured DNA strands by different means *gene meth* >>> DNA hybridization

molecular marker: particular DNA sequences and/or segments that are closely linked to a gene locus and/or a morphological or other characters of a plant; those segments can be detected and visualized by molecular techniques; roughly, three groups of markers can be classified: (1) hybridization-based DNA markers such as restriction fragment length polymorphisms (RFLPs) and oligonucleotide fingerprinting; (2) PCR-based DNA markers such as random amplified polymorphic DNAs (RAPDs), which can also be converted into sequence characterized amplified regions (SCARs), simple sequence repeats (SSRs) or microsatellites, sequence-tagged sites (STS), amplified fragment length polymorphisms (AFLPs), inter-simple sequence repeat amplification (ISA), cleaved amplified polymorphic sequences (CAPs) and amplicon length polymorphisms (ALPs); (3) DNA chip and sequencing-based DNA markers such as single nucleotide polymorphisms (SNPs) *gene biot*

molecular weight (MW): the sum of the atomic weights of all of the atoms in a given molecule *chem*

molecule: that ultimate unit quantity of a compound that exists by itself and retains all the chemical properties of the compound *chem*

molybdenum (Mo): an element that is required in small amounts by plants and is found largely in the enzyme nitrate reductase; deficiency leads to interveinal chlorosis *chem phys*

monad: a meiocyte-derived individual cell instead of a tetrad as a result of meiotic disturbances *cyto*

monoallelic: applied to a polyploid in which all alleles at a particular locus are the same (in a tetraploid—*A1A1A1A1*) as opposed to diallelic (in a tetraploid—*A1A1A1A2*), triallelic (in a tetraploid—*A1A1A2A3*), tetraallelic (in a tetraploid—*A1A2A3A4*), etc. *gene*

monobrachial: a chromosome with a terminal centromere *cyto* >>> telocentric >>> Figures 11, 37

monocarpic: bearing one fruit *bot*

monocentric chromosome: a chromosome with only one centromere *cyto* >>> Figure 11 >>> neocentric >>> polycentric

monocentric crop plant: a crop plant species that has only one center of origin *evol* >>> center of diversity

monoclonal: describing genetically identical cells produced from one clone *gene*

monoclonal antibody: an antibody preparation that contains only a single type of antibody molecule; monoclonal antibodies are produced naturally by myeloma cells; a myeloma is a tumor of the immune system; a clone of cells producing any single antibody type may be prepared by fusing normal lymphocyte cells with myeloma cells to produce a hybridoma *meth sero*

monocot: an abbreviated name for monocotyledon, referring to plants having single-seed leaves; flower parts arranged in threes or multiples thereof, parallel-veined leaves, closed vascular bundles arranged randomly in the stem tissue *bot* >>> Table 32

monocotyledonous: having one cotyledon *bot* >>> monocot

monoculm mutant: a mutant form that shows only one culm instead of normally more (e.g., in wheat) *gene*

monoculture: the growing over a large area of a single crop species or of a single variety of a particular species *agr*

monoecism: the condition in plants that have male and female flowers separated on the same plant (e.g., in maize or pumpkin) *bot* >>> Table 18

monogenic: a trait controlled by the alleles of one particular locus, as opposed to digenic, trigenic oligogenic, or polygenic *gene* >>> Table 33

monogenic resistance: resistance determined by a single gene *phyt*

monogenomatic >>> monohaploid

monogerm(ous): a fruit of, for example, sugarbeet containing only one ovule in contrast to a multigerm fruit, which represents an aggregate fruit containing several ovule units *bot seed* >>> Table 33

monohaploid: a haploid cell or individual possessing only one chromosome set in the nucleus *gene cyto*

monohybrid: a cross between two individuals that are identically heterozygous for the alleles of one particular gene (i.e., *Aa* × *Aa*) *gene*

monohybrid heterosis >>> superdominance

monohybrid segregation: a segregation pattern according to a monogenic inheritance *gene* >>> Table 2

monoisodisomic: a cell or individual showing monosomy for one chromosome but an isochromosome for one of the arms of the missing chromosome *cyto* >>> Figure 37

monoisosomic: a cell or individual showing nullisomy of one chromosome but having an isochromosome for one arm of the missing chromosome pair *cyto* >>> Figure 37

monophyletic: a group of species that share a common ancestry, being derived from a single interbreeding population *bot evol*

monophyllous: showing one leaf *bot*

monoploid: having the basic chromosome number in a polyploid series *cyto*

monoplontic: it refers to a haploid individual or monoploid phase of the life cycle *bot*

monosome: a chromosome that lacks a homologue in a diploid organism *cyto*

monosomic: a genome that is basically diploid but that has only one copy of one particular chromosome type, so that its chromosome number is $2n - 1$ *cyto gene* >>> Figure 37

monosomic analysis: a common method for gene mapping (e.g., in hexaploid wheat); when genes determining phenotypes of interest for which an aneuploid series is not available, crosses can be made to a monosomic series in a variety with a contrasting phenotypic pattern; the monosomics are used as female parents in order to ensure the majority of progeny (~ 72 percent; the transmission of $n - 1$ gametes is about 75 percent by the egg cell, but only about 4 percent by the pollen) will be become monosomic; if the gene involved is both recessive and hemizygous-effective (an uncommon situation), direct phenotypic observations on F1 monosomic progenies enable the researcher to locate the gene on a particular chromosome; only monosomic individuals in the critical cross will exhibit the phenotype, whereas disomic sibs and monosomic (or disomic) progenies from all other crosses will display the dominant phenotype; if the plants of the monosomic series carry the dominant allele of a gene of interest, monosomic individuals in the critical cross exhibit the recessive phenotype, whereas disomic sibs display the dominant phenotype *gene* >>> aneuploid >>> Figure 37 >>> Table 23

monosomy >>> monosomic

monospermic >>> monospermous

monospermous: bearing one seed *bot*

monostand: sometimes it refers to a grass community composed of only one cultivar *agr*

monotelic: a mitotic chromosome with one oriented and one unoriented centromere *cyto* >>> centromere

monotelocentric: a cell or individual lacking one chromosome pair but showing one telocentric chromosome for one arm of the two missing homologues *cyto* >>> Figure 37

monotelodisomic: a cell or individual lacking one chromosome pair but showing two homologous telocentric chromosomes for one arm of the two missing homologues *cyto* >>> Figure 37

monotelomonoisosomic: a cell or individual lacking one chromosome pair but showing a telocentric chromosome for one arm of the missing homologous pair and an isochromosome for the other arm *cyto* >>> Figure 37

monotelotrisomic: a cell or individual showing an additional telocentric chromosome to a certain pair of chromosomes *cyto* >>> Figure 37

mordant >>> mordanting

mordanting: in a broad sense, to produce surface conditions by metal ions in the fixed structures that will enable them to hold the particular stains intended for making them visible *cyto*

MORGAN unit, morgan (M): a unit of relative distance between genes on a chromosome; one morgan (1 M) represents a crossing-over value of 100 percent; a crossing-over value of 10 percent is a decimorgan (dM); 1 percent is a centimorgan (cM) *gene*

morphogenesis: the developmental processes leading to the characteristic mature form of a plant or parts of it *phys*

morphosis: a modification of the morphogenesis of an individual caused by environmental changes *phys*

morphotype >>> habit(us)

mortality: the relative frequency of deaths in a specific population (i.e., death rate) *bio*

mosaic: a pattern of disease symptoms displaying mixed green and lighter colored patches *phyt*

mosaicism: intraindividual variation of chromosome numbers or chromosome structure, usually in different tissues *cyto*

mother cell: special cells in the anther and ovule that give rise to pollen or egg cells *bot*

mother plant: the female ancestor of a hybrid and/or hybrid progeny; in horticulture, a mature plant from which cuttings are taken *meth agr gene hort* >>> donor plant

mottled leaf: disease caused by a zinc deficiency that reduces the size of leaves and fruits (e.g., in citrus plants) *phyt phys*

mould >>> mold

mound layering: a method of propagation whereby a branch or stem is scored and then brought into contact with the soil to spur rooting *meth hort*

MP-PCR >>> microsatellite-primed PCR

mRNA >>> messenger RNA

MS medium: in vitro culture medium named after the description by MURASHIGE and SKOOG *biot*

mtDNA: an abbreviation for mitochondrial DNA *gene*

mucilage: the gummy, sticky complex carbohydrate (consisting principally of polyuronides and galacturonides that chemically resemble the pectic compounds and hemicellulose) substances that cover the seeds, the root tip, bark, or stems of some plants *bot*

mugeinic acid: a chelating agent; it plays an important role in the uptake of heavy metal ions from the soil when it is exudated by roots of some gramineous plants (e.g., of rye) *chem phys* >>> chelate

mulch: a crumbly intimate mixture of organic and mineral material formed mainly by worms *agr*

mule: a plant hybrid that is self-sterile and usually cross-sterile due to infertile pollen or undeveloped pistils *gene*

multidimensional scaling (MDS): a multivariate method of showing similarities and/or dissimilarities of empirical data (e.g., genotypes) within an *n*-dimensional euclidic space (e.g., $n = 2$ or 3) in which the distances between the objects are as best as possible arranged according to the distances of the data matrix *meth stat*

multifactorial >>> polygenic

multigene family >>> multigene variety

multigene variety: a variety that carries a number of specific genes governing resistance to a particular pathogen *phyt*

multigenic: a trait that is controlled by many genes, as opposed to monogenic *gene* >>> polygenic

multigerm: an aggregate fruit containing several ovules *bot* >>> monogerm(ous)

multihybrid: an individual that is heterozygous for more than one gene *gene*

multiline: a cultivar or variety that is composed by many more or less defined lines *seed* >>> multiline variety

multiline variety: a composite (blended) population of several genetically related lines of a self-pollinated crop, but bearing different genes (e.g., for resistance to pathogens) *seed*

multilineal variety >>> multiline variety

multilocation testing: a testing of breeder's strains and varieties on several geographically different sites in order to estimate the adaptive environmental response and/or performance stability *meth*

multi-locus probe: a DNA probe that hybridizes to a number of different sites in the genome of an organism *biot*

multinucleate: describes cells that have more than one nucleus *bot*

multiple alleles: the existence of several known allelic forms of a gene *gene* >>> allelism

multiple cropping: the growing of more than one crop on the same field in one year *agr*

multiple fruit: developed from a cluster of flowers on a common base *bot*

multiple genes: two or more genes at different loci that produce complementary or cumulative effects on a single, quantitative genetic trait *gene* >>> polygenes

multiplication: the increase in number of individuals produced from seed or by vegetative means *meth*

multiply >>> multiplication

multitude of genes >>> polygenes

multivalent: designating and association of more than two chromosomes whose homologous regions are synapsed by pairs (e.g., in autopolyploids or in translocation heterozygotes) *cyto* >>> Figure 15

multivar >>> cultivar mixture

mummy: a dried, shriveled fruit or seed colonized by a fungus or parasite *phyt*

mutability: the ability of a gene to undergo mutation *gene*

mutable genes: a class of genes that frequently spontaneously mutate *gene*

mutable site >>> mutational site

mutagen: an agent that increases the mutation rate within an organism or cell; for example, X-rays, gamma-rays, neutrons, or chemicals (base analogues, such as 5-bromo uracil, 5-bromo deoxyuridine, 2-amino purine, 8-ethoxy caffeine, 1.3.7.9.-tetramethyl-uric acid, maleic hydrazide; antibiotics, such as azaserine, mitomycin C, streptomycin, streptonigrin, actinomycin D; alkylating agents, such as sulfur mustards [ethyl-2-chloroethyl sulfide], nitrogen mustards [2-chloroethyl-dimethyl amine], epoxides [ethylene oxide], ethyleneimines, sulfates, sulfonates, diazoalkanes, nitroso compounds [*N*-ethyl-*N*-nitroso urea]; azide [sodium azide]; hydroxylamine; nitrous acid; acridines [hydrocyclic dyes], such as acridine orange) *gene*

mutagenesis: the process leading to a mutant genotype *gene*

mutagenic: substances and circumstances inducing mutants *gene*

mutagenic agent >>> mutagen

mutagenicity: the potential of agents and circumstances to induce mutations *gene*

mutagenicity testing: the assessment of chemical or physical agents for mutagenicity *gene meth*

mutagenize: treatments that result in mutations *meth*

mutagenized: cells or individuals that were treated with mutagens *gene*

mutant: a plant bearing a mutant gene that expresses itself in the phenotype *gene* >>> Table 35

mutant site: a site on a chromosome at which a mutation can occur or has occurred *gene*

mutant strain: a strain of cells or individuals that, by one or more mutations, is differentiated from the original strain *gene*

mutation: a change in the structure or amount of the genetic material of an organism; in cytogenetics, a gene or a chromosome set that has undergone a structural change *gene* >>> Table 35

mutation breeding: to experimentally introduce or remove a character from a cell or organism by exposure to mutagenic agents followed by screening for the desired attribute; it also refers to several techniques, involving induced mutations, that were utilized (mainly in the 1960s and 1970s) to introduce desirable genes into the plants (e.g., resistance to plant diseases, increased yield, improvements in composition); usually seeds or pollen were soaked in mutation-causing chemicals (mutagens) or via bombardment treated with ionizing radiation followed by screening of the resultant plants and selection of the particular mutation (beneficial trait) *meth* >>> Figure 1 >>> Table 35

mutation map: the frequency of mutations recorded along of chromosomes, represented as a diagrammatic drawing *gene*

mutation pressure: the continued production of an allele by mutations *evol*

mutation rate: the number of mutation events per gene and per unit of time (e.g., per cell generation) *gene*

mutational hot spot: a site within a gene or genome that frequently mutates *gene*

mutational site: the more or less defined position along a gene at which mutations occur *gene*

mutator gene: a gene that may increase the spontaneous mutation rate *gene*

muton: the smallest unit of DNA in which a change can result in a mutation (i.e., the single nucleotide) *gene*

mycelium (mycelia *pl*): a mass of hyphae that form the body of a fungus *bot*

mycobacterium (mycobacteria *pl*): any of several rod-shaped aerobic bacteria of the genus *Mycobacterium bot*

mycology: the science of fungi; the study of mushrooms *bot*

mycoplasma: the smallest free-living microorganism; it lacks a rigid cell wall and is therefore pleomorphic (polymorphic); mycoplasmas cause many diseases in plants; many formerly attributed to viruses are now known to be caused by mycoplasmas *phyt*

mycor(r)hiza: a close physical association between a fungus and the roots (or seedlings) of a plant from which both fungus and plant appear to benefit *bot*

mycotoxin(s): toxic substances produced by fungi or molds on agricultural crops *phys chem phyt*

N

n: the symbol for the haploid chromosome number

N banding: a special chromosome staining method related to C banding, which reveals specific types of heterochromatin; the pattern of bands and interbands along a chromosome serves as a tool for chromosome or chromosome segment identification; the method was successfully applied in wheat *cyto* >>> C banding

naked (barley or oats): mutant varieties that thresh free from their husk in contrast to conventional varieties where the husk is held firmly to the grain *bot*

nanometer: equals 10^{-9} meter equals 1 Ångström *phy*

nanovirus >>> geminivirus

nap: large masses of curled and loosely matted fibers found in raw cotton *agr*

naphthalene acetic acid (NAA): a synthetic auxin *chem phys*

napiform: turniplike in form *bot*

narrow wing: the smaller of the two parts of, for example, the wheat glume, which are separated by the keel *bot*

narrow-leafed >>> angustifoliate

nastic movement >>> nasty

nasty: the response of a plant organ to a nondirectional stimulus (e.g., light); it is facilitated by changes in cell growth or changes in turgor *bot phys*

native >>> indigenous

native breed >>> landrace

native DNA: double-stranded DNA isolated from a cell with its hydrogen bonds between strands intact, as opposed to denatured DNA *gene biot*

natural selection: a complex process in which the total environment determines which members of a species survive to reproduce and so pass on their genes to the next generation *gene*

nature reserve: an area of land set aside for nature conservation and associated scientific research, usually with strong legal protection against other uses *eco*

near-isogenic lines: not fully isogenic, for example, in maize, two distinct composites of F3 lines from a single cross, one consisting of lines homozygous recessive and the other consisting of lines homozygous dominant for a certain gene (i.e., there is same genetic background), however, differing only in being homozygous dominant versus recessive for the genes; in wheat, near-isogenic lines were produced for different *Rht* (reduced height) genes causing different straw length *gene*

neck: the uppermost part of the culm between the flagleaf sheath and the collar *bot*

necrosis: death of plant tissue, which is usually accompanied by discoloration or becoming dark in color; commonly a symptom of fungus infection *phys*

necrotrophic pathogen(e): a fungal pathogen that causes the immediate death of the host cells as it passes through them; a colonizer of dead tissue *phyt*

nectar: a sticky, sometimes sweet, secretion of flowers, which has an attraction for insects *bot*

nectar gland >>> nectarium >>> nectary

nectarine: a variety of peach having a smooth, downless skin *bot hort* >>> peach

nectarium >>> nectary

nectary: refers to a sugar-secreting gland; nectaries are usually situated at the base of a flower, sometimes in a spur, in order to attract pollinators; nectaries can also be extrafloral (e.g., the gland spin of certain cacti where they attract seed dispersal insects, such as ants) *bot*

needle: a linear, commonly pungent leaf *bot*

neighborhood: a partially isolated subpopulation with a certain degree of inbreeding; it may arise when a large population splits into subpopulations by inbreeding effects *eco*

nema >>> eelworm

nematicid: a substance or preparation used for killing nematodes parasitic to plants *phyt*

nematode: a microscopic soil worm that may attack roots or other structures of cereal, sugarbeet, potato, and other plants and cause extensive damage *phyt* >>> eelworm

neocentric: secondary centromeres, that under certain conditions, show movement as the primary centromere; sometimes they are observed on chromosome ends when they move toward the poles during anaphase of meiosis I *cyto*

neocentric activity >>> neocentric

neo-Darwinian evolution: evolutionary theory incorporating Darwinism and Mendelian genetics *bio evol*

neomorph: alleles that show qualitatively different effects compared to the wild-type allele *gene*

neomycin: an aminoglycoside antibiotic produced by *Streptomyces fradiae* that functions by interfering with ribosomal activity and so causing errors in the reading of the mRNA *chem phys gene*

neomycin phosphotransferase (NPT): an enzyme used in gene transfer experiments as a reporter for transient gene expression *chem phys biot*

nervation (of leaf): the arrangement of veins in a leaf *bot*

nervature >>> nervation

nerve: the line, usually raised, on the surface of a lemma or glume marking the presence of conducting tissue below the surface *bot*

nervure >>> leaf vein >>> nervation

nest planting: setting out a number of seedlings or seeds close together in a prepared hole, pit, or spot *meth*

net blotch disease (of barley): an important disease of winter barley *(Pyrenophora teres, syn Helminthosporium teres, syn Drechslera teres);* it is favored by cool, damp conditions; diseased leaves bear short, dark brown stripes or blotches consisting of a fine network of lines, running at right angles to as well as parallel to the veins; the blotches are usually surrounded by a narrow yellow zone *phyt*

netting: covering plots or individual plants with nets in order to prevent bird damage *meth*

niche: the functional position of an organism in its environment, comprising the habitat in which the organism lives, the period of time during which it occurs, and the resources it obtains there *eco*

nick: the two parents for producing hybrid seed when they produce high yields of seed of a highly productive and desirable hybrid; in breeding, synchronization of the receptivity of the female organ to the maximum pollen load of the pollinator for cross-fertilization *meth*; in molecular genetics, a single-strand break of DNA *gene biot* >>> nick translation

nick translation: a technique by which a DNA molecule is radioactively labeled with high specificity; such labeled DNA is used by different DNA hybridization methods as a probe (e.g., Southern transfer); within a double-stranded DNA several single-strand breaks (nicks) are produced by DNase hydrolysis; each break or gap is extended by 5'>3' exonuclease activity of DNA polymerase I; the removed 5' nucleotides are immediately substituted by the polymerase activity of the same exonuclease; however, now ^{32}P-deoxynucleotides are used, which label the newly synthesized DNA *gene biot*

nick-translated >>> nick translation

nicotianamide: a soluble crystal amide of nicotinic acid that is a component of the vitamin B complex *chem phys*

nicotine: a colorless, oily, water-soluble, highly toxic liquid alkaloid, $C_{10}H_{14}N_2$, found in tobacco and valued as an insecticide *chem phys* >>> biological control

***nif* genes:** the genetic designation of genes participating in the process of nitrogen fixation; about 17 genes are organized in the *nif* operon; by the *nif* genes produced proteins the atmospheric nitrogen (N_2) will be fixed as NH_4^+ and NO_3^- ions; many soil bacteria may fix atmospheric nitrogen; there are many research activities dealing with the transfer of the bacterial system of nitrogen fixation into crop plants other than legumes *gene biot*

nitrification: the oxidation of ammonia to nitrite and nitrite to nitrate by chemolithotrophic bacteria *chem phys*

nitrocellulose: a nitrated derivative of cellulose; it is used in the form of a membrane as a filter for macromolecules in blotting techniques *prep meth*

nitrocellulose filter: a filter made up of nitrocellulose membrane *prep* >>> nitrocellulose

nitrogen (N): an element that is essential to all plants; it is found reduced and covalently bound in many organic compounds and its chemical properties are especially important in the structure of proteins and nucleic acids; deficiency causes chlorosis and etiolation *chem phys*

nitrogen (N) consumers: a crop plant in the crop rotation that takes nitrogen up from soil, as opposed to plants (e.g., legumes) that provide nitrogen to the soil by their nitrogen fixation activity *agr* >>> nitrogen fixation

nitrogen (N) fixation: the reduction of gaseous molecular nitrogen and its incorporation into nitrogenous compounds; it is facilitated by lighting, photochemical fixation in the atmosphere, and by the action of nitrogen-fixing microorganisms (bacteria) *chem phys agr*

nitrogenase: an enzyme complex that catalyzes the reduction of molecular nitrogen in the nitrogen-fixation process in which dinitrogen is reduced to ammonia *chem phys*

NMS >>> nuclear male sterility

nobilization: a term used in the breeding of sugarcane to indicate repeated matings (backcrossing) to the "noble" canes (i.e., restoring intergeneric *Saccharum* hybrids to the phenotype of *Saccharum officinarum*) *agr*

nodal bud: the lateral shoot bud located within the root ring at the node *bot*

node: a slightly enlarged portion of a stem where leaves and buds arise, and where branches originate *bot*

nodulation: in legumes, species of *Rhizobium* bacteria fixing nitrogen of the air in association with the roots on which they are the cause of swellings (nodules) *bot agr*

nodule: a small, hard lump or swelling; root nodules are characteristic of *Rhizobium* infection and nitrogen fixation in legumes *bot*

nodule bacteria: it refers to several species of nitrogen-fixing *Rhizobium* bacteria, which form ball-like nodules along legume roots *bot agr*

nodus >>> node

nominal scale: a scale for scoring quantitative data using a series of predefined values (e.g., flower color) *stat*

noncoding DNA: a certain portion of DNA that obviously does not determine a gene product, such as a protein and/or character *gene*

nonconjunction: the failure of metaphase chromosome pairing during meiosis *cyto*

nondemanding variety >>> low-input variety

nondisjunction: the failure of separation of paired chromosomes at metaphase, resulting in one daughter cell receiving both and the other daughter cell none of the chromosomes in question; it can occur both in meiosis and mitosis *cyto*

nonhomologous association: pairing of chromosomes, which obviously are not homologous, however, it is presumed that cryptically homologous segments allow the chromosome association *cyto*

nonhomologous chromosomes: the different chromosomes of a haploid chromosome set, which usually cannot pair with another *cyto*

nonhost resistance: inability of a pathogen to infect a plant because the plant is not a host of the pathogen due to lack of something in the plant that the pathogen needs or to the presence of substances incompatible with the pathogen *phyt*

nonimmune >>> susceptible

noninfectious disease: a disease that is caused by an abiotic agent (i.e., by an environmental factor, not by a pathogen) *phyt*

non-Mendelian inheritance: an unusual ratio of progeny phenotypes that does not reflect the simple operation of Mendel's law, for example, mutant: wild-type ratios of 3 : 5, 5 : 3, 6 : 2, or 2 : 6, indicate that gene conversion has occurred; in general, it refers to extrachromosomal and/or nonchromosomal inheritance *gene*

nonparametric tests: these are tests that do not make distributional assumptions, particularly the usual distributional assumptions of the normal-theory-based tests; nonparametric tests usually drop the assumption that the data come from normally distributed populations *stat*

nonpreference: a term used to describe a resistance mechanism where parasites prefer to be on some host genotypes more than others; the less preferred genotypes are resistant *phyt* >>> antixenosis

nonrandom mating: a mating system in which the frequencies of the various kinds of matings with respect to some trait or traits are different from those expected according to chance *gene*

nonrecurrent apomixis: it refers to occasional apomixis, usually caused by haploid parthenogenesis *bot*

nonrecurrent parent: a parent that is not involved in a backcross *meth*

nonsense codon: a codon that does not determine an amino acid; it may terminate the translation *gene* >>> stop codon

nonsense mutation: a mutation that alters a gene so that a nonsense codon is inserted; such a codon is one for which no normal tRNA molecule exists, therefore it does not code for an amino acid; usually nonsense codons cause the termination of translation; several nonsense codons are recognized (e.g., amber, ochre, opal) *gene*

nonsense suppressor: a mutation in tRNA that leads to the insertion of an amino acid at the position of a "stop" codon and often restores enzyme activity in nonsense mutations *biot*

nonsibling chromatid: a chromatid that derives from the other homologue of the two homologous chromosomes or from nonhomologous chromosomes of the complement *cyto*

nonsister chromatid >>> nonsibling chromatid

nonspecific resistance >>> horizontal resistance

nontill rotation: a method of planting crops that involves no seedbed preparation in the rotation other than opening small areas in the soil for placing seed at the intended depth; moreover, there is no cultivation during crop production, despite chemicals that are used for vegetation control *agr*

nonuniform resistance >>> resistance

nopaline: a rare derivate of an amino acid; it is produced in some crown galls of plants; the controlling genes are part of the T-DNA of Ti plasmids *gene phys*

norm of reaction >>> range of reaction

normal curve >>> normal distribution

normal distribution: the most commonly used probability distribution in statistics; in nature, a vast number of continuous distributions are normally distributed; a continuous symmetrical bell-shaped frequency distribution *stat*

normalizing selection: the removal of genes and/or alleles that produce deviations from the normal phenotype of a population *meth*

Northern blot >>> Northern blotting

Northern blotting: a procedure analogous to Southern blotting, but transferring RNA from a gel to a carrier (like nitrocellulose) instead of DNA *gene meth*

novel plant >>> intergeneric cross

noxious: injurious (e.g., a noxious weed is one that crowds out desirable crops, robs them of plant food and moisture and causes extra labor in cultivation) *agr*

noxious weeds: undesirable plants that infest either land or water resources and cause physical and economic damage *agr* >>> noxious

NPK analysis: the analysis of the ratio of nitrogen (N), phosphorus (P), and potassium (K) in an organic soil amendment *agr phys*

N-type: in sugarbeet breeding, varieties with normal sugar content and normal yielding capacity (N = Normal) *seed*

nucellar embryony: a way of parthenogenesis in which the embryo arises directly from the nucellus *bot*

nucellus: the mass of tissue in the ovule of a plant that contains the embryo sac; size and shape can be diagnostic for species *bot*

nuclear division: the division of the cell nucleus by mitosis, meiosis, or amitosis *cyto*

nuclear division cycle: the sequence of stages of the division of the nucleus *cyto*

nuclear dye >>> nuclear stain

nuclear envelope >>> nuclear membrane

nuclear fragmentation: the degeneration of the nucleus by partition of the nucleus into more or less different parts *cyto*

nuclear gene: a gene that is located on the chromosome of the nucleus *gene*

nuclear male sterility (NMS): refers to male sterility that is determined by nuclear genes, as opposed to cytoplasmic male sterility *gene seed*

nuclear membrane: the structure that separates the nucleus of eukaryotic cells from the cytoplasm; it comprises two unit membranes each 10 nm thick, separated by a perinuclear space of 10-40 nm; at intervals, the two membranes are fused around the edges of circular pores, which allow for the selective passage of materials into and out of the nucleus *bot*

nuclear pore: it allows the selective passage of materials into and out of the nucleus *bot* >>> nuclear membrane

nuclear sap: a nonstaining or slightly stainable liquid or semiliquid substance of the interphase nucleus *cyto*

nuclear stain: usually, basic dyes such as methylene blue, methylgreen, crystal violet, greenpuron, or azure B bromide that bind preferentially to the cell nuclei (i.e., chromosomes and nucleolus) with slight cytoplasmic effect *micr meth cyto*

nuclear staining >>> nuclear stain

nuclear transplantation: the transfer of a nucleus into the cytoplasm of another cell *biot*

nuclease: any enzyme that degrades DNA and RNA *gene* >>> endonuclease >>> exonuclease

nucleic acid: nucleotide polymers with high relative molecular mass, produced by living cells and found in both the nucleus and cytoplasm; they occur in two forms (DNA and RNA) and may be double- or single-stranded; DNA embodies the genetic code of a cell or organelle, while various forms of RNA function in the transcriptional and translational aspects of protein synthesis *gene*

nucleolar chromosome: the chromosome that carries the nucleolus organizer (region); it may also be called a satellite chromosome *cyto*

nucleolar constriction: that region of a chromosome that carries the nucleolus organizer; besides the centromeric region it is observed as a secondary constriction along particular chromosomes; it is not stained by the standard chromosome techniques; by the nucleolar constriction the chromosome arm appears divided into two parts; the terminal part is called satellite, the whole chromosome is called the satellite chromosome *cyto*

nucleolar dominance >>> amphiplasty

nucleolar organizer: a region on a loop of DNA emanating from a chromosome in the nucleolus and around which rRNA genes are clustered; it is involved in the regulation of chromosome behavior *cyto*

nucleolar zone *syn* **nucleolus organizer region** (NOR): a chromosome region that is associated with the formation of the nucleolus during telophase *cyto* >>> nucleolar constriction

nucleolin: one of the nonribosomal proteins; it is considered to play a key role in regulation of rDNA transcription, perisomal synthesis, ribosomal assembly, and maturation; it influences the nucleolar chromatin structure through its interaction with DNA and histones; it is involved in cytoplasmic-nucleolar transport of preribosomal particles *phys* >>> ribonucleoprotein

nucleolus: a clearly defined, often spherical area of the eukaryotic nucleus, composed of densely packed fibrils and granules; its composition is similar to that of chromatin, except that it is very rich in RNA and protein; it is the site of the synthesis of ribosomal RNA; the assembly of ribosomes starts in the nucleolus but is completed in the cytoplasm *cyto*

nucleolus organizer >>> nucleolar organizer

nucleolus organizer region (NOR) >>> nucleolar zone

nucleoprotein: a conjugated protein, composed of a histone or protamine bound to a nucleic acid as the nonprotein portion *cyto*

nucleosid(e): a glycoside that is composed of ribose or deoxyribose sugar bound to a purine or pyrimidine base *chem gene*

nucleosome >>> karyosome

nucleotide: a nucleoside that is bound to a phosphate group through one of the hydroxyl groups of the sugar; it is the unit structure of nucleic acids *chem gene*

nucleotide pair: a pair of nucleotides joined by hydrogen bond that are present on complementary strands of DNA *gene*

nucleotide sequence: the order of nucleotides along a DNA or RNA strand *gene*

nucleotide synthesis >>> biotechnology

nucleus: the double-membrane-bound organelle containing the chromosomes that is found in most nondividing eukaryotic cells; it disappears temporarily during cell division; within the nucleus several independent approaches point to the compartmentalization of particular activities such as transcription, RNA processing, and replication; chromosomes are revealed to occupy defined domains and to represent highly differentiated structures; the numerous activities that use DNA and RNA as a template occur with a defined spatial and temporal relationship (e.g., compartmentalization of nuclear functions is particularly seen with replication); DNA moves through a fixed architecture containing the molecular machines directing replication *cyto* >>> Figure 25

null allele: a "silent" allele without an obvious expression *gene*

null hypothesis: a hypothesis that there is no discrepancy between observation and expectation based on some sets of postulates *stat*

null hypothesis test: the standard hypothesis used in testing the statistical significance of the difference between the means of samples drawn from two populations; the null hypothesis states that there is no difference between the populations from which the samples are drawn; one then determines the probability that one will find a difference equal to or greater than the one actually observed; if this probability is 0.05 or less, the null hypothesis is rejected and the difference is said to be statistically significant *stat*

null mutation: a mutation that eliminates all enzymatic activity, usually deletion mutations *biot*

nulli-haploid: a cell or individual that possesses a haploid chromosome set plus a missing single chromosome *cyto* >>> Figure 37

nulliplex type: the condition in which a polyploid carries a recessive gene at a particular locus in all homologues; simplex denotes that the dominant gene is represented one, duplex two, triplex three, quadruplex four times, etc. *gene* >>> autotetraploid >>> Table 3

nullisome: a plant lacking both members of one specific pair of chromosomes *cyto* >>> Figure 37

nullisomic >>> nullisome

nullisomic analysis: in nullisomic analysis, observations are made for phenotypic or other differences between the nullisomic for each chromosome and the disomic condition within the same variety; the method is applied for localization of genes within a given genome; the method can only be used in polyploids, while diploids commonly do not tolerate the loss of both homologous chromosomes *gene meth* >>> Figure 37

nulli(somic)-tetrasomic line >>> nulli-tetrasomic

nullisomy >>> nullisome

nulli-tetrasomic: a cell or individual, usually an allopolyploid, that possesses one lacking pair of chromosomes, which is partially compensated by a tetrasomic (four-fold dosage) of another, usually homoeologous, chromosome; a whole series of nulli-tetrasomics was produce in hexaploid wheat, and successfully used in numerous genetic and molecular studies *cyto* >>> Figure 37

numeric aberration: the variation of the number of genomes or chromosomes, for example, ploidy variation, aneuploids (nullisomics, monosomics, trisomics, tetrasomics, etc.), substitutions, or additions *cyto* >>> Figure 37

numeric constancy of chromosomes: the constant inheritance of the same number of chromosomes from generation to generation, which is facilitated by the mitotic and meiotic mechanisms *evol*

numerical aperture (NA): relationship between the objective light collection angle (a) and the refractive index (h) of the medium between the objective and specimen (NA = h × sin q) *micr*

numerical taxonomy: the classification of related organisms using a multitude of characteristics, each one of which is given equal weight; the degree of similarity between them is calculated using a digital computer, which treats the data collected for all characters and determines the similarities taking all possible pairs *tax*

nurse crop >>> companion crop

nurse culture: a culture technique or the callus upon which a filter paper is placed separating single cells from the callus in the paper raft technique; the callus (i.e., the nurse tissue) releases growth factors and nutrients that induce growth in the single cells supported by the filter paper and sharing the communal environment *biot*

nursery: a place where young trees or other plants are raised, either for propagation or for testing and observations *agr*

nurseryman's tape >>> grafting tape

nut: a dry, indehiscent, woody fruit *bot*

nutation: the turning of a plant or plant organ toward light *bot*

nutlet: a little nut (e.g., in strawberry) *bot* >>> nut

nutrient: a nutritive substance or ingredient, such as major and minor mineral elements, necessary for plant growth and development as well as the organic addenda such as sugars, vitamins, amino acids, and others employed in plant tissue culture media *phys biot* >>> Table 33

nutrient-enhanced varieties (crops): plants that have been modified to possess novel traits that make those plants more economically valuable for nutritional uses (e.g., higher than normal protein content in feedgrains; high-zinc content or high-glutenin in wheat; high-amylose, high-lysine, high-methionine high-oil in maize; high-phytase in maize and soybeans; high-oleic oil, high-stearate, high-sucrose in soybeans) *biot seed*

nutritional mutant >>> nutrient-enhanced varieties

nutritive substance >>> nutrient

nutritive value index: a measure for daily digestible amount of forage per unit of metabolic body size relative to a standard forage; it is used as a selection criterion in forage crop breeding *meth agr* >>> Table 33

nyctanthous: flowering by night *bot*

nymph: the immature stage of certain insects whose growing young resemble the parents in body form *zoo phyt*

O

objective: the lens or combination of lenses that first receives the rays from an observed object, forming its image in an optical device, as a microscope *micr*

objective lens: the lower lens in a microscope that is closest to what is being looked at *micr*

obligate: restricted to a particular way of life *phyt*

obligate apomixis: seed apomixis with maternal offspring to 100 percent *bot*

obligate parasite: an organism that cannot live in the absence of its host *phyt*

observation: a data collection strategy in which the activities of subjects are visually examined; the observer attempts to keep his presence from interfering in or influencing any behaviors *stat meth*

observation tube >>> eyepiece

ocDNA: an open circular DNA molecule that has at least one nick; it cannot be supercoiled and has the same density as linear DNA in $CsCl_2$/EB density gradients *biot*

ochre codon: UAA stop codon *gene biot* >>> nonsense mutation

octopine: a rare derivate of an amino acid; it is produced in some crown galls of plants; the controlling genes are part of the T-DNA of Ti plasmids *chem phys*

octoploid: having eight chromosome sets of identical or different complements *cyto*

octovalent: chromosome configuration consisting of eight chromosomes *cyto*

ocular >>> eyepiece

OECD: Organization for Economic Cooperation and Development

off-grade: postharvest removal of pathogen-infested or damaged fruit, seeds, or plants by screening procedures; the culled or off-graded material can later be individually analyzed or discarded *meth*

offset: a young plant produced by the parent, usually as its base (offshoot) or a small bulb at the base of a mother bulb *bot*

offspring >>> progeny

off-type: an individual differing from the population norm in morphological or other traits; the term also includes escapes and contaminants (e.g., seeds that do not conform to the characteristics of a variety, uncontrolled self-pollination during production of hybrid seed, segregates from plants, etc.) *gene agr*

Ohio method >>> ear-to-row selection

oil crops: plants that are grown for oil or oil-like products; main oil crops are: castor, peanut, rapeseed, safflower, sesame, soybean, sunflower, crambe, niger, jojoba, and poppy *agr*

oil immersion: the oil that is placed between the lens of a microscope and the coverslip above a microscopical preparation *micr*

oil legumes >>> oil crops

oil poppy >>> oil crops

oil-immersion objective: the objective lens system used for highest resolution with the light microscope; the space between the coverslip over the object to be examined and the lens is filled with a drop of oil of the same refractive index as the glass *micr*

oilseed >>> oil crops

OKAYAMA-BERG procedure: a method of cDNA cloning using specialized vectors favoring the generation of full-size clones *biot*

olein: a colorless to yellowish, oily, water-insoluble liquid, $C_{57}H_{104}O_6$, the triglyceride of oleic acid, present in many vegetable oils *chem phys*

oligogene: a gene that produces a pronounced phenotypic effect on characters that show normal inheritance *gene*

oligogenic: inheritance due to a small number of genes with discernible effects *gene*

oligogenic resistance: resistance controlled by one or a few genes *phyt*

oligomer: a protein composed of two or a few identical polypeptide subunits *chem phys*

oligonucleotide (oligos): a small piece of ssDNA or ssRNA; oligos are synthesized by chemically linking together a number of specific nucleotides; they are used as synthetic genes and DNA probes or in site-directed mutagenesis *biot*

oligopeptide: a small protein composed of 5-20 amino acids *chem phys*

oligos >>> oligonucleotide

oligospermous: showing only few seeds *bot*

omnipotency >>> totipotency

omnivorous: of parasites, or attacking a number of different hosts *phyt*

once blooming: it refers to plants or varieties (e.g., in roses) that bloom once a year *hort*

one-gene-one-enzyme (polypeptide) hypothesis: the hypothesis that a large class of structural genes exists in which each gene encodes a single polypeptide that may function either independently or as a subunit of a more complex protein; originally it was thought that each gene encoded the whole of a single enzyme, but it has since been found that some enzymes and other proteins derive from more than one polypeptide and hence from more than one gene *gene bio*

one-point crossover: in genetic algorithms, a breeding technique using one randomly chosen point, interchanging the portions of the two breeding individuals to the right of that point *stat*

ontogenesis: the course of growth and development of an individual from zygote formation to maturity *phys*

ontogeny >>> ontogenesis

oogonium: a primordial germ cell that gives rise, by mitosis, to oocytes, from which the ovum and polar bodies develop by meiosis *bot*

opal: a UAG stop codon *biot* >>> nonsense mutation

opaque: partially pervious to light *bot*

opaque (–2) maize (mutant): a mutant form that produces proteins rich in lysine and higher in content of calcium, magnesium, iron, zinc, and manganese *gene*

open continuous culture: a cell culture in which inflow of fresh medium is balanced by outflow of a corresponding volume of culture *biot* >>> closed continuous culture >>> continuous culture

open pollination: natural, cross, or random pollination; a free gene flow *bot*

open storage: storage with free access to normal atmospheric conditions *seed*

open-pollinated crossing group >>> crossing group(s)

open-reading frame (ORF): the mRNA region between the start and stop codon *gene*

operator (gene): a region of DNA at one end of an operon that acts as the binding site for a specific repressor protein and so controls the functioning of adjacent cistrons *gene*

operon: a set of adjacent structural genes whose mRNA is synthesized in one piece, together with the adjacent regulatory genes that affect the transcription of the structural genes; it is under the control of an operator gene, lying at one end of it *gene*

opposite: applied to the leaf arrangement in which leaves arise in pairs, one pair at each node *bot*

optic chiasma: it refers to a visible chiasma on meiotic chromosomes through the microscope *cyto*

optical density (OD): a logarithmic unit of transmission; OD = –logT (transmission), for example, a change of the optical density from 1 to 2 represents a tenfold increase in absorption *micr*

optimal sampling strategy: a sampling strategy that ensures that the genetic diversity of a species is represented in the samples *stat meth*

opuntia: within the subtribe Opuntioideae there are several species used as crop and horticultural plants; edible fruits and fleshy parts of the plant; special use for production of the stain "carmine red" by the ecto-parasite Cochenille *(Dactylopius coccus) bot*

orcein: a dye used in cytology; it is dissolved in acetic acid and used for staining of squash preparations of chromosomes *cyto meth prep*

orchard: an area of land devoted to the cultivation of (fruit or nut) trees *hort*

ordinal scale: a scale for scoring quantitative data using a series of predefined intervals arranged in a logical sequence (e.g., a typical ordinal scale may involve responses of "very good," "good," "satisfactory," "poor," or "very poor") *stat*

organ asymmetry: in many plants the left and right halves of their organs have distinct shapes (e.g., the leaves of *Begonia, Tilia* (lime tree), *Ulmus* (elm) and petals of *Anhirrhinum* (snapdragon) or *Pisum* (pea) flowers; they can occur in two mirror-image forms, left handed and right-handed; in many cases these two forms occur in equal numbers on the plant, either being located opposite each other or alternating along the stem; asymmetry of each organ traces back to a meristem with a single plane of symmetry (bilateral symmetry), such that mirror-image organs arise from opposite halves of the meristem *bot*

organ culture: the growth in aseptic culture of plant organs, such as roots or shoots, beginning with organ primordia or segments and maintaining the characteristics of the organ *biot*

organelle: within a cell, a persistent structure that has a specialized function; mostly separated from the rest of the cell by selective membranes *bot*

organic soil: a soil that is composed predominantly of organic matter; it usually refers to peat *agr*

organogenesis: the initiation and growth of an organ from cells or tissue *phys*

organogenetic: it refers to cells or tissue able to form organs *biot*

organogenic >>> organogenetic

ornamental plant: a plant that is grown for visual display *hort*

ornithogamy >>> ornithophily

ornithophilous >>> ornithophily

ornithophily: pollination by birds *bot*

Orobanchaceae: a family of totally parasitic herbs; often, specific root parasites of particular angiosperms and/or crop plants (e.g., sunflower) *bot phyt*

orphan gene: a gene identified by sequencing; its function is unknown *gene*

ortet: the original plant from which a clone is started through rooted cuttings, grafting, tissue culture, or other means of vegetative propagation (e.g., the original plus tree used to start a grafted clone for inclusion in a seed orchard) *meth hort fore*

orthodox seed: seed that can be dried and stored for long periods at reduced temperatures and under low humidity *seed*

orthologous genes: homologous genes that have become differentiated in different species derived from a common ancestral species, as opposed paralogous genes *gene*

orthoploid >>> euploid

oryzalin: an agent that is efficient for chromosome doubling of haploid apple shoots *in vitro meth cyto*

oryzinin >>> glutenin

osmolarity: the total molar concentration of the solutes affecting the osmotic potential of a solution or nutrient medium *phys*

osmolyte: osmolytes are osmolytic active, neutral organic compounds, such as sugars (polyols), certain amino acids, and quaternary ammonium compounds; proline is the most widely distributed compatible osmolyte; there is a strong correlation between increased cellular proline levels and the capacity to survive both water deficit and the effects of high environmental salinity *phys*

osmosis: the net movement of water or of another solvent from a region of low solute concentration to one of higher concentration through a semipermeable membrane *phys*

other crop seed: one of the four components of a purity test; the total percentage (by weight) of seed of all crop species, each comprising less than 5 percent of the seed lot *seed*

O-type: a maintainer plant in sugarbeet breeding; it carries the same sterility genes as the male sterile plants but having the normal cytoplasm—(N)xxzz; this genotype exists at low frequencies (3-5 percent) in most sugarbeet populations; it can be identified only by test-crossing prospective O-types with CMS plants; if all the offspring from a test cross are male sterile, the test-crossed pollinator plant is of the O-type genotype; by repeated selfing of an identified O-type, and simultaneous repeated backcrossing to a CMS line, inbred O-type lines and their equivalent inbred CMS lines can be developed *seed*

ounce (oz): equals 31.1030 g

outbreeding: the crossing of plants that are not closely related genetically, in contrast to inbreeding, in which the individuals are closely related *bot*

outclassed: it refers to a crop variety that is taken away from registration *seed*

outcrossing: cross pollination between plants of different genotypes *bot gene;* in biotechnology, the transfer of a given gene or genes (e.g., one synthesized by humans and inserted into a plant *via* genetic engineering) from a domesticated organism (e.g., crop plant) to wild type (relative of crop) *biot eco*

outgrades (in potato): outgrades are tubers considered unmarketable because of size, disease, greening, second growth, or slug or mechanical damage *seed*

outlier: an individual that occurs naturally some distance away from the principal area in which its population is found; they are anomalous values in the data and can be due to recording errors, which may be correctable or they may be due to the sample not being entirely from the same population *stat*

outplant: a seedling, transplant, or cutting ready to be established on a certain site *agr fore hort*

ovary: the part of the flower that develops into the grain in grasses; the ovary has one or more ovules, each containing an embryo sac *bot* >>> Figure 35 >>> Table 8

ovary wall >>> pericarp

ovate: egg-shaped; having an outline like that of an egg, with the broader end basal *bot* >>> ovoid

overall resistance: resistance to disease expressed at all plant growth stages *phyt*

overdominance: the phenomenon in which the character of the heterozygotes is expressed more markedly in the phenotype than in that of either homozygote; usually the heterozygote is fitter than the two homozygotes; this can give rise to monohybrid heterosis when the hybrid vigor obtained by crossing parents differing in a single specified pair of allelic genes *gene* >>> Tables 20, 21

overhang: 3' and 5' ssDNA overhangs of dsDNA; overhangs may also be called extensions or sticky ends *biot*

overlapping code >>> overlapping DNA (segments)

overlapping DNA (segments): a special type of gene organization; one DNA sequence may code for different proteins; it is performed by two open reading frames, which subsequently act *gene*

overseeding: seeding into an existing crop stand or turf *agr*

overstored seeds >>> method of overstored seeds

oversummering: the survival through the summer and/or to keep alive through summer *agr*

overwintering: the survival through the winter and/or to keep alive through winter *agr*

ovoid: egg-shaped *bot* >>> ovate

ovule: a structure in angiosperms and gymnosperms that, after fertilization, develops into a seed *bot*

ovule primordium: meristematic tissue of the ovary wall from which the seeds of angiosperms originate *bot*

ovum >>> egg

oxalate oxidase: an enzyme detected in, for example, barley seedling roots soon after germination and in the leaves of mature plants, and in response to powdery mildew infection; the enzyme contains manganese; the enzyme shows almost identical structure to the wheat protein germin *phys*

oxalic acid: a white, crystalline, water-soluble, poisonous acid, $H_2C_2O_4 \cdot 2H_2O$, used for bleaching and as a laboratory reagent *chem phys*

oxidase: an enzyme that catalyzes reactions involving the oxidation of a substrate using molecular oxygen as an electron acceptor *chem phys*

oxidation: a reaction in which atoms or molecules gain oxygen or lose hydrogen or electrons *chem phys*

P

P value: probability value; a decimal fraction showing, for example, the number of times an event will occur in a given number of trials *stat*

P1: parental generation; P1 individuals are the parents of the F1 generation *gene meth* >>> parental generation

P1 generation >>> parental generation

pachnema >>> pachytene

pachytene: the state of the prophase of first meiosis, when the homologous chromosomes are completely paired; crossing-over occurs between the nonsister chromatids of homologous chromosomes; the recombination that occurs during crossing-over in a heterozygous individual is responsible for part of the segregation observed in the progeny *cyto*

packer: it compacts loose soil to help the seeds to germinate *agr*

pairing block: a definite chromosome segment that acts as functional unit in meiotic chromosome pairing *cyto* >>> zygomere

palatability: the characteristics of feed that will affect the intake by animals; it has to be considered in forage crop breeding *agr*

pale (palea, paleae *pl*): the upper or inner of the two bracts of the floret, which covers the ventral surface of the caryopsis in grasses *bot*

palea >>> pale

paleontology: the study of the fossil record of past geological periods and of the phylogenetic relationships between ancient and contemporary plant and animal species *bio*

palindrome: adjacent inverted DNA repeats; the identical base sequences are on the opposite strands; long (>130 bp) uninterrupted palindromes occur in eukaryotic DNA but are lethal in bacteria *biot*

palisade layer: in leaves, a somewhat compacted layer of elongated cells that underlie the upper epidermis with the long axis perpendicular to the leaf surface; in seeds, the term is used interchangeably with Malphigian layer *bot*

palmate: it refers to leaves; a compound with two or more leaflets arising from the top of a stalk or rachis; the main nerves of a leaf may palmate, a situation in which several more or less equally large ones diverge along the blade from an origin at the top of the petiole *bot*

palmitic acid: a white, crystalline, water-insoluble solid, $C_{16}H_{32}O_2$ *chem phys*

pangenesis: recapitulation of certain ancestral traits during embryogenesis *evol*

panicle: the branched inflorescence of, for example, oats and other grasses *bot*

***Panicum* type:** apospory in which there is no initial polarization in the embryo sac; the spindle at first meiosis lies crosswise at the micropylar end and a second mitosis leads to a mature four-nucleate, monopolar, and unreduced embryo sac *bot* >>> Figures 28, 35

panmictic: a random interbreeding population; the individuals mate at random (e.g., in rye) *gene* >>> panmixis

panmictic population >>> panmictic

panmixia >>> panmixis

panmixis: random mating; a mode of sexual reproduction in which male and female gametes encounter each other incidentally (e.g., each mating between two individuals of a population show the same probability, as in a rye field) *bot*

paper-piercing test: a stress test for seedling vigor utilizing sand covered by filter paper, through which the seedlings must emerge to be considered vigorous *seed*

pappus: a tuft of delicate fibers or bristles at the tip of a tiny fruit, such as the feathery structure of the ripe dandelion seed that is easily blown from the head *bot*

paracasein >>> casein

paracentric: around the centromere *cyto*

paracentric inversion >>> pericentric inversion

paraffin section: a section of tissue cut by a microtome after embedding in paraffin wax *micr*

Parafilm: the brand name for a stretchable, waxed adherent used as a glassware closure or for other purposes in scientific work *prep*

parallel mutation: a mutation that causes similar phenotypes but in different species *gene*

paralogous genes: homologous genes that have arisen through gene duplication and that have evolved in parallel with the same organism, as opposed to orthologous genes *gene*

parameter: a numerical quantity that specifies a population in respect to some characteristics *stat*

paramutable allele: an unstable allele where the phenotypic consequences are enhanced by the presence of a paramutagenic allele in a heterozygote *gene*

paramutagenic allele: an allele possessing the ability to cause paramutation *gene*

paramutation: a mutation in which one allele in the heterozygous condition permanently changes the partner allele *gene*

parapatric: applied to species whose habits are separate but adjoining *bot* >>> allopatric >>> sympatric

parasexual: genetic systems that achieve genetic recombination by ameiotic exchange mechanisms *bot*

parasite: an organism living in or on a host organism, from which it obtains food and other support *phyt*

parasitic plant >>> parasite

parasitization *syn* **parasitize** *syn* **parasitized** >>> parasite

paravariation: modification that is contrary to the idiovariation and mixovariation *gene*

parenchym(a): a tissue composed of specialized plant cells with a system of air spaces running between them; the cells are regarded as the basic cells from which all other cell types have evolved *bot*

parent: pertaining to an organism, cell, or structure that produces another gene >>> P1

parental generation: the generation comprising the immediate parents of the F1 generation; the symbols P2 and P3 may be used to designate grandparental and great-grandparental generations, respectively *gene* >>> P1

parental type: an association of genetic markers, found among the progeny of a cross, that is identical to an association of markers present in a parent *gene*

parent-offspring analysis: the linear regression of the performance of offspring on that of the parents was proposed as a method of estimating heritability *stat*

parthenocarp >>> parthenocarpy

parthenocarpy: production of fruit without seeds, as in bananas and some grapes *bot*

parthenogenesis: the development of an individual from an egg without fertilization; it occurs in some plants *bot* >>> Figure 28

partial diallel >>> incomplete diallel

partial dominance *syn* **semidominance:** incomplete dominance; the production of an intermediate phenotype in individuals heterozygous for the gene concerned; it is generally considered to be a type of incomplete dominance, with the heterozygote resembling one homozygote more than the other *gene* >>> Tables 20, 21

partial pedigree: some ancestors of a particular genotype are known, usually the female parents; partial pedigrees are most common where open-pollinated or polycross seed is used for progeny testing *gene*

partial resistance: resistance that is expressed by the slower development of fewer pustules or lesions, compared with normally susceptible varieties *phyt*

partial seeding: seeding confined to limited areas (drills, strips, patches, or nests), generally according to a regular spatial pattern *fore*

partial self-incompatibility: self-incompatibility in some species results in a lower percentage of seed set than occurs with cross-pollination *gene*

partially balanced incomplete block design (PBIB): an experimental design where the residual variance of the difference between the candidates may adopt one out of only two different values *stat meth*

participatory plant breeding: plant breeding (crossing and subsequent selection in the heterogeneous progeny) involving farmers *agr*

particle bombardment >>> gene transfer

particle gun >>> gene transfer

particulate inheritance: the model proposing that genetic information is transmitted from one generation to the next in discrete units (particles) so that the character of the offspring is not a smooth blend of factors from the parents *gene* >>> MENDEL's laws of inheritance

partitioning: the physiological process by which assimilates are distributed among competing sink tissues *phys*

parts per million (ppm): designates the quantity of a substance contained in a million parts of a mixture or solution in a carrier, such as air or water *meth*

passage number: the number of times the cells in the in vitro culture have been subcultured *biot*

passage time: the interval between successive subcultures or the culture period *prep*

passive immunity: the immunity against a given disease produced by conditioning *phyt*

passive resistance >>> disease avoidance

passport data: information about the origin of an accession and any other relevant information, including descriptors, which assist in the identification of the accession *seed*

pasture: an area of vegetation that is used for grazing and/or devoted to the production of forage for harvest primarily by grazing *agr*

path diagram: pedigree arrangement showing only the direct line of descent from common ancestors *gene meth*

path-coefficient analysis: a method for analyzing regular and irregular breeding systems; the path coefficient is a measure of the importance of a given path of influence from cause to effect; it is the ratio of the standard deviation of the effect to the total standard deviation *stat*

pathogen: a (micro)organism (e.g., bacteria, viruses, fungi) that causes disease in a particular host or range of hosts *phyt*

pathogenesis: the complete sequence of events starting with the arrival of the pathogen at the host surface to the completion of the disease cycle *phyt*

pathogenic: producing disease or toxic symptoms *phyt*

pathogenicity: the ability of an organism to survive at the expense of its host and to incite disease *phyt*

pathologist: a specialist for origin, nature, and course of diseases *phyt*

pathotoxin: microbial metabolite with a causal role in disease; it induces all disease symptoms and exhibits the same host specificity as the pathogen of origin *phyt*

pathotype: a subspecific classification of a pathogen distinguished from others of the species by its pathogenicity on a specific host *phyt* >>> pathogenicity

patroclinous inheritance: an inheritance in which all offspring have the nucleus-based phenotype of the father *gene*

pBR322 vector: the most-used plasmid cloning vector; size: 4.3 kb, about 20 copies per chromosome, it carries ampicillin and tetracycline resistance genes *biot*

PCN >>> potato cyst nematode

PCR >>> polymerase chain reaction technique

peat: an accumulation of dead plant material often forming a very deep layer; it shows various stages of decomposition and is completely waterlogged; it has a low pH value and may fix metal ions, such as copper or iron *agr*

peat pot: a preformed plant pot (of different sizes and shapes) manufactured from peat and other supplements, usually used for seedling cultivation *meth*

pectin: one of a group of homopolysaccharides, but that are especially rich in galacturonic acid; they form a kind of cement, so contributing to the structure of plant cell walls *chem phys*

pectinase: an enzyme that degrades pectin (the adhesive material that cements cells together); it is used alone or with enzymes to digest the polygalacturonic acid of plant cell walls to sugar and galacturonic acid in protoplast production *chem phys biot*

pedicel: the small laterals of the panicle branch in, for example, oats, which bear the spikelets on their swollen tips *bot*

pedigree: the record of the ancestry of an individual, genetic line or variety *gene* >>> full pedigree >>> partial pedigree

pedigree breeding: a system of breeding in which individual plants are selected in the segregating generations from a cross on the basis of their desirability judged individually and on the basis of a pedigree record; the advantages are (1) if selection is effective, inferior genotypes may be discarded before lines further evaluated, (2) selection in generation involves a different environment, which provides a good genetic variability, and (3) the genetic relationship of lines is estimated and can be used to maximize genetic variability *meth* >>> Figure 7 >>> Tables 5, 28, 35

pedigreed seed: seed of a named cultivar that is produced under the supervision of a certification agency to ensure genotype and purity *seed*

peduncle: flower stalk; the inflorescence stalk of a plant supporting either a cluster or a solitary flower *bot*

pedunculate: possessing or pertaining to a peduncle; stalked *bot*

pegging: the burying of, for example, peanut fruits *agr*

pellet: a small, roundish mass of matter *biot*

pelleted seeds: seed that are commercially prepared for precision planting by pelleting them inside a special preparation in order to make them more uniform in size; sometimes special nutrient or growth-promoting substances are placed in the pellets to aid in seed germination and growth *seed*

pelleting drum: used for pelleting seeds of different crops *seed* >>> pelleted seeds

pendulous: drooping or hanging downward *bot*

penetrance (of genes): the proportion of individuals of a specified genotype who manifest that genotype in the phenotype under a defined set of environmental conditions *gene*

penetration: the phenotypic expression of an allele may depend on the growing condition or other factors, such as age or sex; sometimes only a fraction of individuals with a certain genotype shows the "expected" phenotype; this portion is sometimes called penetration *gene*

penetration peg: a minute protuberance from a hypha, germ tube or appressorium that effects penetration of the host plant surface *phyt bot*

pentaploid: having five homologous or inhomologous genomes *gene*

pentasomic: the presence of five homologous chromosomes in a complement *cyto*

pentosan: any of a class of polysaccharides that occur in plants, and form pentoses upon hydrolysis *chem phys*

pentose: a monosaccharide that consists of five carbon atoms *chem phys*

pepo: a fruit that has a hard rind (e.g., watermelon, cantaloupe, squash, cucumber) but without internal separations or septa *bot*

peptidase: any of the class of enzymes that catalyze the hydrolysis of peptides or peptones to amino acids *phys*

peptide: a linear molecule that consists of two or more amino acids linked by peptide bonds *chem phys*

peptide bond: a covalent bond formed between the NH_2 group of one amino acid and the COOH group of another, with the elimination of H_2O *chem*

perennation: the survival of plants from growing season to growing season with a period of reduced activity in between *bot*

perennial: a plant that normally lives for more than two seasons and that, after an initial period, produces flowers annually *bot*

perennialism: a plant that continues to grow from year to year *bot*

perfect flower: having both functional pistil and stamens *bot* >>> Table 18

perfect state: the state of the life cycle of a fungus in which spores are formed after nuclear fusion or by parthenogenesis *phyt*

perianth: flower envelope; the outer covering of a flower composed of the floral leaves, usually an outer greenish calyx and an inner, brightly colored carolla *bot*

pericarp: the mature ovary wall, which is fused with the testa in the caryopsis *bot* >>> carycpsis

pericarp layer: it comprises several layers of cells; in the mature grain, the innermost layers form a solidified, dense, protective coat, which prevents water entering and moulds from infecting the grain; the outer layers form a coarse coat over the grain surface; these outer layers can become infected by moulds such as *Alternaria* and *Cladosporium* by harvest time, giving a dull color to the grain *bot prep*

pericentric: an intrachromosomal inversion that includes the centromere, as opposed to paracentric *cyto*

pericentric inversion: the inversion of a chromosome segment containing a block of genes that involves the centromere *cyto*

periclinal: referring to a layer of cells running parallel to the surface of a plant part *bot*

periclinal chimera: a plant made up of two genetically different tissues, one surrounding the other *bot*

periodicity: repetition of events at fairly regular intervals *bot*

perisperm: it may derive from the nucellus in some plant species (e.g., sugarbeet, leafy spurge); it contributes substantially to the storage tissue *bot*

perithecium: a rounded or flasklike ascocarp from which ascospores formed inside; spores are discharged to the exterior via a small pore *bot*

permaculture: a contraction of permanent agriculture where the inputs equal the outputs *agr*

permanent preparation: preparing a microscopic slide (e.g., chromosome preparation) for long-term storage by using special recines or other substances *prep micr cyto*

permanent slide >>> permanent preparation

permanent wilting (point) >>> wilting point

permeability: a property of a membrane, or other barrier, being the ease with which a substance will diffuse or pass through it *phys*

permutation: changing the order of a set of elements arranged in a particular way *stat*

peroxidase (Per): an enzyme that catalyzes the oxidation of certain organic compounds using hydrogen peroxide as an electron acceptor *chem phys*

peroxisome: a cytoplasmic organelle characterized by the association of peroxide-generating oxidases with a catalase *cyto*

persistence: the act or fact of persisting *phyt*

persistent >>> persistence

persistent modification: nonheritable morphological or physiological changes over a more or less long period of generation cycles induced by varying abiotic or biotic influences *gene*

pest: any form of plant or animal life, or any pathogenic agent, injurious or potentially injurious of plants or plant products *phyt*

pest resistance: resistance to any form of plant or animal life or any pathogenic agent *phyt*

pesticide: a chemical preparation for destroying plant, fungal, or animal pests *phyt*

petal: one of the inner floral leaves, usually brightly colored and borne in a tight spiral or whorled carolla *bot*

petiole: the stalk by which a leaf is attached *bot*

PETRI dish: a covered glass container in which cells, organs, seeds, or microorganisms are cultured *prep*

***Ph* locus:** a gene that controls homoeologous chromosome pairing in wheat and, similarly, in other allopolyploid plants; the *Ph* gene restricts pairing between homologous chromosomes in a polyploid *gene*

pH value: the negative logarithm of the hydrogen-ion activity, expressed in terms of the pH scale from 0-14 *chem*

phage: the abbreviation for bacteriophage (i.e., a virus that infects bacteria) *bio*

phase-contrast microscopy: light rays passing through an object of high refractive index that will be retarded in comparison with light rays passing through a surround medium with a lower refractive index; the retardation or phase change for a given light ray is a function of the thickness and the index of refraction of the material through which it passes; thus in a given unstained specimen, transparent regions of different refractive indexes retard the light rays passing through them to differing degrees; such phase variations in the light focused on the image plane of the light microscope are not visible to the observer; the phase contrast microscope is an optical system that converts such phase variations into visible variations in light intensity or contrast; it allows observation of the cell and structures (even living) without staining and degrading treatments *micr*

phasmid: a cloning vector that has the possibilities to replicate as a plasmid or as a phage; the two modes of replication are usually functional in different bacterial species *biot*

phellogen: a layer of plant tissue outside of the true cambium, giving rise to cork tissue *bot*

phene: the phenotype of the plant, which is a product of the gene (gene > DNA > transcription > RNA > processing > translation > protein) and the interaction with the environment *gene* >>> phenotype

phenocopy: a nonhereditary phenotypic change that is environmentally induced during a limited developmental phase of an organism; it may mimic the effect of a known genetic mutation *gene*

phenogenetics: a branch of genetics that studies the interaction of the genotype and its manifestation *gene*

phenol: a chemical used to remove proteins from DNA preparations *biot*

phenol test (reaction): the color produced in the grain of, for example, wheat and barley, by treatment with a one percent solution of phenol in water *seed meth*

phenology: the study of the impact of climate on the seasonal occurrence of flora *eco*

phenotype: the observable manifestation of a specific genotype (i.e., those properties on an organism, produced by the genotype in conjunction with the environment) *gene*

phenotypic expression: the manifestation of a particular gene resulting in a particular phenotype *gene*

phenotypic plasticity: the capability of a genotype to assume different phenotypes *gene*

phenotypic segregation: the phenotypic differentiation patterns of cells or individuals in segregating populations, as opposed to genetic segregation *gene* >>> Table 2

phenotypic selection: development of a variety based on its physical appearance without regard to its genetic constitution *gene*

phenotypic variance (VP): the total variance observed in a character; it includes experimental error, genotype × environment interaction and the genotypic variance *stat*

phenylalanine (Phe): an aromatic, nonpolar amino acid *chem phys*

pheromone: a chemical exchanged between members of the same animal species that effects behavior (sex attractants, alarm substances, aggregation-promotion substances, trail substances, etc.) *phyt bio*

phloem: a tissue comprising various types of cells that transports dissolved organic and inorganic materials over long distances within vascular plants *bot*

phosphatase: an enzyme that catalyzes reaction involving the hydrolysis of esters of phosphoric acid *chem phys*

phosphorus (P): an element that is required by plants in the oxidized form; it is utilized in reactions in which energy is transferred, often involving ATP *chem phys* >>> mitochondrion

phosphorylation: the addition of a phosphate group to a compound, involving the formation of an ester bond between the reactants *phys*

photobleaching: photochemical reaction of fluorophores, light, and oxygen that causes the intensity of the fluorescence emission to decrease with time *micr*

photoinhibition: the slowing or stopping of a plant process by light (e.g., the germination of some seeds) *phys*

photo-insensitive plants >>> daylength insensitivity

photometry: the measurement of the intensity of light or of relative illuminating power *meth micr*

photomorphogenesis: changes in plant growth due to light; there is a main plant regulator protein Cop1 that suppresses genes controlling the photomorphogenesis; for example, when a seedling gets exposed to light the Cop1 protein is reduced in the nucleus and photosynthesis is initiated—the seedling becomes green; cryptochromes may interact with photoreceptor-proteins, which can recognize blue light; thus the interaction can "switch-off" the Cop1 protein *phys*

photon: a packet of light energy *phy*

photonasty >>> nasty

photoperiod: the relative length of the periods of light and darkness associated with day and night; in many species, floral induction occurs in response to daylength; species have been categorized according to their daylength requirements as short-day, long-day, intermediate-day, or day-neutral *phys*

photoperiodism: the response of a plant to periodic, often rhythmic, changes in either the intensity of light or to the relative length of day *phys*

photoreceptor: a pigment that absorbs the light used in various metabolic plant processes that require light *phys*

photosynthesis: the series of metabolic reactions that occur in certain autotrophs, whereby organic compounds are synthesized by the reduction of carbon dioxide using energy absorbed by chlorophyll from light *phys* >>> Table 33

phyletic: the evolution by which a race or line is progressively transformed from its ancestral form without branching or separating into related parts *evol*

phyletic gradualism: the process of gradual evolutionary change over time *evol*

phyletic series >>> phyletic

phyllode: an expanded petiole resembling and having the function of a leaf, but without a true blade *bot*

phyllody: the condition in which parts of a flower are replaced by leaf-like structures; often it is a symptom of certain diseases *bot*

phyllosphere: the surface of a living leaf *bot*

phyllotaxis: the active arrangement of leaves on a stem or axis *bot*

phyllotaxy >>> phyllotaxis

phylloxera: a fatal vine pest that destroys the soft vine roots of *Vitis vinifera* cultivars; the only remedy is to replant on phylloxera-resistant rootstocks; roots of most American hybrid vines are immune to the effects of the pest; phylloxera will generally not inhabit soils that are 80 percent sand; in all other soil textures, cultivars should be grafted onto phylloxera-resistant rootstocks *hort* >>> vine-louse

phylogenesis >>> phylogeny

phylogenetic: of, pertaining to, or based on phylogeny *evol*

phylogenetic tree: a diagram showing evolutionary lineages of organisms *evol*

phylogeny: evolutionary relationships within and between taxonomic levels, particularly the patterns of lines of descent, often branching, from one organism to another *evol* >>> Figure 10 >>> Tables 1, 14, 17

phylum (phyla *pl*): an important group of organisms *bot*

physical map: representation of a DNA molecule showing the location of restriction sites or genes *gene* >>> mapping

physiologic specialization: the existence of a number of races or forms of one species of pathogen based on their pathogenicity to different cultivars of a host *phyt*

physiological maturity: the maturity of a seed when it reaches its maximum dry weight; this usually occurs prior to the normal harvest date *seed*

physiological race: pathogens of the same species with similar or identical morphology but differing pathogenic capabilities *phyt*

physiologist: a specialist dealing with the functions and activities of living organisms and their parts *phys*

phytase: an enzyme that catalyzes the breakdown of phytin, the source of inorganic phosphorus in seed metabolism *chem phys*

phytoalexin(e): an antifungal substance that is produced by a plant in response to damage or infection *phyt phys*

phytochrome: a photoreversible pigment that occurs in every major taxonomic group of plants; it exists in two interchangeable forms with respect to absorption, a red and a far-red form; Pr phytochrome is receptive to orange-red light (600-680 nm) and inhibits flowering; Pf-r phytochrome is receptive to far-red light (700-760 nm) and induced flowering *phys*

phytogenetics: synonymous to plant genetics, dealing with inheritance in plants *gene*

phytohormone >>> plant hormone

phytoncide >>> herbicide

phytopathology: the study of plant diseases *phyt*

phytosanitary certificate: a certificate from a recognized plant quarantine service that indicates that a sample is substantially free from diseases or pests *seed*

phytosiderophore(s): nonproteinogenic amino acids developed by plants under conditions of mineral deficiency (especially under iron and zinc deficiency); the production and exudation of phytosiderophores is controlled by several genes; there are crop plants, such as rye, showing a high level of phytosiderophore production and/or exudation toward the rhizosphere *phys*

phytotoxic: being poisonous to plants *phyt*

phytotoxin >>> pathotoxin

phytotron: a group of rooms or a room for growing plants under controlled and reproducible environmental conditions *prep meth*

pick-up reel: a special device on some harvesters for taking up lodging straw *agr*

pigment: an organic compound that produces color in the tissue of the plant *bot*

pileorhiza >>> root cap

piliferous layer >>> root hair

pilose: hairy *bot*

pilosum: used for describing a hairy ventral furrow, for example, barley grain *bot*

pilot test: preliminary test or study of the program or evaluation activities to try out procedures and make any needed changes or adjustments *stat meth*

pin flower: flowers with long styles and short stames *bot* >>> Table 18

pincers: a gripping tool consisting of two pivoted limbs forming a pair of jaws and a pair of handles *prep*

pinching (pinching back, pinching out) >>> disbud

pinna: one of the leaflets of a pinnate leaf *bot*

pinnate: feathered *bot*

pinnate leaf >>> pinnate

pinninervate: pinnate-veined; feather-veined *bot*

pipette: a slender, graduated tube for measuring liquids or transferring them from one container to another *prep*

pistil: the gynoecium of a syncarpous flower; a pistil includes an ovary, style, and stigma; the stigma is the receptor of the pollen *bot* >>> Figure 35

pistillate flower: designating a flower having one or more pistils and no stamens *bot*

pistillody: the conversion of any organ of a flower into carpels (e.g., stamens into pistils or pistillike structures) *bot* >>> pistillate flower

pistillum >>> pistil

pit: in botany, a term used for the widening in the center of the ventral furrow in some wheat and barley grains *bot*

pith: a tissue that occupies the central part of a stem (composed of parenchyma cells) *bot*

pixel: a picture (**pix**) element (**el**); it refers to points of information used to map images; pixels exist in arrays with specific x- and y-coordinates *phy meth*

placenta: the part of the ovary wall formed from the fused margins of the carpel or carpels on which the ovules are carried *bot*

placentation: the position of the placenta within the carpel; it may be parietal (on the walls), axile (on the axis), basal (on the basis), or free-central *bot*

plan apochromatic objective lens (plan apo): a modern, high-resolution microscope objective lens designed with high degrees of corrections for various aberrations; it is corrected for (1) spherical aberration in four wavelengths (dark blue, blue, green, red), (2) for chromatic aberration in more than these four wavelengths, and (3) for flatness of field; a single plan apo may contain as many as 11 lens elements *micr*

plant breeder: a person or organization actively engaged in the breeding and maintenance of varieties of plants, applying a wide range of methods of different scientific disciplines

plant breeding: the application of genetic principles and practices to development of individuals or cultivars more suited to the needs of humans; it uses knowledge from agronomy, botany, genetics, cytogenetics, molecular genetics, physiology, pathology, entomology, biochemistry, statistics, etc.

plant density: the rate at which seed or vegetative propagules are placed in a field or experimental planting *meth agr*

plant gall >>> cecidium

plant hair >>> trichome

plant hormone: a compound that is synthesized by a plant, but is not a nutrient, coenzyme, or detoxification product, and which regulates growth, differentiation or other specific physiological processes *phys*

plant label: plastic, wood, or other stakes for gardens or experiments to indicate what seeds or material are planted where until they appear, which varieties are included, what sort of evaluation is carried out, etc.; in plant conservation, paper forms to include in drying plant samples, with formal printed forms as permanent labels on herbarium specimens; the minimum information includes the name of the collector, the location collected, the date collected, and the correct identification of the specimen *meth agr hort*

plant passport: an official seed label used for forthcoming marketing; it shows the crop, crop class, e.g., European community grade, inspections, etc. *seed* >>> passport data

plant pathogenesis-related proteins (PPRP): groups of proteins with different chemical properties produced in a cell within minutes or hours following inoculation, but all being more or less toxic to pathogens *phyt*

plant pathology >>> phytopathology >>> biological control

plant protection >>> phytopathology >>> biological control

plant variety protection >>> plant variety rights

Plant Variety Protection Act: enacted in 1970 in the United States to provide patentlike protection for seed plant species; prior to 1970, breeders in the industry worked primarily with maize and sorghum, with some efforts directed to alfalfa, cotton, sugarbeet, and certain vegetables; the purpose of the PVPA was to stimulate private plant breeding research and to provide better seed cultivars to farmers and gardeners

plant variety rights: the legal rights of a plant breeder (not necessarily as a person) who has developed a new variety *seed*

plantation: a closely set stand of trees or special crops that has been planted by humans *agr hort*

plantibodies: antibodies produced in transgenic plants expressing the antibody-producing gene(s) of an animal *biot*

planting cord: a string of different manufacturing and length used for marking experimental plots, paths, margins between landmarks, or applied to mark planting rows *prep*

plantlet: a stage of in vitro culture; the stage after torpedo stage and usually one of the last before a whole plant is generated *biot*

plantling >>> plantlet

plaque: a circular zone of lysis produced by bacteriophage in a colony of bacteria on an artificial medium *meth*

plasmagene: an extranuclear hereditary determinant showing no Mendelian inheritance *gene*

plasmalemma >>> cell membrane

plasmatic: all functions, processes, or properties of the cytoplasm *bot* >>> cytoplasm

plasmic >>> plasmatic

plasmid: any extrachromosomal hereditary determinant of bacteria; such ring-shaped structures are intensively used for amplification of DNA segment in recent bioengineering approaches; it is capable of self-replication; it can carry genes into a host cell *gene* >>> Figure 27

plasmid vector: a plasmid or plasmidlike structure used as a carrier for alien DNA segments or genes *biot* >>> Figure 27

plasmodesm(a) (plasmodesmata *pl*): cytoplasmic bridges, lined with a plasma membrane that connect adjacent cells; they provide major pathways of communication and transport between cells *bot*

plasmodium: in cellular slime molds, a vegetative structure consisting of a noncellular, mobile mass of naked protoplasm containing many nuclei *bot*

plasmogamy: the fusion of the cytoplasm of two or more cells after karyogamy during the process of fertilization *cyto*

plasmolysis: the result of placing plant cells in a hypertonic solution so that water is drawn out of the cell; the cytoplasm shrinks and the cell membrane is pulled away from the cell wall *phys*

plasmon: all the extrachromosomal hereditary determinants *gene* >>> plasmotype

plasmon mutation: a mutation that genetically changes the cytoplasm and/or its hereditary determinants *gene*

plasmotype: the sum of the extrachromosomal hereditary determinants *gene*

plastid: one of a group of double-membrane-bound plant-cell organelles that vary in their structure and function (e.g., chloroplasts, leucoplasts, amyloplasts) *bot*

plastid DNA: organelle DNA that is present in a plastid *gene*

plastid inheritance: non-Mendelian inheritance that is caused by hereditary factors present in the plastids *gene*

plastidome >>> plastome

plastidome mutation: mutations that genetically change the DNA of the plastids of a plant cell *gene*

plastogene: the hereditary determinants located in the plastid *gene*

plastom mutation >>> plastidome mutation

plastome: a term usually used for the plastids of a cell or for the genetic information of the plastid DNA *bot gene*

plate: to place on or in special media in a culture dish *prep*

plating efficiency: an estimate of the percentage of viable cell colonies developing on an agar plate relative to the total number of cells spread onto the plate; the plating efficiency is a function of the tissue, medium composition, plating density, and the phase of the stock culture *biot*

pleiotropic: an allele or gene that affects several traits at the same time *gene*

pleiotropy: the phenomenon of a single gene being responsible for a number of different phenotypic effects that are apparently unrelated *gene*

plesiomorphy: ancestral state of a particular character *bot*

ploidy: the number of complete chromosome sets in the cell nucleus (e.g., diploid, tetraploid, etc.) *cyto*

plot: in field experiments, more or less large pieces of land used for planting and evaluation; in forestry, a group of trees, all from the same entry (family, clone, provenance) planted together; a five-tree-plot row is the most common design for forest genetics experiments *meth agr fore hort*

plot size: in field experiments, the size, number, and distribution of plots are essential elements; an efficient combination of plot size and plot number is required; larger plots offer smaller between-plot variance, long measurement time per plot, shorter walking time between plots, less edge, and less statistical error; small plots require a higher number of plots in order to achieve the same level of precision of estimates; plot size and shape vary with crops and stage of testing, and among characters under selection; plots are generally small at the initial stage of testing and reach a maximum size during the second and third years of replicated trials; unbordered plots with few replications at one or few locations are used for traits with low heritability; row spacing and plant populations are chosen to be similar to commercial production of crop; seedbed preparation, fertilization, weed control practices, etc., are generally the same as those used for commercial production; the mechanization and computerization of most plant breeding programs have greatly increased a breeder's ability to handle more plots, populations, etc. *meth stat*

plum pox virus: a disease of plum and related trees whose symptoms are variable according to the plant species, but which usually include the appearance of pale or dark rings or spots on leaves and fruits *phyt*

plumula: the undeveloped shoot consisting of unexpanded leaves and the growing point (i.e., the terminal bud of developed embryo) *bot*

plurannual >>> perennial

plus tree: a tree phenotype judged (but not proved by testing) to be unusually superior in some qualities (e.g., growth rate relative to site, growth habit, high wood quality, resistance to disease and insect attack, or to other adverse local factors) *fore*

pneumatophore: a specialized root in certain aquatic plants that performs respiratory functions *bot*

pod: a fruit that dehisces down both sides into two separate valves that are most typically dry and somewhat woody; they are characteristic fruits of the Leguminosae *bot* >>> legume

pod drop: losses due to the premature drop of pods; *phys agr*

pod shattering: seed losses due to the premature shattering of pods; for example, in oilseed rape it can be as great as 50 percent of the potential yield in some seasons; average losses are around 10-15 percent, equivalent to 500 kg/ha or ten times the sowing rate *agr*

poikiloploid >>> mixoploid

point mutation: a mutation that can be mapped to one specific locus; it is caused by the substitution of one nucleotide for another; it may also be caused by deletion and inversion *gene*

poison >>> toxicity

poisonous >>> toxicity

POISSON distribution: the basis of a method whereby the distribution of a particular attribute in a population can be calculated from its mean occurrence in a random sample of the population, provided that the population is large and the probability that the attribute will occur is less than 0.1 *stat*

polar mutation: a mutation of one gene that affects the expression of the adjacent nonmutant gene on one side, but not of that on the other side *gene*

polar nuclei >>> pole nucleus

polar plate: the functional center for meiotic division of chromosomes; spindle fibers converge on it *cyto*

polarization: restriction of the orientation of the vibration of electro-magnetic waves of light *phy micr* >>> polarization microscope

polarization microscope: a compound light microscope used for studying the anisotropic properties of objects and for rendering objects visible because of their optical anisotropy *micr* >>> polarization

pole: one of the two ends of the cell spindle toward which chromosomes move during mitotic and meiotic anaphase *cyto*

pole nucleus (pole nuclei *pl*): the two haploid nuclei present in the center of the embryo sac after division of the megaspore; they may fuse to form a diploid definitive nucleus before fusing with the male gamete to form the triploid primary endosperm nucleus prior to double fertilization *bot* >>> Figure 25

pollard: a tree cut back almost to the trunk in order to form a thick head of spreading branches, which are cut for basket-making and kindling (e.g., poplars and willows can be pollarded) *bot*

pollen: collectively, the mass of microspores or pollen grains produced within the anthers of a flowering plant; it is a highly specialized tissue whose function is the production of two sperm cells and their subsequent delivery through the style and ovary to the embryo sac cells where the double fertilization takes place; it is a highly reduced structure consisting of only three cells; the processes of pollen development, germination, and fertilization involve the specific expression of a large number of genes *bot* >>> Figure 35

pollen analysis: a method to study pollen grains, particularly their size, shape, and surface; since those characters are highly specific for species the method is used for taxonomic classifications *bot*

pollen barrier: in seed production, the separation of the varieties, lines, etc. (mostly in allogamous crops) in order to prevent intercrossing; it is done by different means, such as strips of other plants (e.g., hemp or maize), which prevent the free flow of pollen, or by isolation walls of tissue *seed*

pollen culture >>> microspore culture

pollen embryoids: embryoids that derive from anther culture *biot*

pollen grain: a microspore in flowering plants that germinates to form the male gametophyte, a structure made up of the pollen grain plus a pollen tube *bot* >>> Figure 35 >>> pollen

pollen grain mitosis: in microsporogenesis, a nuclear division that occurs in the pollen grain after the formation of tetrads; it results in a smaller generative nucleus and a larger vegetative nucleus *bot*

pollen mixing: cross-incompatibility in interspecific crosses is associated with proteins of the pistil that interact with proteins of the pollen to prevent normal pollen tube germination and growth; this unfavorable reaction may be avoided in certain combinations by mixing pollen from a compatible species with pollen from an incompatible parent *meth*

pollen mother cell (PMC): the microsporocyte, which undergoes two meiotic divisions to produce four microspores; each microspore becomes a pollen grain *bot*

pollen parent: the parent that furnishes the pollen and that fertilizes the ovules of the other parent in the production of seed *meth*

pollen sac: a sac within the anther of a stamen within which microspores are produced *bot*

pollen sterility: pollen that is not able to fertilize an egg cell *phys*

pollen transfer: refers to the kind of pollen transfer from the male to female organs (e.g., mediated by wind, insects, by hand, etc.) *bot*

pollen tube: the tube formed from a germinating pollen grain and down which the two male gametes pass to the ovum *bot* >>> Figure 35

pollen tube competition >>> certation

pollen-shedding: the status or process when the pollen grains are released from the anthers *bot*

pollinarium (pollinaria *pl*): a functional unit in orchid pollination that consists of two or more pollinia, stalk, or stipe, and a viscidium *bot*

pollinate: to transfer pollen from the anther to the receptive surface of the stigma of the same or another flower in angiosperms and from male to female in gymnosperms; this process usually requires a vector in outbreeding plants *meth* >>> Figure 35

pollination: the transfer of pollen grains from the anther to the stigma of a flowering plant; pollination can also be done in artificial culture (i. e., in vitro pollination) *bot* >>> Figure 35

pollinator: that parental individual, line, variety, or species, which is used as a donor of pollen in a cross *meth*

polyacrylamide gel: a gel used to separate biological molecules (proteins); it is prepared by mixing a monomer (acrylamide) with a cross-linking agent (*N,N'*-methylene-bisacrylamide) in the presence of a polymerizing agent; it leads to the formation of an insoluble three-dimensional network of monomer chains, which become hydrated in water prep

polyacrylamide gel electrophoresis (PAGE): a method for separation of proteins and amino acids on the basis of their molecule size; the molecules move through the gel under the influence of an electric field *meth*

polyadenylation: posttranscriptional addition of 50-200 adenine residues to the 3' end of eukaryotic mRNA; the poly-A tail can be used to separate eukaryotic mRNA from other RNA species with oligo T cellulose *biot*

polyandrous: having many stamens *bot*

polyandry: the state of having more than one male mate at one time; in fertilization, the fusion of one female and two or more male pronuclei within an egg cell *bot*

polycarpellary: composed of several carpels *bot*

polycarpic: in general, bearing many fruits (i.e., producing fruit many times or indefinitely); in botany, having a gynoecium forming two or more distinct ovaries or carpels *bot*

polycarpous >>> polycarpic

polycentric: a chromosome that shows more than two centromeres *cyto* >>> centromere

polycistronic RNA: mRNA that codes for more than one polypeptide *gene*

polycotyledonous embryo(s): embryos having more than two cotyledons or seed leaves (e.g., in pines and conifers) *bot*

polycross: open pollination of a group of genotypes (generally selected) in isolation from other compatible genotypes in such a way as to promote random mating inter se; it is a widely used procedure for intercrossing parents by natural hybridization *meth* >>> Table 35

polyembryonic >>> polyembryony

polyembryony: the condition in which an ovule has more than one embryo like in certain grasses or cereals; in the past, the phenomenon was used for haploid selection among the embryos, which show often different ploidy levels *bot meth* >>> Figure 17

polyethene >>> polyethylene

polyethylene: a plastic polymer of ethylene *prep chem*

polyethylene glycol (PEG): a polymeric substance of molecular weight between 1,000-6,000; it is used for stimulation of protoplast fusion *chem biot*

polygamous: plants that show male, female, and hermaphrodite flowers on the same or different plants *bot*

polygenes: one of a group of genes that together controls a quantitative character; individually each gene has little effect on the resulting phenotype, which instead requires the interaction of many genes *gene* >>> Table 33

polygenic: of traits determined by many genes, each having only a slight effect on the expression of the trait *gene* >>> multifactorial >>> Table 33

polygeny: a trait that is controlled by many genes *gene* >>> Table 33

polygynoecial: having a number of pistils joined together, as in aggregate fruits (e.g., raspberry) *bot*

polygyny: the state of having more than one female mate at one time; in fertilization, the fusion of one male and two or more female pronuclei within an egg cell *bot*

polyhaploid: haploid plant derived from a polyploid individual *cyto*

polyhybrid: individuals that are heterozygous with respect to the alleles of many gene loci or of crosses involving parents that differ with respect to the alleles of more loci *gene*

polykaryotic: cells showing many nuclei *cyto*

polylinker: synthetic oligonucleotide with recognition sites for several restriction endonucleases *biot*

polymerase: an enzyme that catalyzes the replication and repair of nucleic acids *gene*

polymerase chain reaction (PCR) technique: a technique for continuous amplification of DNA and/or DNA fragments in vitro; the DNA sequence must be known so that oligonucleotides can be synthesized that are complementary to the extremes of the fragment that is to be amplified; heat stable DNA polymerase (e.g., from *Thermus aquaticus*) is used for DNA synthesis *gene*

polymeric genes: genes with equal effects but cumulative action *gene*

polymerize: the act or process of forming a polymer or polymeric compound *chem*

polymery: the production of a trait by cooperation among several polymeric genes *gene*

polymorphic: occurring in several different forms *bot gene*

polymorphism: the existence of two or more forms that are genetically distinct from one another but contained within the same interbreeding population *gene*

polynemic chromosome: describes metaphase chromosome and/or chromatids with more than two DNA helices *cyto*

polynucleotide: a sequence of many nucleotides *gene*

polypeptide: a linear polymer that consists of ten or more amino acids linked by peptide bonds *chem*

polypheny >>> pleiotropy

polyphylesis: originating from several lines of descent *bot*

polyphyletic: designating a group of species arbitrarily classified together, some of the members of which have distinct evolutionary histories, not being descended from a common ancestor *evol*

polyploid: an individual carries more than two complete sets of homologous chromosomes *cyto*

polyploidization: the spontaneous or induced multiplication of a haploid or diploid genome of a cell or individual *cyto* >>> Table 35

polyploidy: the condition in which an individual possesses one or more sets of homologous chromosomes in excess of the normal two sets found in diploids; it can be produced in nature by somatic doubling due to irregular mitosis in the meristematic cell and by unreduced gametes due to irregular reductional division during meiosis; for example, approximately 70 percent of grass species and 20 percent of legumes are polyploid *cyto* >>> Tables 17, 35

polysaccharide: a molecule composed of chains of sugar units *phys*

polysome: a polyribosome, consisting of two or more ribosomes bound together by their simultaneous translation of a single mRNA molecule *gene phys*

polysomy: the reduplication of some but not all of the chromosomes of a set beyond the normal diploid number *cyto*

polyspermy: the entry of more than one sperm cell into an egg cell, irrespective of whether or not the additional sperm cells fertilize *bot*

polytene chromosome: a chromosome that is formed by repeated reduplication of single chromatids; sections may appear to puff or swell due to differential gene activation; it is visible through the light microscope *cyto* >>> puff >>> lampbrush chromosome

polyteny >>> polytene chromosome

pomaceous fruit >>> pome

pome: a fruit (e.g., apple, pear, quince) in which the seeds are protected by a tough carpel wall and the entire fruit is embedded in a fleshy receptacle *bot*

pomology: the science or study of growing fruit *hort*

population: a community of individuals, which share a common gene pool *bot gene*; in statistics, a hypothetical (often infinitely large) series of potential observations among which observations actually made constitute a sample *stat* >>> Figure 38 >>> Table 35

population, closed: a group of interbreeding plants (occurring in a certain area or in an experimental design) or a group of plants originating from one or more common ancestors, where there is no immigration of plants or pollen *stat eco*

population density: the number of individuals of a population per unit area of a particular habitat *bot eco*; in biotechnology, the cell number per unit area or volume of a medium *biot* >>> plant density

population genetics: the study of inherited variation in populations and its modulation in time and space; it relates the heritable changes in populations to the underlying individual processes of inheritance and development *gene*

population, Mendelian: a group of (potentially) interbreeding plants (cross-fertilizing crops), which may occur in a certain area or in an experimental design *gene stat*

population size: the number of individuals in a population that are included in reproduction during a certain generation *meth stat*

population waves: irregular or rhythmic changes of the number of individuals in a population that are included in reproduction during certain generations *meth stat*

porosity: the proportion, as the percentage volume, of the total bulk volume of a body of rock or soil occupied by pore space *agr*

position (positional) effect: the change in the expression of a gene with respect to neighboring genes *gene*

positional cloning: a process of molecular cloning of a gene with reference to its position on the genetic or physical map *biot*

postemergence herbicide: a herbicide that affects the weeds after emergence *phyt*

postgamous incompatibility >>> crossbreeding barrier

postzygotic incompatibility: a condition where, in the case of incompatible or wide crosses, the zygote fails to develop, often for nutritional reasons; in some cases embryo culture can be used to rescue the embryo *bot*

pot feet: supports placed under pots and planters to raise them off the ground for better drainage and air circulation *hort*

potassium (K): an element that is required for plant growth; deficiency leads to reduced growth and to dark or blue-green coloration in the leaves *chem phys*

potassium iodide >>> chemical desiccation

potato cyst nematode (PCN): a major pest of potato (*Globodera rostochiensis* [golden PCN] + *G. pallida* [white PCN]) causing severe yield loss if populations are allowed to reach high levels *zoo phyt*

potato leaf roll virus (PLRV): a virus spread by aphids that can cause severe reduction in yield *phyt*

potato virus Y: a virus spread by aphids that can cause severe reduction in yield *phyt*

pot-bound: used when a plant has an overly extensive root system in a too-small container *hort*

pound (lb): equals 373.2420 g

power (of statistical test): it is the probability that a statistical test will detect a defined pattern in data and declare the extent of the pattern as showing statistical significance *stat*

PPA: Plant Patent Act; in 1930, the PPA was enacted into law in the United States; the plant patent grants the breeder the exclusive right, for 17 years, to propagate the patented plant by asexual reproduction; the purpose of the

PPA was to encourage research investment in asexually reproduced plant species; since 1930, more than 6,000 plant patents have been issued by the Patent and Trademark Office, primarily for fruit trees, flowers, ornamental trees, grape, and other horticultural species

ppm >>> parts per million

PPRP >>> plant pathogenesis-related proteins

preadaptation: an adaptation evolved in one adaptive zone, which proves especially advantageous in an adjacent zone and so allows the organism to radiate into it; in breeding, the pretreatment of plants under moderate climatic (light, temperature, or nutrient) conditions in order to gradually accustom them to stress conditions *phys evol*

prebasic seed >>> breeder('s) seed

prebloom: the stage or period immediately preceding blooming *agr*

prebreeding: all research and screening activities before a plant material enters the directed breeding process (e.g., the development of germplasm to a state where it is viable for breeder's use); primarily, it involves the evaluation of traits from exotic material and their introduction into more cultivated backgrounds *meth*

prechilling: the practice of exposing imbibed seeds to cool (+5 to +10°C) temperature conditions for a few days prior to germination at warmer conditions *seed* >>> stratification

precipitant: a substance or process that causes precipitation *chem*

precision seed: calibrated or pelleted sorts of seeds used for precision drilling *agr*

precleaning: the process for removing the bulk of foreign materials grossly different in size from the harvested seed or other crop products *seed*

precocious embryo development: asexual development of the embryo before the flower opens and anthers dehisce *bot*

predecessor: within the crop rotation, the crop before the recent cropping *agr*

predisposition: an increase in susceptibility resulting from the influence of environment on the suspect *phyt*

preferential pairing: chromosome pairing in allopolyploids in which the most structurally similar chromosomes preferentially pair with another *cyto*

prefoliation >>> vernation

presoaking: presoaking of seeds in water has been suggested as a means to speed up germination *meth*

prevalence: the observed frequency of a trait or disease in a population, often at a particular age or time *gene*

prezygotic incompatibility: in the case of incompatible or wide crosses the inhibition of pollen germination or the prevention of pollen tube growth, among other possible barriers of plant fertilization; in some cases the barrier can be overcome by in vitro pollination *bot*

pricking off: a method of transplanting tiny seedlings; the blade of a knife or plant marker is used to remove each plant from one spot and move it to another *meth hort*

pricking out: a method of thinning seedlings by cutting them off at soil level so as not to disturb the roots of the other plants *hort meth*

pricking-out peg: an adjusted peg used for thinning or transplanting seedlings by cutting them off at soil level *meth hort*

prickle pollination >>> tripping mechanism

prickly: having thorns *bot*

primary constriction: the centromeric region of a chromosome *cyto*

primary culture: a culture resulting from cells, tissues, or organs taken from an organism *biot*

primary fluorescence: fluorescence originating from the specimen itself *micr*

primary gene pool: it includes the cultivated species of a crop and related species from which useful genes can be most readily obtained for breeding; in general, it is the total sum of all the genetic variation in the breeding population of a species and closely related species that commonly interbreed with, or can be routinely crossed with, the species *evol*

primary host >>> host

primary infection: the first infection of a plant by the overwintering or oversummering pathogen *phyt*

primary inoculum: spores or fragments of a mycelium capable of initiating a disease *phyt*

primary leaf: one of the first pair of leaves to emerge above the cotyledon during the development of a seedling; it is often morphologically distinct from subsequent leaves *bot*

primary sex ratio >>> sex ratio

primer: an oligonucleotide that forms a double strand by using a complementary segment; the primer is prolonged till the double strand is completed *biot*

priming: the treatment of seeds with an osmotic solution (e.g., polyethylene glycol, which allows controlled hydration); the seed embryo develops to the point of germination and then is dried; it is applied for more uniform and rapid germination of certain vegetable seeds *seed*

primitive form: in phylogeny, seedless vascular plants with underground rhizomes; they grow in tropical to subtropical areas and are terrestrial or epiphytic; in botany and breeding, as compared to cultivated crops, plants still show wild characteristics, such as brittle rachis, seed shattering, and others *bot*

primordium (primordia *pl*): the early cells that serve as the precursors of an organ to which they later give rise by mitotic development *bot*

principal host >>> host

probe: a radioactively or nonradioactively labeled and defined nucleic acid sequence that can be used in molecular hybridization; usually it is used for *ex situ* or in situ identification of specific, complementary nucleic acid sequences *biot*

processing of seeds: the complex of measures in order to clean, calibrate, disinfect, store, and pack seeds *seed*

prochromosome: a heterochromatin block that is seen during the interphase of cell division and which is related to the number of chromosomes per complement or less *cyto*

procumbent: trailing or laying flat on the ground *bot*

proembryo: the young embryo in its early stages of development after zygote formation *bot* >>> direct embryogensis

progenitor: the original, ancestral, or parental cell, individual, or species *bot gene evol*

progenitor cell: undifferentiated cell (i.e., immature cell) which will go on to develop into any cell type *biot*

progeny: offspring; plants grown from the seeds produced by parental plants *bot gene* >>> Figure 19

progeny selection: selection based on progeny performance *meth*

progeny test(ing): a test of the value of a genotype based on the performance of its offspring produced in some definite system of mating *meth* >>> Figure 19

prokaryote: the class of organisms that does not have discrete cell nuclei in a nuclear envelope, including bacteria, but shows single, circular DNA molecules within the cytoplasm *bio*

prokaryotic >>> prokaryote

prolamin: a protein; it is soluble in 70-90 percent ethyl alcohol but not in water; it is found only in cereal seeds (e.g., gliadins in wheat and rye, zein in maize); upon hydrolysis they yield proline, glutamic acid, and ammonia *chem phys* >>> Table 15

proliferation: successive development of new parts, organs, etc. *phys*

proline (Pro): a heterocyclic, nonpolar imino acid, which is present in all proteins; the major pathway for proline synthesis, which takes place in the cytoplasm, is from glutamate, through gamma-glutamyl phosphate and glutamyl-gamma-semialdehyde, a two-step reaction that is catalyzed by a single enzyme, D1-pyrroline-5-carboxylate synthetase *chem phys* >>> osmolyte

prometaphase: the stage in mitosis between the dissolution of the nuclear membrane and the organization of the chromosomes on the metaphase plate *cyto*

promoter: a nucleotide sequence within an operon, lying between the operator and the structural gene or genes, that serves as a recognition site and point of attachment for the RNA polymerase; it is the starting point for transcription of the structural gene(s) in the operon, but is not itself transcribed;

the promoter controls where (e.g., which portion of a plant) and when (e.g., which stage in the lifetime) the gene is expressed; for example, the promoter Bce4 is seed-specific *gene biot* >>> cauliflower mosaic virus (CaMV)

promoter sequence >>> promoter

propagate: to reproduce or cause to multiply or breed *agr*

propagating bench: a stationary, shallow box (sometimes covered by a glass pan or other means); it is usually filled with fine sand or certain soil (often sterilized), which is kept moist; cuttings, slips, or shoots after in vitro culture are inserted into the growing medium until they form roots *meth hort*

propagating case >>> propagating bench

propagation: various methods by which plants are increased (e.g., seeds, division, separation, softwood cuttings, slips, grafting, budding, or layering) *meth hort agr*

propagule: any type of plant to be used for reproduction (e.g., seedling, a rooted or unrooted cutting, a graft, a tissue-cultured plantlet, etc.) *meth hort*

prophage: the noninfectious form of a temperate bacteriophage in which the phage DNA has become incorporated into the lysogenic, host bacterial DNA *biot*

prophase: the first phase of mitosis and meiosis *cyto*

prophyll(um): the first leaf or protective scale of a lateral shoot *bot*

proplastid: a colorless, double-membrane-bound organelle with little internal structure that acts as a precursor in the development of all plastids *bot*

protandry: the maturation of anthers before carpels (e.g., in sugarbeet, sunflower, or carrot) *bot* >>> Table 18

protease: an enzyme that catalyzes the hydrolysis of the peptide bonds of proteins and peptides *chem phys* >>> proteinase

protected variety: a variety that is released and granted a certificate of plant variety protection under the legal statutes of a given country; the owner of a protected variety has the right of selling, offering, reproduction, import, export, or using for hybrid seed production *seed*

protein: a polymer that has a high relative molecular mass of amino acids; it has many functions in the living cell *chem phys* >>> Tables 15, 16, 33

protein engineering: production of altered proteins by site-directed mutagenesis *biot* >>> biotechnology

proteinase: an enzyme that hydrolyzes protein molecules *chem phys*

proteome: the complete set of proteins detectable in a tissue *phys*

proteome analysis: the basis of proteome analysis is an electrophoretic separation of the proteins on a two-dimensional protein gel, silver staining of proteins, followed by an image analysis of the stained gel; interesting protein spots identified on the gel can be excised; the critical protein can be extracted from the excised spot and further analysis of the protein (amino acid composition, partial amino acid sequence, isoelectric point, molecular weight) may result in protein identification *meth phys biot*

protoclonal variation: variability of somatic cells derived from protoplast culture *biot*

protoclone: a plant regenerated from a protoplast culture *biot*

protogyny: a condition in which the female parts develop first (e.g., in rapeseed) *bot* >>> Table 18

protoplasm: a complex, translucent, colorless, colloidal substance within each cell, including the cell membrane, but excluding the large vacuoles, masses of secretions, ingested material, etc. *gene*

protoplast: that part of a cell that is actively engaged in metabolic processes or a cell without a cell wall; protoplasts are produced by enzymes, which digest the wall; they are used for production of hybrid cells by protoplast fusion or for injection of foreign DNA *bot biot*

protoplast culture: the isolation and culture of plant protoplasts by mechanical means or by enzymatic digestion of plant tissue, organs, or cultures derived from these; protoplasts are utilized for selection or hybridization at the cellular level and for a variety of other purposes *biot*

protoplast fusion: a technique used in somatic hybridization experiments; it is used for overcoming crossing barriers; protoplasts are placed together and induced to fuse, applying fusogenic agents, such as polyethylene glycol or physical means; subsequent regeneration of the cell wall allows the propagation and regeneration of a somatic hybrid plant *biot*

prototroph: a strain of organisms capable of growth on a defined minimal medium from which they can synthesize all of the more complex biological molecules they require, as opposed auxotroph *phys*

provenance: the geographical source and/or place of origin of a given lot of seed, propagules, or pollen *fore*

provenance test: an experiment, usually replicated, comparing trees grown from seed or cuttings that were collected from many geographical regions of a species distribution *fore*

proximal: toward or nearer to the place of attachment *cyto* >>> centromere >>> chromosome arm

prune >>> pruning

pruning: trimming branches or parts of trees and shrubs in order to trim a plant or to bring the plant into a desired shape; in addition, the removal of the growing point from a plant frequently causes the initiation of tillering or branching; the onset of flowering on the new vegetative growth may be delayed compared with that of unpruned plants; the method may be also used for synchronization of flowering prior to hybridization *hort meth*

pseudoallele: genes that behave as alleles in the allelism test but that can be separated by crossing-over *gene*

pseudoallelism >>> pseudoallele

pseudobivalent: a bivalent-like association of two mitotic chromosomes due to reciprocal chromatid exchange *cyto*

pseudocarp: a fruit consisting of one or more ripened ovules attached or fused to modified bracts or other nonfloral structures *bot*

pseudocompatibility: the occurrence of fertilization that normally is prevented by incompatibility mechanisms; it is caused by specific environmental or genotypic conditions; for example, in rye, by high temperature (about +30°C) self-incompatibility can be broken so that seeds are set *bot*

pseudodominance: the apparent dominance of a recessive gene (allele), owing to a deletion of the corresponding gene in the homologous chromosome *gene*

pseudodominant >>> pseudodominance

pseudofertility >>> pseudocompatibility

pseudogamous heterosis: increased vigor of maternal offspring due to male parent influence on the endosperm *bot gene*

pseudogamy: a type of apomixis in which the diploid egg cell develops into the embryo without fertilization of the egg cell, although only after fertilization of the polar nuclei with one of the sperm cells from the male gamete to form a normal triploid endosperm *bot*

pseudogene: genes that have been switched off in evolution and no longer have any function; they are entirely neutral and evolve at a constant rate *gene*

pseudoheterosis: luxuriance; it designates hybrids between species, varieties, or lines that exceed the parents in some traits, however, neither by sheltering deleterious genes nor by balanced gene combinations *gene* >>> heterosis

pseudoisochromosome: a chromosome that shows only equal ends as a result of interchanges *cyto*

pseudovivipary: vegetative proliferation of plantlets in the inflorescence axes *bot*

pubescent: covered with soft hairs, downy *bot*

pUC18 vector: a plasmid cloning vector; size: 2.7 kb; about 100 copies per chromosome; it shows ampicillin resistance for selection and alpha complementing fragment of beta-galactosidase with in-frame polylinker for cloning *biot*

puff: a structural modified region of a polytene chromosome; it originates from the despiralization of deoxyribonucleoprotein *cyto* >>> polytene chromosome >>> lampbrush chromosome

puffing >>> puff

pulp: the soft, succulent part of a fruit, usually composed of mesocarp and/or the pith of a stem *bot*

pulse: the edible seeds of any leguminous crop *bot* >>> legume(s)

pulsed-field gel electrophoresis: an electrophoretic technique in which the gel is subjected to electrical fields alternating between different angles, allowing very large DNA fragments to move through the gel, and hence permitting efficient separation of mixtures of such large fragments *meth biot*

PUNNETT square: a diagrammatic representation of a particular cross used to predict the progeny of the cross; a grid is used as a graphic representation of the progeny zygotes resulting from different gamete fusions in a specific cross *meth*

pure bred: derived from a line subjected to inbreeding *gene*

pure line: a number of individuals of a successive, self-pollinated crop, which derives from a single plant; a strain homozygous at all loci *gene meth*

pure line breeding >>> true breeding

pure-live seed: the percentage of the content of a seed lot that is pure and viable; it is determined by multiplying the percentage of pure seed by the percentage of viable seed (germination percentage) and dividing by 100 *seed*

purity testing: determination of the degree of contamination of seed lots with genetically nonidentical, damaged, or pest-infected seeds *seed*

puroindoline proteins: small, basic, cysteine-rich proteins found in bread wheat *(Triticum aestivum);* by engineered introduction of a puroindoline gene the hard textured grain of durum wheat *(Triticum durum)* and other cereals, including maize and barley, can be converted into soft texture (Morris and Geroux, 2000) *phys gene*

pustule: a blisterlike spot or spore mass developing below the epidermis, which usually breaks through at maturity *phyt*

PVPA >>> Plant Variety Protection Act

pycnidium (pycnidia *pl*): a flask-shaped fungal receptacle bearing asexual spores (i.e., pycniospores) *bot*

pycniospore: a spore from a pycnidium *bot phyt*

pycnotic: the concentration of the nucleus into a compact, strongly stained mass, taking place as the cell dies *cyto*

pyriform: pear-shaped *bot*

pyrimidine: a basic, six-membered heterocyclic compound; the principal pyrimidines uracil, thymine, and cytosine are important constituents of nucleic acids *chem phys*

Q

Q banding: a chromosomal staining technique using the fluorescence dye quinacrin mustard; under UV light a characteristic light and dark banding is induced *cyto meth*

QTL >>> quantitative trait locus

quadriduplex type >>> autotetraploid >>> nulliplex type

quadrivalent: a chromosome association of four members *cyto* >>> Figure 15

quadruplex type >>> autotetraploid

qualitative character: a character in which variation is discontinuous *gene* >>> Table 33

qualitative inheritance: an inheritance of a character that differs markedly in its expression amongst individuals of a species; variation is discontinuous; such characters are usually under the control of major genes *gene* >>> Table 33

qualitative resistance >>> vertical resistance

quality-declared seed: a terminology introduced by the FAO, for a seed system in which a proposed 10 percent of the seed produced and distributed is checked by an autonomous seed control agency and the rest by the seed-producing organization *seed*

quantitative character: a character in which variation is continuous so that classification into discrete categories is arbitrary *gene* >>> metric character >>> Table 33

quantitative genetics: a branch of genetics that deals with the inheritance of quantitative traits; sometimes it is also called biometrical or statistical genetics *gene*

quantitative inheritance: an inheritance of a character that depends upon the cumulative action of many genes, each of which produces only a small effect; the character shows continuous variation (i.e., a gradation from one extreme to the other) *gene* >>> Table 33

quantitative resistance >>> horizontal resistance

quantitative trait locus (QTL): a locus or DNA segment that carries more genes coding for an agronomic or other traits *gene* >>> Table 33

quantum speciation: the rapid rise of a new species, usually in small isolates, with the "founder effect" and random genetic drift; it is also called saltational speciation *tax evol*

quarantine: the official confinement of plants subject to phytosanitary regulations for observation and research or for further inspection and/or testing; more general, a legal ban on the export or import of certain noxious weeds or insects that may be attached to the plants *meth seed agr*

quarternary hybrid: a hybrid derived from four different grandparental individuals *meth*

quartet: the four nuclei and/or cells produced during meiosis *bot* >>> tetrad

quereitron >>> glucoside

quick test (of seed testing): a type of test for evaluating seed quality, usually germination, more rapidly than standard laboratory tests *seed*

quiescence >>> dormancy

quincunx planting: planting four young plants to form the corners of a square with a fifth plant at its center *meth*

quinone >>> growth inhibitor

R

R line >>> restorer

R1, **R2**, **R3**, etc.: the first, second, third, etc. generation following any type of irradiation in mutation breeding *meth*

RABL configuration: in plants and insects, it refers to chromosomes that are spatially organized within the nucleus, with centromeres clustered on the nuclear membrane at one pole and telomeres attached to the nuclear membrane of the other hemisphere *cyto*

race: a genetically and, as a general rule, geographically distinct interbreeding division of a species; in other contexts, a population or group of populations distinguishable from other such populations of the same species by the frequencies of genes, chromosomal rearrangements, or hereditary phenotypic characteristics; a race that has received a taxonomic name is a subspecies *tax*

raceme: an inflorescence in which the main axis continues to grow, producing flowers laterally, such that the youngest ones are apical or at the center *bot*

racemose: an indeterminate, unbranched inflorescence in which the flowers are borne on pedicels of about equal length, along an elongated axis *bot*

race-nonspecific type: host-plant resistance that is operational against all races of a pathogen species; it is variable, sensitive to environmental changes, and usually polygenetically controlled *phyt*

race-specific type: host-plant resistance that is operational against one or a few races of a pathogen species; generally produces an immune or hypersensitive reaction and is controlled by one or a few genes *phyt*

rachilla: the spikelet axis; it is also applied to the segment of the rachilla that remains attached to the oat grain *bot*

rachis: the main axis of the ear and of the panicle of grasses *bot*

rad: an abbreviation for "radiation absorbed dose;" a measure of the amount of any ionizing radiation that is absorbed by the tissue; one rad is equivalent to 100 ergs of energy absorbed per gram tissue *phy meth* >>> radioactive

radiation absorbed dose >>> rad

radicle: the root of the embryo, which develops into the primary root of the seedling *bot*

radicule >>> radicle

radioactive: a substance when a constituent chemical element is undergoing the process of changing into another element through the emission of radiant energy; radioactivity is used as a tool in research to tag or trace the movement of compounds; the presence of a compound containing a radioactive element is revealed by instruments that measure the radiant energy emitted or by radio-sensitive films *phy meth*

radioactive tracer >>> isotopic tracer

radiocarbon dating *syn* **^{14}C dating:** a dating method for organic material that is applicable to about the last 70,000 years; it relies on the assumed constancy over time of atmospheric ^{14}C:^{12}C ratios and the known rate of decay of radioactive carbon, of which half is lost in a period of about every 5.730 years *meth*

raffinose: the raffinose family of oligosaccharides; it is believed to play an important role in the resistance of plants to environmental stress by protecting membrane-bound proteins; it is relevant when seeds mature because they rapidly lose moisture as they dry out *phys*

rain forest: a tropical forest of tall, densely growing, broad-leaved evergreen trees in an area of high annual rainfall *eco*

rain gauge: an instrument for measuring rainfall *prep*

ramet: an individual that belongs to a clone *bot*

ramification: the act or process of ramifying (i.e., the repeated division of branches into secondary branches) *bot*

random amplified polymorphic DNA (RAPD) technique: a comparative study (among individuals, populations, or species) of the DNA fragment length produced in controlled DNA synthesis reactions started with short sequences of DNA (primers); as a genetic mapping methodology, it utilizes as its basis the fact that specific DNA sequences (polymorphic DNA) are repeated (i.e., appear in sequence) with a gene of interest; thus, the polymorphic DNA sequences are linked to that specific gene; their linked presence serves to facilitate genetic mapping within a genome *biot meth*

random effects: effects of a treatment in an experiment in which the treatments are considered in relation to a whole range of possible treatment effects *stat*

random genetic drift: variation in gene frequency from one generation to another due to chance fluctuations *gene evol*

random lines >>> isogenic lines

random mating: for a given population, where an individual of one sex has an equal probability of mating with any individual of the opposite sex, or insofar as the genotypes with respect to given genes are concerned *gene*

random sample: a sample of a population selected so that all items in the population are equally likely to be included in the sample *stat*

random sampling: a sample drawn from a population in such a way that every individual of the population has an equal chance of appearing in the sample; it ensures that the sample is representative and provides the necessary basis for virtually all forms of inference from sample to population, including the informal inference, which is characteristic of rerandomization statistics *stat*

randomization: the process of making assignments at random; in field trials, randomization of entries is required to obtain a valid estimate of experimental error; each entry must have an equal chance of being assigned to any plot in a replication and an independent randomization is required for each replication *stat meth* >>> Table 25

randomized block: the entries to be tested are assigned at random to the plots within the block *stat meth* >>> Table 25

randomized block design: a randomized block analysis of variance design (e.g., one-way blocked ANOVA) is created by first grouping the experimental subjects into blocks; the subjects in each block are as similar as possible (e.g., littermates); there are as many subjects in each block as there are levels of the factor of interest; randomly assigning a different level of the factor to each member of the block, such that each level occurs once and only once per block; the blocks are assumed not to interact with the factor *stat* >>> Table 25 >>> randomized block >>> randomized complete block

randomized complete block (RCB): each block contains a plot for each candidate to be tested; in that case the classification of the data according to the blocks and the classification according the candidates are orthogonal *stat meth* >>> Table 25

range of reaction: the range of all possible phenotypes that may develop by interaction with various environments from a given genotype; it is also called "norm of reaction" *gene*

range pole >>> landmark

RAPD >>> random amplified polymorphic DNA technique

raphanobrassica: an intergeneric hybrid between *Raphanus* and *Brassica* species; a first hybrid was already reported in 1826; about 100 years later

KARPECHENKO produced a hybrid between *Raphanus sativus* ($2n = 2x = 18$, RR) x *Brassica oleracea* ($2n = 2x = 18$, CC) >>> $2n = 4x = 36$, RRCC

raphe: a ridge, sometimes visible on the seed surface, which is the axis along which the ovule stalk joins the ovule *bot*

rate of response to selection >>> selection progress

ratooning: a sprout or shoot from the root of a plant (e.g., in sugarcane) after it has been cropped *bot*

ratooning crop: obtaining a second crop from the same plant (e.g., in sugarcane) *agr*

ray: a pedicel in an umbellate inflorescence *bot*

ray flower: a flower head with outer ray flowers forming petallike structures surrounding the inner disc flowers (e.g., in the *Asteraceae* or in sunflower) *bot*

RCB >>> randomized complete block

rDNA >>> ribosomal DNA

reading frame: the mechanism that moves a ribosome, one codon at a time, from a designated start sequence during genetic translation; a shift in the reading frame by any number of nucleotides other than three or multiples of three will cause an entirely new sequence of codons *gene*

reading mistake: the placement of an incorrect amino acid into a polypeptide chain during protein synthesis *gene* >>> reading frame

reagent: a substance that, because of the reactions it causes, is used in analysis and synthesis *meth chem*

realized genetic gain: it refers to the observed difference between the mean phenotypic value of the offspring of the selected parents and the phenotypic value of the parental generation before selection *meth*

realized heritability: it refers to heritability measured by a response to selection; it is the ratio of the single-generation progress of selection to the selection differential of the parents *stat*

reaper: a machine for cutting standing grain *agr*

reaper-binder: an implement that cuts and ties hay into bundles *agr*

rearrangement: all chromosome mutations that result in modified karyotypes *cyto* >>> translocation >>> interchange >>> chromosome mutation

reassociation >>> anneal

recalcitrant seed: seed that does not survive drying and freezing; in particular, seed that cannot withstand either drying or temperatures of less than +10°C and, therefore, cannot be stored for long periods, as compared to orthodox seeds *seed*

receptacle: the part of the stem from which all parts of the flower arise; in Compositae, the flattened tip of the stem that bears the bracts and florets *bot*

receptaculum >>> receptacle

receptor site: in molecular genetics, a set of reactive chemical groups in the cell wall of a bacterium that are complementary to a similar set in the tailpiece of bacteriophage *gene*

recessive: a gene and/or allele whose phenotypic effect is expressed in the homozygous state but masked in the presence of the dominant allele; usually the dominant gene and/or allele produces a functional product whereas the recessive gene and/or allele does not; both one and two doses per nucleus of the dominant allele may lead to expression of its phenotypes, whereas the recessive allele is observed only in the complete absence of the dominant allele *gene* >>> Table 6

recessive allele >>> recessive

recessive epistasis: the effect of a recessive allele of a gene suppressing the phenotype manifestation of another gene *gene* >>> Table 6

recessiveness >>> recessive

recipient: one that receives; receiver *gene meth*

recipient cell >>> recipient

reciprocal cross: one of a pair of crosses in which the two opposite mating types are each coupled with each of two different genotypes and mated with the reciprocal combination; for example, male of genotype A x female of genotype B (first cross) and male of genotype B x female of genotype A (reciprocal cross); such crosses are used (1) to detect sex linkage, (2) maternal inheritance, or (3) cytoplasmic inheritance *meth*

reciprocal full-sib selection: a method of interpopulation improvement for species in which the commercial product is hybrid seed; a cycle of selection is completed in the fewest number of seasons by use of plants from which both selfed and hybrid seed can be obtained *meth*

reciprocal genes: nonallelic genes that reciprocate or complement one another *gene*

reciprocal half-sib selection >>> reciprocal recurrent selection

reciprocal recurrent selection: a breeding method used to achieve an accumulation of genes that are valuable for specific traits but also for combining ability; in practice, two populations form the basis of selection; reciprocally, one population serves as a tester for the investigation of the selections deriving from the other population; from the first population, usually a greater number of plants is selfed and at the same time crossed with several plants of the second population; the same procedure is realized with the second population; during the second year, the progenies of the test-crosses are subjected to performance trials; the progenies of the cross of one female parent, derived from population 1, are combined with several male parents from population 2; based on the performance testing, progenies from selfed seed, produced by best plants, are grown during the third year; in the same year, the best progenies of selfings are crossed in many combinations; seeds obtained from those crosses form the improved populations for growing during the fourth year; within such populations individual plants may be selfed again and crossed for utilization in a new cycle *meth* >>> recurrent selection >>> Figure 21

reciprocal translocation: a translocation that involves an exchange of chromosomes segments between two nonhomologous chromosomes *cyto* >>> translocation >>> Robertsonian translocation >>> chromosome mutation

recognition sequence (site): a nucleotide sequence composed typically of 4, 6, or 8 nucleotides; it is recognized by a restriction endonuclease; so-called type II enzymes cut (and their corresponding modification enzymes methylate) within or very near the recognition sequence *biot gene*

recombinant: an individual or cell with a genotype produced by recombination (i.e., with combinations of genes other than those carried in the parents); they result from independent assortment or crossing over *gene*

recombinant DNA molecule: DNA molecule created by ligating together two not normally contiguous DNA molecules *biot*

recombinant DNA technology: DNA molecules constructed by joining, outside the cell, natural or synthetic DNA segments to DNA molecules capable of replication in living cells *biot* >>> biotechnology

recombinant protein: a protein synthesized from a cloned gene *biot*

recombinant type: an association of genetic markers, found among the progeny of a cross, that is different from any association of markers present in the parents *gene*

recombination: the process whereby new combinations of parental characters may arise in the progeny, caused by exchange of genetic material of different parental lines *gene* >>> Figure 24 >>> Table 22

recombination frequency: the number of recombinants divided by the total number of progeny, expressed as a percentage or fraction; such frequencies indicate relative distances between loci on a genetic map *gene* >>> mapping >>> Table 22

recombination nodule: swellings along DNA strands and associated proteins during prophase pairing of chromosomes; the nodules can be identified under the electron microscope; it is suggested that recombination nodules are the sites of genetic crossing over *cyto*

recombination system: all factors that mediate and control the process of genetic recombination *gene*

recombination unit >>> MORGAN unit

recon: the smallest unit of DNA capable of recombination *gene*

recurrent full-sib selection: a method of intrapopulation improvement that involves the testing of paired-plant crosses; it is the only method of recurrent selection in which the seeds from two individuals, rather than one, are used for testing and to form the new population *meth*

recurrent half-sib selection: a method of intrapopulation improvement that includes the evaluation of individuals through the use of their half-sib progeny; the general procedure for a cycle of selection is (1) to cross the plants being evaluated to a common tester, (2) evaluate the half-sib progeny from each plant, and (3) intercross the elected individuals to form a new population *meth*

recurrent parent: the parent to which a hybrid is crossed in a backcross; it replaces the dragged alleles step by step with the alleles of the original variety *meth*

recurrent (backcross) parent >>> recurrent parent

recurrent reciprocal selection: a recurrent selection breeding system in which genetically different groups are maintained and, in each selection cycle, individuals are mated from the different groups to test for combining ability *meth* >>> recurrent selection >>> Figure 21

recurrent selection: a method designed to concentrate favorable genes scattered among a number of individuals; it is performed by repeated selection in each generation among the progeny produced by matings inter se of the selected individuals of the previous generation; in practice, plants from a population are selfed and, after the yield of the selfed seeds, the progenies of the phenotypically best individuals are grown in the second year; the best progenies are then crossed in as many combinations as possible and the seeds received hereby are grown in the third year as a population; within the already improved population, selection and selfing can be carried out again; with this population a second cycle of recurrent selection can be started *meth* >>> facilitated recurrent selection >>> Figures 4, 20, 21

red rust: uredospore state of rusts, particularly of cereals *phyt*

rediploidization: in anther culture, the haploidization of the genome by culturing pollen grains to haploid plantlets and its rediploidization after spontaneous or induced doubling of the chromosome set *biot* >>> Figures 17, 26 >>> Table 7

reductase: an enzyme responsible for reduction in an oxidation-reduction reaction *chem phys*

reduction division: the two nuclear divisions in meiosis that produce daughter nuclei, each of which has half as many chromosomes as the parental nucleus *bot* >>> meiosis

reductional division >>> reduction division >>> meiosis

redundant DNA: DNA that does not appear to be genetically active and hence is not translated or transcribed; it often consists of repeated sequences *gene*

redundant gene: a gene that is present in many functional copies, so that one copy can complement the loss of another copy *gene*

reduplication: doubling of the genetic matter of a haploid chromosome set *cyto* >>> rediploidization

reed: the straight stalk of any of various tall grasses growing in marshy places *agr eco*

reel (at a harvester): it draws the cut crop into the intake auger, which carries it to the center of the cutting table *agr*

reforest: to replant trees on land denuded by cutting or fire *eco fore*

refraction: the change of direction of a ray of light in passing obliquely from one medium into another in which its wave velocity is different *micr* >>> refraction index

refraction index: a number indicating the speed of light in a given medium, as the ratio of the speed of light in a vacuum or in air to that in the given medium; for example, distilled water 1.336, liquid paraffin 1.343, glycerine 1.473, Euparal 1.483, xylene 1.497, cedarwood oil 1.520, or balsam 1.524 *micr* >>> refraction

refractometer: an instrument for determining the refractive index of substance *prep* >>> refraction >>> refraction index

regenerable >>> regeneration

regenerant: an entire plant grown from a single cell *phys*

regenerate >>> regenerant

regeneration: the replacement by a plant of tissue or organs that have been lost *hort*; in biotechnology, forming a new, entire plant from a clump of cells and/or from a single cell *biot* >>> Figure 27

registered seed: a class of certified seed that is produced from breeder; it also can be selected or foundation seed planted to produce certified seed; it is handled under procedures acceptable to the certifying agency to maintain satisfactory genetic purity and identity *seed* >>> Table 28

registration >>> release of variety

regression coefficient: the rate change of the dependent variable with respect to the independent variable *stat*

regression line: a line that defines how much an increase or decrease in one factor may be expected from a unit increase in another *stat*

regulator(y) gene: in the operon theory of gene regulation, a gene that is involved in switching on or off the transcription of structural genes; when transcribed, the regulator gene produces a repressor protein, which switches off an operator gene and hence the operon that this controls; the regulator gene is not part of the operon and may even be on a different chromosome *gene*

regulatory region: stretches of the DNA sequence, which control the activity of genes *biot* >>> regulator(y) gene

regulatory system >>> regulator(y) gene

reiterated (DNA): nucleotide sequences that occur many times within a genome *gene* >>> redundant DNA

rejuvenation: synonymous with de-differentiation or treatment that leads to culture invigoration or revival *biot;* in horticulture (e.g., in fruit tree planting) reversion from adult to juvenile by restoration of juvenile vigor (growth) on a mature entity *hort*

rejuvenation (of seed samples): the restoration of viability of seeds by new propagation of the material *seed*

release of variety: a crop variety or germplasm that is released and designated to be reproduced, marketed and made available as seed for public use *seed*

REMI >>> restriction-enzyme-mediated integration

remote hybridization >>> distant hybridization

RENNER complex: a specific gametic chromosome combination in evening primrose (*Oenothera* spp.) *cyto*

renovation: usually, it refers to the mechanical removal of plants from a very dense, unproductive, or sodbound stand for the purpose of revitalizing of its productivity *seed*

repeat: small tandem duplication or a nucleotide sequence that occurs many times within a DNA molecule *gene*

repeated sequence: a DNA sequence that occurs in many copies *gene* >>> repetitive DNA

repellent: a material or substance that animals try to avoid *phyt*

repetitive (repetitious) DNA: a type of DNA that constitutes a significant fraction of the total DNA; it is characterized by its large number of copies of repeated nucleotide sequences; some may be redundant DNA of unknown function, but mRNA, tRNA, 5S-RNA and histones are coded by repetitive sequences *gene*

replicate: a more or less exact duplication or repetition of a test, an experiment, or an experimental single plot to assure or to increase confidence in the resulting data *stat* >>> randomization

replication: in cytology, the synthesis of new daughter molecules of nucleic acid from a parent molecule, which acts as a template *cyto;* in a experimental field design, it allows not only estimation of the error variance, and consequently application of statistical tests, but it also promotes the accuracy of the estimation of genotypic values of the entries tested *stat meth* >>> randomization >>> Figures 5, 9 >>> Table 25

replication error: any modification that prevents or disturbs the DNA replication process *gene*

replication unit >>> replicon

replicon: a structural gene that controls the synthesis of a specific initiator along with a replicator locus upon which the corresponding initiator acts *gene*

reporter gene: in DNA or gene transfer experiments, the linkage of a gene that (transient) expression can easily be detected with a target DNA sequence or gene *biot*

repressor: a protein produced by a regulatory gene that inhibits the activity of an operator gene, and hence switches off an operon *gene*

repressor gene >>> repressor

reproduce: to create another individual of the parental type that will in turn produce another *meth* >>> reproduction >>> Table 35

reproducible >>> reproduction

reproduction: the process of forming new individuals of a species by sexual or asexual ways *bot gene* >>> Table 35

reproductive isolating mechanism: any biological property of an organism that interferes with its interbreeding with organisms of other species *gene eco* >>> Table 35

reproductive isolation: the absence of interbreeding between members of different species *eco*

reproductive meristem >>> generative meristem

reproductive organ: usually, it refers to the sexual organ *bot*

repulsion: the linkage phase of a double heterozygote for two linked gene pairs, which has received one dominant factor from each parent and the alternative recessive factor from each parent (e.g., for genes and/or alleles *A, a* and *B, b* the repulsion heterozygote receives *Ab* from one parent and *aB* from the other, where *A* and *B* are dominant, and *a* and *b* are recessive) *gene*

research: diligent and systematic inquiry into a subject in order to discover or revise facts, theories, etc. *meth*

residue seed method >>> method of overstored seeds

resilience: the ability of a population to persist in a given environment despite disturbance or reduced population size; based upon the ability of individuals within the population to survive (fitness) and reproduce (fecundity) in a changed environment *gene eco*

resin: an exudate of tree wood or bark, liquid but becoming solid on exposure to air, consisting of a complex of terpenes and similar compounds *bot*

resistance: inherent capacity of a host plant to prevent or retard the development of an infectious disease; there are different types of resistance: (1) hypersensitivity (infection by the pathogen is prevented by the plant), (2) specific resistance (specific races of the pathogen cannot infect the plant), (3) nonuniform resistance (the host prevents the establishment of certain races), (4) major gene resistance (races of the pathogen are controlled by major genes in the host), (5) vertical resistance (host resistance controls one or a certain number of races), (6) field resistance (severe injury in the laboratory, but resistance under normal field conditions), (7) general resistance (the host is able to resist the development of all races of the pathogen), (8) nonspecific resistance (host resistance is not limited to specific races of the pathogen), (9) uniform resistance (host resistance is comparable for all races of the pathogen, rather than being good for some races), (10) minor gene resistance (host resistance is controlled by a number of genes with small effects), (11) horizontal resistance (variation in host resistance is primarily due to differences between varieties and between isolates, rather than to specific variety x isolate interactions) *phyt* >>> biological control >>> Table 33

resistance breeding: special crossing and selection methods in order to improve the inherent capacity of a crop plant to prevent or retard the development of an infectious disease *phyt meth* >>> resistance >>> Table 33

resistant >>> resistance

resolution: the smallest distance by which two objects can be separated and still be resolved as separate objects *micr*

resolving power >>> resolution

respiration: oxidative reactions in cellular metabolism involving the sequential degradation of food substances and the use of molecular oxygen as a final hydrogen acceptor *phys*

rest: a condition of a plant in which growth cannot occur, even though temperatures and other environmental factors are favorable for growth *phys* >>> dormancy

rest period >>> rest >>> dormancy

resting bud >>> hibernaculum

resting spore: a spore germinating after a resting period (frequently after overwintering), as does an oospore or a teliospore *phyt*

restitution nucleus: a nucleus with an unreduced chromosome number *cyto*

restorer: an inbred line that permits restoration of fertility to the progeny of male sterile lines to which it is crossed *seed* >>> Figure 23

restorer gene: a gene and/or allele that is able to restore fertility of a sterile genotype; while genes for sterility frequently belong to the mitochondrial genome (i.e., cytoplasmic), the restorer genes are very often found to belong to the nuclear complement; they are used in hybrid variety production *gene* >>> Figure 2

restorer line (R line): a pollen parent line; it contains the restorer gene or genes, which restores cytoplasmic male sterile plants to pollen fertility; it is crossed with an A line in the production of hybrid seeds *seed* >>> Figures 2, 23

restoring gene >>> restorer gene

restriction analysis: determination of the number and size of the DNA fragments produced when a particular DNA molecule is cut with restriction endonucleases *biot* >>> restriction enzyme

restriction endonuclease >>> restriction enzyme

restriction enzyme: an enzyme that functions in a bacterial modification-restriction system and recognizes specific nucleotide sequences and breaks the DNA chain at these sites; there is a great number of them, each with different recognition and/or cutting sites; they are intensively used as a tool in molecular genetics and also in producing chromosomal banding patterns in cytogenetics *gene* >>> restriction analysis

restriction fragment length polymorphism (RFLP): a comparative study (in individuals, populations, or species) of the DNA fragment lengths produced by particular restriction enzymes; by using a DNA hybridization technique, restriction fragments can be identified if they are complementary to a specific DNA probe; each mutation that produces or eliminates a restriction site in a homologous region leads to a change of length of the restriction fragment, which has to be detected; it is used to infer genomic relationships; RFLPs represent an important tool in detecting variability; they are free of secondary effects due to pleiotropic action and they are frequently associated with the segregation of alleles affecting morpho-physiological traits; the advantages are as follows: (1) they are everywhere present in the genome and in living organisms, (2) they show Mendelian inheritance, (3) they show codominant expression, (4) they have no pleiotropic effects, (5) they are independent of environmental effects, (6) they are available at each developmental and/or physiological stage, (7) different loci within the genome can be identified by one DNA probe, (8) heterologous genes may also be used as probes, (9) any number of DNA probes can be established, (10) probes are available for coding and silent genes (DNA sequences), (11) probes show also the variability of flanking DNA sequences, and (12) several traits can be screened in the same experimental sample *meth* >>> Table 29

restriction map: representation of DNA with the position of restriction sites indicated *gene* >>> restriction analysis

restriction site: a certain nucleotide sequence within the double-stranded DNA; it is recognized by a restriction endonuclease; the enzyme cuts the double strand within the recognition sequence; the restriction sites are usually composed of four to six base pairs and are bilaterally symmetric; both strands are cut either on exactly opposite positions (blunt ends) or alternated ones (sticky ends); the type of cutting depends on the enzyme used *gene*

restriction-enzyme-mediated integration (REMI): a method of transformation that generates tagged mutations *biot*

resynthesis: the artificial production of autopolyploids or allopolyploids of naturally occurring autopolyploid or allopolyploid plants by utilization of the presumable parental species (e.g., it was done in wheat and rapeseed) *meth*

retrotransposon: retrotransposons are a ubiquitous and major component of plant genomes; those with long terminal DNA repeats (LTRs, Ty1-copia-like family) are widely distributed over the chromosomes of many plant species *gene*

reverse genetics: using linkage analysis and polymorphic markers to isolate a disease gene in the absence of a known metabolic defect, then using the DNA sequence of the cloned gene to predict the amino acid sequence of its encoded protein; in general, a technology aiming at isolating mutants of a given sequence; it is also applied for identification of gene function *gene biot*

reverse mutation: the production by further mutation of a premutation gene from a mutant gene; it restores the ability of the gene to produce a functional protein; strictly, reversion is the correction of a mutation (i.e., it occurs at the same site) *gene*

reverse transcriptase: an enzyme from retroviruses for the synthesis of a DNA complementary to a RNA molecule (i.e., cDNA); it is used for (1) filling-in reactions, (2) for DNA sequencing, and (3) for cDNA synthesis *biot*

reversion >>> reverse mutation

revertant: an allele that undergoes reverse mutation or a plant bearing such an allele *gene* >>> reverse mutation

RFLP >>> restriction fragment length polymorphism

rhizome: a horizontally creeping underground stem that bears roots and leaves and usually persists from season to season *bot*

rhizosphere: the soil near a living root *agr*

rhodanese: the enzyme is defined biochemically by its ability to transfer sulfur from thiosulfate to cyanide, yielding thiocyanate; it is found in plants, animals, and bacteria *phys*

***Rht* gene** >>> short-straw mutant

***Rhynchosporium* leaf blotch** (of barley and rye): it frequently occurs in wet seasons and in high humidity; symptoms first appear as irregular or diamond-shaped blue-gray water-soaked lesions on the leaves and leaf sheathes; as the lesions mature, they become pale brown with a dark purple margin and co-alesce to form large areas of dead tissue; ears may also be infected *phyt*

rhytidome >>> bark

rhytmicity >>> periodicity

rib: a primary or prominent vein of a leaf *bot*

ribonuclease >>> RNase

ribonucleic acid (RNA): a polymer composed of nucleotides that contain the sugar ribose and one of the four bases adenine, cytosine, guanine, and uracil *gene*

ribonucleoprotein: a protein composed of pre-rRNAs and ribosomal as well as nonribosomal protein components; one of the nonribosomal proteins, the nucleolin, is considered to play a key role in regulation of rDNA transcription, perisomal synthesis, ribosomal assembly, and maturation *phys*

ribosomal DNA: DNA as components of ribosomes *gene*

ribosomal RNA: the RNA molecules that are structural parts of ribosomes (i.e., 5S, 16S, and 23S RNAs in prokaryotes and 5S, 18S, and 28S RNAs in eukaryotes) *gene*

ribosome: one of the ribonucleoprotein particles, which are the sites of translation; it consists of two unequal units bound together by magnesium ions *gene*

ridge tillage: a type of soil conserving tillage in which the soil is formed into ridges and the seeds are planted on the tops of the ridges; the soil and the crop residue between the rows remain largely undisturbed; the practice offers opportunities to reduce crop production costs by banding fertilizers and pesticides and reducing the need for field trips *agr*

rifamycin(s): a group of antibiotics that inhibit initiation of transcription in bacteria *biot phys*

rind: a thick and firm outer coat or covering (e.g., in watermelon, orange, etc. or the bark of a tree) *bot*

ring bivalent: an association of two chromosomes with terminal chiasmata on both arms *cyto* >>> Figure 15

ring chromosome: a (sometimes aberrant) chromosome with no ends (e.g., the chromosome of bacteria); an isochromosome may also form a ring in MI of meiosis *cyto*

RINGER solution: a physiological saline containing sodium, potassium, and calcium chlorides used in physiological experiments for temporarily maintaining cells or organs alive in vitro *phys*

ringspot: a circular area of chlorosis with a green center; a symptom of many virus diseases *phyt*

ripe >>> mature

RNA >>> ribonucleic acid

RNA transcriptase: the enzyme responsible for transcribing the information encoded in DNA into RNA; it is also called transcriptase or RNA polymerase *biot*

RNase: an enzyme hydrolyzing RNA *gene*

Robertsonian translocation: a chromosomal mutation due to centric fusion or centric fission (i.e., a reciprocal translocation with breakpoints within the centromeric regions) *cyto* >>> translocation

rod (rd): equals 5.03 m

rolled paper toweling: adjusted filter paper or paper towels are used for this method in order to germinate seeds inside and/or between the layers of paper; after germination and growth the viability and/or germability are determined *seed*

roller: a device that compacts the soil to produce a firm seedbed, like a packer *agr* >>> packer

root: the lower part of a plant, usually underground, by which the plant is anchored and through which water and minerals enter the plant *bot*

root ball: the roots and soil or soil mix that they are growing in when lifted from the open ground *hort*

root cap: a cap of cells covering the apex of the growing point of a root and protecting it as it is forced through soil *bot*

root crop: a crop, such as beets, turnips, or sweet potatoes, grown for its large, edible roots *agr*

root culture: the in vitro growth of roots (e.g., root tips or root meristem on a synthetic medium) *biot*

root cuttings: root cuttings are made by cutting off pieces of root and planting them under suitable conditions; in this way some plants species or varieties can be easily propagated *meth agr hort*

root grafting: the process of grafting scions (shoots) directly on a small part of the root of some appropriate stock, the grafted root then being potted *hort*

root hair: a tabular outgrowth of an epidermal cell of a root, which functions to absorb water and nutrients from the soil *bot*

root knot nematode: a nematode *(Meloidogyne naasi)* that induces small, gall-like growth on the roots of certain types of plants *phyt*

root nodule: a small, gall-like growth on the roots of certain types of plants (legumes); the nodules develop as a result of infection of the root by bacteria *bio agr*

root pruning: cutting the roots of large plants, mainly trees and shrubs, to force more vigorous growth or to prepare the plant for transplantation or transportation *meth hort*

root sucker: a shoot arising adventitiously from a root of a plant; mostly at some distance from the main trunk *bot*

rooting: the natural or induced process of root formation *phys bot*

rooting compound: a powdery substance into which fresh cuttings are dipped before inserting in soil or medium, containing hormones, such as kinetins, to encourage root growth *meth prep hort*

rootstock: synonymous with "rhizome"; in horticulture, the bottom or supporting root used to receive a scion in grafting *hort* >>> rhizome

rootstock variety: in horticulture, there are special (fruit) tree varieties (often of wild-type character) that serve as rootstock for graftings; usually they show good root formation, resistance traits, and compatibility with the scion *hort*

rosarium: since Roman times, a rose garden and breeding site of roses *hort*

rosel(l)ate >>> rosette

rosette: an arrangement of leaves radiating from a root crown near the earth *bot*

rosette plant >>> rosette

rosular >>> rosette

rot: to deteriorate, disintegrate, fall, or become weak due to decay *agr*

rotary hoe: an implement that breaks the soil with a circular motion *agr*

rotation >>> crop rotation

rotation of crops >>> crop rotation

rouge: a noun referring to an off-type plant; when used as a verb it refers to the act of removing, to uproot or destroy such plants that do not conform to a desired standard or are diseased *seed*

rouging: a manual removal of infected or inferior specimens from an otherwise healthy crop of plants *seed*

row spacing: the distance between rows of crop plants; it depends on needs for optimal plant growth, plant density, weed control, and harvest technology *agr*

rRNA >>> ribosomal RNA

rub >>> rubbing

rubbed seeds >>> rubbing

rubbing: smooth the surface of multigerm seed of sugarbeet *seed*

rubisco: a CO_2-fixing enzyme; the key enzyme in photosynthesis; it is the most frequent protein on earth; it has a unique double function of being both a carboxylase and an oxygenase; when acting as an oxygenase, it catalyzes the light-dependent uptake of O_2 and the formation of CO_2 in a complicated process (photorespiration), which takes place concomitantly in three organelles—chloroplasts, peroxisomes, and mitochondria; there are projects to manipulate the enzyme in order to create an artificial plant that can contribute to reduction of CO_2 content in the atmosphere (i.e., decreasing the so-called greenhouse effect) *phys biot*

ruderal plant: a plant that is associated with human dwellings or agriculture, or one that colonizes waste ground *eco*

rudiment >>> rudimentary

rudimentary: incompletely developed *bot*

run out: separation of nucleic acid or protein molecules by gel electrophoresis *prep*

runner: a procumbent shoot that takes root, forming a new plant that eventually is freed from connection with the parent by decay of the runner; it serves as a vegetative propagule (e.g., in strawberry) *bot* >>> stolon

rush >>> reed

russet: a brownish roughened area on the skin of fruits as a result of cork formation *phyt*

rust: a plant disease caused by a fungus of the class Urediniomycetes; the characteristic symptom is the development of spots or pustules bearing masses of powdery spores that are usually rust-colored, yellow, or brown *phyt*

S

S0: a symbol used to designate the original selfed plant *meth*

S1, S2, S3, etc.: the representation for continued selfing (self-fertilization) of plants; S1 designates the generation obtained by selfing the parent plant, S2 the generation obtained by selfing the S1 plant, etc. *meth*

S1 nuclease: a nuclease that cuts single-stranded DNA and RNA; used for S1 protection experiments in transcript mapping *biot*

saccharide: an alternative term for sugar *chem phys*

saccharose: a sweet, crystalline substance, $C_{12}H_{22}O_{11}$, obtained from the juice or sap of many plants (e.g., from sugarcane and sugarbeet) *chem phys*

sacrificial crop: crop planted to distract pests safely *agr*

safranin >>> GRAM's stain

SAGE >>> serial analysis of gene expression

S-allele: an allele of a gene controlling incompatibility in many allogamous plants; alleles present in both style and pollen are referred to as matching S-alleles; S-alleles usually belong to a series of multiple alleles *gene*

sagittate: arrow-shaped *bot*

salicin >>> glucoside

saline soil: a soil containing enough salts to reduce plant fertility *agr* >>> crop rotation

Salmon procedure: a method for producing haploids in hexaploid wheat; "Salmon" is a name of an alloplasmic wheat variety carrying a 1RS.1BL chromosome translocation together with cytoplasm of *Aegilops kotchyi;* the interaction of cytoplasmically genetic determinants with, possibly, a parthenogenesis-inducing gene on chromosome arm 1RS of rye results in haploid progeny *meth* >>> Figures 17, 26 >>> Table 7

saltation: a mutation occurring in the asexual state of fungal growth, especially one occurring in vitro culture *phyt*

saltational speciation >>> quantum speciation

samara: a fruit similar to an achene except that the entire seed coat is tightly fused with the pericarp (e.g., ash, elm, tree of heaven, etc.) *bot* >>> winged fruit

sample: a finite series of observations taken from a population *meth stat*

sample size: the number of experimental units on which observations are considered; it may be less than the number of observations in a data-set, due to the possible multiplying effects of multiple variables and/or repeated measures within the experimental design *stat*

sampling: the method by which a representative sample is taken from a seed lot or something else *meth*

sampling error: variability due to the limited size of the sample *stat*

sanitation: plant disease control involving removal and burning of infected plant parts and decontamination of tools, equipment, etc. *phyt*

sap: the exudate from ruptured tissues emanating from the vascular system or parenchyma *bot*

saponin: any member of a class of glycosides that form colloidal solutions in water and foam when shaken; it occurs in many different plant species; in cereals, only oats are known to produce these compounds; in oats, the resistance to infection by the take-all fungus, *Gaeumannomyces graminis* var. *tritici,* has been attributed to the family of antifungal saponins known as avenacins, which are present in roots *chem phys*

saprophyte: an organism that lives on or in dead or decaying organic matter *bot*

sapwood: the living, softer part of the wood between the inner bark and the heartwood *bot fore*

sarment: a slender running stem *bot* >>> runner

sarmentous plant >>> runner

SAT chromosome >>> satellite(d) chromosome

satellite: a distal segment of a chromosome that is separated from the rest of the chromosome by a chromatic filament *cyto*

satellite(d) chromosome: each chromosome with a secondary constriction; this constriction divides a satellite part from the rest of the chromosome (arm) *cyto*

satellite DNA: a highly repetitive DNA, composed of repeated hepta- to deca-nucleotide sequences; DNA of different buoyant density; a minor DNA fraction that has sufficiently different base composition from the bulk of the DNA in order to separate distinctly during cesium chloride density gradient centrifugation; it can derive from nucleus, plastid, or mitochondrial DNA *biot*

SAT-zone: the secondary constriction of a satellite chromosome *cyto* >>> nucleolar zone

savanna: a plain characterized by coarse grasses and scattered tree growth *eco*

sawflies: larvae of various insects that feed on leaves *phyt*

scab: a general term for any unrelated plant disease in which the symptoms include the formation of dry, corky scabs *phyt*

scald: a necrotic condition in which tissue is usually bleached and has the appearance of having been exposed to high temperatures *phyt*

scale: any thin, scarious body, usually a degenerated leaf *bot*

scalping: the removal of material larger than the crop seed during the processing of seeds *seed*

scanning electron microscopy: a microscope used to examine the surface structure of biological specimens; a three-dimensional screen image is acquired through focusing secondary electrons emitted from a sample surface bombarded by an electron beam *micr*

scarification: the process of mechanically abrading a seed coat to make it more permeable to water; this process may also be accomplished by brief exposure to strong acids (sulfuric acid); it may enhance germination *seed*

scatter diagram: a diagram in which observations are plotted as points on a grid of x- and y-coordinates to see if there is any correlation *stat*

SCE >>> sister chromatid exchange

SCHIFF's reagent: a reagent consisting of fuchsin bleached by sulfurous acid that produces a red color upon reaction with an aldehyde; it is used for chromosome staining *micr*

schizocarp: a dry, two-seeded fruit of some plants that separates at maturity along a midline into two mericarpes; each mericarp has a dry, indehiscent pericarp enclosing a loose-fitting ovule (e.g., carrot) *bot*

scion: a portion of a shoot or a bud on one plant that is grafted onto a stock of another *hort*

scion rooting: covering a low graft with soil so that the plant develops roots directly from both the rootstock and the scion *meth hort*

scission >>> fission

sclerenchyma: tissue composed of cells with thickened and hardened walls *bot*

sclerophyllous: having leaves stiffened by sclerenchyma *bot*

Sclerotinia (of rape or clover): a soil-borne disease (*Sclerotinia sclerotiorum* in rape, *Sclerotinia trifoliorum* in clover) that infects a wide range of crops; symptoms appear from May onward as bleached areas of the stem with black sclerotia within the infected stem *phyt*

sclerotium (sclerotia *pl*): a dense, compact mycelial mass capable of remaining dormant for extended periods *bot*

scorch: "burning" of leaf margins as a result of infection or unfavorable environmental conditions *phyt*

scorpiod cyme: a determinated inflorescence in which the lateral buds on one side are suppressed during growth, resulting in a curved or coiled arrangement *bot*

screening: examining the properties, performance responses of individuals, lines, genotype, or other taxa under an assortment of conditions in order to evaluate the individuals or groups; a routine testing for particular properties *meth*

scutellum: a shield-shaped organ of the embryo of grasses; it is often viewed as a highly modified cotyledon in monocots *bot*

scutiform: platter-shaped *bot*

SDS >>> sodium dodecyl sulfate-polyacrylamide

SDS gel electrophoresis (sodium dodecyl sulfate-polyacrylamide) (SDS-PAGE): in SDS-PAGE, SDS masks protein charge and separations depending only on size as compared to common gel electrophoresis *meth* >>> gel electrophoresis

SE >>> standard error

sealed storage: storage in a sealed (airtight) container (e.g., in gene banks *seed*

secalotricum: a cross combination of rye *(Secale)* and wheat *(Triticum)* in which rye serves as the donor of the cytoplasm (mother plant), as opposed to triticale *bot agr* >>> triticale

second division: second meiotic division, which is a mitotic division of chromosomes *cyto*

secondary constriction >>> SAT-zone

secondary crop: a crop that originated as a weed of a primary crop (e.g., rye) *evol;* in agronomy, a crop grown after a primary crop *agr*

secondary gene pool: species in the secondary gene pool include those from which genes can be transferred to the cultivated species, however with more difficulties as compared to species of a primary gene pool *evol*

secondary infection: any infection caused by inoculum produced as a result of a primary or a subsequent infection and/or an infection caused by secondary inoculum *phyt*

secondary pairing: the association of bivalents in polyploids due to genetic, evolutionary, or structural factors; by any reason those bivalents appear in groups; sometimes it seems that the bivalents of a certain genome are closer together than at random or the bivalents of a certain genome occupy certain domains (spatial order) within the meiotic cell (prometaphase and metaphase) *cyto*

secondary root >>> lateral root

secondary tiller: it can arise from the prophyll node and leaf node of the primary tillers in cereals; in the same manner, tertiary tillers may occasionally be produced by secondary tillers; the primary tillers are usually the smallest of the tillers that emerge *bot*

section cutting: sections are cut from a block of wax around a plant material, usually by microtome *cyto prep*

sectorial chimera: a chimera in which the distinct meristem is cross-sectional present, like sectors of a circle *bot*

sedimentation test >>> ZELENY test

seed: a mature ovule consisting of an embryonic plant together with a store of food, all surrounded by a protective coat *bot* >>> Table 13

seed bank: a place or storage in which seeds of rare plants or obsolete varieties are kept, usually vacuum-packed and under cold conditions in order to prolong their viability *meth fore agr hort* >>> gene bank

seed breeder's rights: national and international rules and laws that provide plant breeders a legal means to apply for proprietary rights to cultivated plant varieties they have bred; "breeder"means the person who bred or discovered and developed a variety; "variety" means a plant grouping within a single botanical taxon of the lowest known rank, whose grouping, irrespective of whether the conditions for the grant of a breeder's right are fully met, can be defined by the expression of the characteristics resulting from a given genotype or combination of genotypes, distinguished from any other plant grouping by the expression of at least one of the said characteristics and considered as a unit with regard to its suitability for being propagated unchanged *agr*

seed certification: a procedure developed as a means of assuring that the seeds have a high standard of purity and quality *seed meth*

seed coat: the protective covering of a seed usually composed of the inner and outer integuments *bot*

seed conditioning: for marketing, seeds are usually cleaned, sized, treated with fungicides, insecticides, or inoculant, and finally bagged *seed* >>> Table 11

seed divider: a device that divides a seed lot and puts subsamples directly into a various number of planting envelopes *seed* >>> Table 11

seed dormancy >>> dormancy

seed drill >>> drill

seed flat >>> flat

seed incompatibility: a postgamous sterility due to failure of tissue development involved in the formation of the seed *seed*

seed increase >>> increase

seed index: the 100-g-weight of seeds *seed* >>> thousand-grain weight

seed leaf >>> cotyledon

seed lot: seeds of a particular crop gathered at one time and likely to have similar germination rates and other characteristics *seed agr hort*

seed mixture: either seed of more than one kind of cultivar or a combination of seed of two or more species *seed*

seed multiplication: all methods required to grow plants to maturity and produce seeds, including those practices necessary for harvesting, processing, and preparing seeds for subsequent plantings *seed*

seed orchard: plantation of fruit or forest trees, assumed or proven genetically to be superior; it is isolated in order to reduce pollination from genetically inferior outside sources; it is managed to improve the plants and produce frequent, abundant, and easily harvestable seeds *hort fore*

seed parent: the strain from which seed is harvested in the hybrid seed field; also commonly used to designate the female parent in any cross-fertilization *meth*

seed plant: an individual plant that is or was used for seed production and/or maintaining the genotype *seed*

seed plants >>> sperma(to)phyta

seed potato: a potato tuber that is used for the next growing season in order to produce next generation for selection and experimental testing *seed*

seed processing: the operations involved in preparing harvested seed for multiplication or marketing *seed* >>> Table 11

seed production area: a forest or other tree stand identified as a good source of seed and in which individual trees are evaluated for desired characteristics; seeds are collected periodically *fore hort*

seed quality control: control of physiological, sanitary, and genetic seed quality characteristics *seed*

seed regulation: the total set of rules and protocols related to variety development and release, seed production, quality control, and delivery *seed*

seed set: the process of producing seeds after flowering *bot agr*

seed source: the location where a seed lot was collected; usually defined on an eco-geographic basis by distance, elevation, precipitation, latitude, etc. *fore* >>> provenance

seed spacing >>> population density

seed stack: the erect stem on a plant that produces flowers and seed; it is particularly applied to root crops and leafy vegetable crops that produce seed after the desired product (root, head, leaves) has fully developed *agr*

seed stand: any stand used as a source of seed *fore hort*

seed stock: seed used as a source of germplasm for maintaining and increasing seed of crop varieties *seed* >>> stock seeds

seed trap: a device for catching the seeds falling on a small area of ground, from trees or shrubs; it is set for determining the amount of seedfall and the time, period, rate, and distance of dissemination *fore hort*

seed vessel: the pericarp (wall of the ripened ovary), which contains the seeds *bot*

seed viability testing: all methods to determine the potential for rapid uniform emergence and development of normal seedlings under both favorable and stress conditions *meth seed*

seed vigor: seed properties that determine the potential for rapid uniform emergence and development of normal seedlings under both favorable and stress conditions *seed*

seedbed: a plot of ground prepared for seeds or seedlings *agr*

seed-borne pathogens: carried on or in seeds; for example, in wheat, the streak mosaic virus of barley, the fungi, such as snow mold *(Fusarium nivale), Septoria* spike blotch *(Septoria nodorum), Helminthosporum* leaf blotch or spot blotch *(Cochiobolus sativus, Helminthosporum sativum, syn Bipolaris sorokiana, Drechslera sorokiana),* loose smut (*Ustilago nuda* or *U. tritici*), common smut or stinking smut *(Tiletia caries),* dwarf smut *(Tiletia controversa);* in barley, the stripe mosaic virus, the fungi, such as snow mold *(Fusarium nivale),* leaf stripe disease *(Pyrenophora graminea), Helminthosporum* leaf blotch *(Helminthosporum gramineum, syn Drechslera graminea),* net blotch disease *(Pyrenophora teres, syn Helminthosporum teres, syn Drechslera teres),* loose smut *(Ustilago nuda),* black smut *(Ustilago nigra),* hard smut *(Ustilago hordei);* in rye, the streak mosaic virus of barley, the fungi, such as snow mold *(Fusarium nivale, syn Griphosphaeria nivalis), Septoria* spike blotch *(Septoria nodorum),* stalk bunt *(Urocystis occulata syn Tuburcinia occulata),* ergot *(Claviceps purpurea);* in oats, the streak mosaic virus of barley, the fungi, such as loose smut *(Ustilago avenea);* and in maize, the fungi, such as common smut *(Ustilago maydis),* seed rots (*Fusarium* spp., *Penicillium* spp.), seedling rots (*Pythium* spp., *Fusarium* spp., *Helminthosporum* spp., *Penicillium* spp., *Rhizopus* spp., *Rhizoctonia* spp., *Deploida* spp.) *phyt*

seeding lath: commonly a wooden device for obtaining uniformly spaced drills in a seedbed and aiding the even distribution of hand-sown seed in them *meth*

seeding machine >>> drill

seedling: a young plant grown from seed *bot*

seedling guard: a row cover to protect seeds indoors or out *meth hort*

seed-tree method: a method of regenerating a forest stand in which all trees are removed from the area except for a small number of seed-bearing trees that are left singly or in small groups *fore*

segmental allopolyploids: a partial homology or so-called homoeology of chromosome sets combined in an allopolyploid *cyto*

segregate >>> segregation

segregation: the separation of alleles during meiosis so that each gamete contains only one member of each pair of alleles *gene* >>> Figure 6 >>> Tables 2, 3, 4, 6, 7, 8, 19, 20, 21

segregation distortion: the distortion of the 1:1 segregation ratio produced by a heterozygote; it can arise because of abnormalities of meiosis, which results in an *Aa* individual producing an unequal number of *A* and *a* bearing gametes or it may arise from *A* and *a* bearing gametes being unequally effective in producing zygotes *gene*

selectable marker: a physiological or morphological character, which may easily be determined as marker for its own selection or for selection of other traits closely linked to that marker *gene*

selection: the process determining the relative share allotted individuals of different genotypes in the propagation of a population; natural selection occurs if zygotic genotypes differ with regard to fitness *meth evol* >>> clonal selection >>> differential selection >>> index selection >>> mass selection >>> recurrent selection >>> tandem selection >>> Figures 38, 39 >>> Table 7

selection after flowering: a selection that is only possible after flowering since the critical selective characters are expressed after flowering time, seed and/or fruit formation period (e.g., grain size, spike length, fruit color, etc.) *meth*

selection coefficient: a measure of the disadvantage of a given genotype in a population *stat* >>> Figure 38

selection criteria: the specific characters and plant reactions on which the selection is focused during the breeding cycles *meth*

selection differential: in artificial selection, the difference in mean phenotypic value between the individuals selected as parents of the following generation and the whole population *meth stat* >>> Figure 38

selection gain: in artificial selection, the difference in mean phenotypic value between the progeny of the selected parents and the parental generation *gene meth* >>> Figure 38

selection intensity: the ratio of the number of genotypes selected divided by the number of genotypes tested *meth gene stat* >>> Figure 38

selection limit: the exhaustion of genetic variance in a population, so that no further selection response can be expected *gene*

selection pressure: the effectiveness of natural selection in altering the genetic composition of a population over a series of generations *stat meth* >>> Figure 38

selection prior to flowering: a selection that is possible before flowering since the critical selective characters are already expressed (e.g., seedling resistance, tillering capacity, head size in cabbage, etc.) *meth*

selection response: the difference between the mean of the individuals selected to be parents and the mean of their offspring; it is expressed by the formula $R = h^2 \times S$ (h^2 = heritability, S = selection coefficient = phenotypic difference between the mean of all selected fractions and the mean of total population) *gene stat* >>> Figure 38

selective advantage: an advantage for survival of a genotype in a population and for production of viable progeny as compared to other genotypes, which may show a selective disadvantage with respect to fitness and viability *meth stat* >>> Figure 38

selective agent: an environmental or chemical agent that imposes a lethal or sublethal stress on growing plants, or portions thereof in culture, enabling selection of resistant or tolerant individuals *biot*

selective culture medium >>> selective agent

selective disadvantage: inferior fitness of one genotype compared to others in the population *meth stat* >>> selective advantage

selective fertilization: the nonrandom participation of male or female gametes or different genotypes in the formation of zygotes and/or hybrids *bot*

selective gametocide: a treatment that inactivates certain gametes, such as one that produces male sterility but does not affect the female gametes *meth seed*

selective herbicide: a herbicide that acts against either monocots or dicots, against weeds and not against crop plants or even against species weeds *phyt*

selective medium >>> selective agent

selective neutrality: the situation in which different alleles of a certain gene confer equal fitness *gene*

selective system: any experimental method that enhances the recovery of specific genotypes *meth*

selective value: a measure of the fitness of a gene within a genotype or of a genotype within a population; it is proportional to the probability of that gene or genotype surviving and it is a function of gene or genotype frequencies *gene stat* >>> Figure 38

self: an individual plant produced by self-fertilization, as opposed to crossbred *gene*

self-compatible: a plant that can be self-fertilized *bot* >>> Table 35

self-fertile: capable of producing seed upon self-fertilization *bot* >>> Table 35

self-fertility >>> autogamy

self-fertilizing: the fusion of male and female gametes from the same individual *bot* >>> Table 35

self-fertilizing crop >>> self-fertilizing

self-incompatibility: controlled physiological hindrance to self-fertilization; inability to set seed from application of pollen produced on the same plant; there are several mechanisms responsible for self-incompatibility in higher plants: (1) pollen may fail to germinate on the stigma, (2) pollen tube growth in the style may be inhibited to the extent that pollen fails to reach the ovary, (3) pollen tubes of sufficient length may fail to penetrate the ovule, (4) a male gamete that enters the embryo sac may fail to unite with the egg cell; the relative length of stamens and style in a bisexual flower is associated with incompatibility; most plant species have homomorphic flowers in which the stamens and styles attain comparable length; self-incompatibility in homomorphic flowers can be gametophytic or sporophytic; some species exhibit heteromorphic flowers in which the stamens and styles attain different heights; the presence of either pin or thrum flowers is termed distyly; there are also possibilities to avoid or break the self-incompatibility though (e.g., by high temperatures in rye) *gene* >>> Table 35

selfing: when a pistil is fertilized with pollen from the same plant that bears the pistil; also applied to seed resulting from such fertilization *meth* >>> self-pollination >>> Table 35

selfish DNA: a segment of the genome with no apparent function other than to ensure its own replication *gene*

self-pollination: the transfer of pollen from anther to stigma of the same plant (e.g., in barley, chickpea, clover cowpea, crambe, crotelaria, guar, field bean, field pea, linseed, jute lentil, lespedeza, millet, mungbean, oats, peanut, potato, rice, sesame, soybean, tobacco, tomato, vetch, wheat, or wheatgrass) *bot* >>> Table 35

self-seed: a plant that releases ripe seeds that sprout into new plants without human help *agr hort*

self-seeding >>> self-seed

self-sown cereals >>> self-seed

self-sterility: the inability of some hermaphrodites to form viable offspring by self-fertilization *bot* >>> Table 35

semiallele: mutant allele(s) that are allelic with respect to the function but not with the structure of the allele (i.e., they are not present on the same chromosome position) *gene*

semiarid: of a region or land characterized by very little annual rainfall, usually from 250 to 5,000 mm *eco*

semiconservative replication: replication of DNA in which the molecule divides longitudinally, each half being conserved and acting as a template for the formation of a new strand *gene*

semidesert: an extremely dry area characterized by sparse vegetation *eco*

semidomesticated: the incompleted breeding of species in order to accommodate human needs *evol*

semidominance >>> partial dominance

semidominant >>> semidominance

semidwarf: a common term used in wheat breeding; it designates individuals or a variety showing intermediate stem length compared to very short mutant lines; during the 1970s semidwarf wheat genotypes triggered the so-called "green revolution" in agriculture, mainly of third-world countries; several alleles of *Rht* (reduced height) loci contributed to shorter straw length and thus better lodging resistance as well as to higher spikelet fertility and thus yield; several dwarf *(Rht)* mutants show a decrease of the amount of certain gibberellins within the meristematic cells, which reduce plant height, not only in wheat but also in rice, maize, and *Arabidopsis* spp.; the genes modifying those gibberellic acids are producing several proteins regulating different cell activities; some of the *Rht* genes that show a specific gibberellic acid insensitivity are sequenced and available for use in genetic engineering *agr* >>> gibberellic acid (GA) >>> short-straw mutant

semigamy >>> hemigamy

seminoferous: seed-bearing (e.g., a cone consisting of seed-bearing, overlapping scales surrounding a central axis; a seed-bearing spike of a grass; a pistil as a seed-bearing organ of a flower) *bot*

semisterile >>> semisterility

semisterility: a situation in which half or more of all zygotes are inviable *gene bot*

semolina: a granular, milled product of durum wheat, used in the making of pasta *agr*

senescence: the phase of plant growth that extends from full maturity to death *bot*

sense strand DNA: the DNA strand with the same sequence as mRNA *gene biot*

sepal: a floral part of the outer whorl, referred to collectively as the calyx *bot*

***Septoria* seedling disease** (of wheat): the same pathogen *(Septoria nodorum, syn Leptosphaeria nodorum)* that causes *Septoria* glume blotch in wheat *phyt*

septum (septa *pl*): a cross-wall or partition *bot*

sequence-tagged-site (STS) marker: a unique (single-copy) DNA sequence used as a mapping landmark on a chromosome *biot gene*

sequencing: the determination of nucleotides and their order along a DNA or RNA molecule or the determination of the amino acids and their order in a protein molecule *gene biot*

serial analysis of gene expression (SAGE): a technique applied to study transcriptomes; most useful for fully sequenced genomes or for organisms having a quasi-complete collection of ESTs; it is based on cloning of concatemeres of very small tags (9-11 bases), each representing a transcript; sequencing of the concatemers results in a list of tags present in a population; the frequency of each tag is proportional to the abundance of the transcript it represents *biot*

serology: the study of the nature, production, and interactions of antibodies and antigens *phyt sero*

serotype: a subdivision of virus strains distinguished by protein or a protein component that determines its antigenic specificity *phyt*

serrate: with sharp teeth pointed toward apex or forward *bot*

sessile: without stalk of any kind *bot*

set: a short piece of a stem or a plantlet used for propagation *hort agr*

set (of seeds, etc.): the development of fruit and/or seed following pollination; it refers also to transplant as seedlings, to apply as a graft, a young bulb, tuber, or other type of vegetative propagule ready for planting *meth hort*

seta (setae *pl*): a bristle *bot*

setting >>> set

severe mosaic virus of potato >>> potato virus Y

sex: contrasting and complementary traits shown by female and male individuals within a given species *bot*

sex chromatin: a condensed mass of chromatin representing an inactivated X chromosome *cyto*

sex chromosome: a chromosome whose presence or absence is linked with the sex of the bearer; it plays a role in sex determination (e.g., in asparagus, hops, hemp, etc.) *cyto* >>> heterochromosome

sex determination: the mechanism by which the sex is determined *bot*

sex dimorphism: the different morphology of individuals within a species caused by their sexual constitution (e.g., in *Melandrium alba,* hops, hemp, etc.) *bot*

sex expression >>> sex-controlled

sex inheritance >>> sex linkage

sex linkage: genes located and inherited on a sex chromosome *gene*

sex ratio: the number of males divided by the number of females (also given in percent) at fertilization (equals primary sex ratio) *gene*

sex-controlled: a trait whose appearance is controlled by the type of sex of the individual (e.g., in asparagus, hops, hemp, etc.) *gene*

sexfoil: a leaf with six leaflets *bot*

sex-influenced >>> sex-controlled

sex-limited: pertaining to genetically controlled characters that are phenotypically expressed in only one sex *gene*

sexual: processes in which meiosis and fertilization is included leading to genetic recombination *gene*

sexual dimorphism >>> sex dimorphism

sexual partner >>> mate

sexual reproduction: reproduction involving the union of gametes that are haploid and derive from two sexes *gene*

sexual selection: contributes to the sex dimorphism; it bases on the male competition or on the female choice *gene*

sexuality: sexual character; possession of the structural and functional traits of sex *gene*

SGE >>> starch gel electrophoresis >>> gel electrophoresis

shade cloth: any of various fabrics used in the summer to lower soil temperatures, accelerate germination of cool-season autumn crops, prevent bolting, or protect against drying *meth agr hort*

shattering: the opening or disintegrating of the seed coat, fruit, or husk before harvesting; the consequence is loss of seeds in crop plants (e.g., seeds drop to the ground prior to harvest) *agr;* in viticulture, the physiological stage following bloom when impotent flowers and small green berries begin to fall from the cluster *hort*

sheath: of leaves, the base of a blade or stalk that encloses the stem *bot* >>> Table 30

shedding: the release of seeds from seed-bearing organs, such as spikes in cereals, usually before harvest *agr*

shelf life: the term or period during which a stored commodity (vegetables or food) remains effective, useful, or suitable for consumption *agr*

shifting dominance >>> alternating dominance >>> dominance

shoot: a stem or branch and its leaves; a new young growth *bot*

shoot apex: the tip of a shoot; the apical or lateral shoot meristematic dome together with the leaf primordia, from which emerge the leaves and subadjacent stem tissue *bot*

shoot bud >>> bud

shoot culture >>> shoot-tip culture

shoot meristem: a meristem located at the apex of shoots, and from which the aerial parts of a plant are derived *bot*

shoot regeneration: in tissue culture, the phenomenon of in vitro shoot formation from callus *biot*

shoot tip >>> shoot apex

shoot-tip culture: the in vitro culture of shoot apical meristem plus one to several primordial leaves *biot*

short-day conditions: short days and corresponding long nights *phys*

short-day plant: a plant in which flowering is favored by short days and corresponding long nights (e.g., *Chrysanthemum* spp.) *phys*

short-lived perennial: for example, several grasses normally expected to live only 2-4 years *bot agr*

short-straw mutant: a mutant genotype of monocots showing reduced plant height by reduced length of internodes; several genes are known to cause the shortening of plants; some of those genes (*Rht* and *Gai* genes) attained great attention in cereal breeding *gene* >>> semidwarf

shot hole: a leaf spot disease characterized by holes made by the dead parts dropping out; for example, caused by *Heteropatella antirrhini* in snapdragon, by *Stigmina carpophila* in peach, or by *Pseudomonas mors-prunorum* in plum *phyt*

shotgun method: a way preparing DNA for genetic studies; random-cutted fragments of DNA are cloned into a vector; the fragments resemble a clone library; out of this collection of clones several molecular techniques are applicable *biot* >>> sibling technique

shoulder: the upper edge of the broad wing of the glume (e.g., in wheat) *bot*

shrub: a perennial woody plant, less than 10 m tall, which branches below or near ground level into several main stems *bot*

shrubby: shrublike, bushy, with many stems rather than a single trunk; a form of lichen that appears bushy or hairlike *bot*

shuck: an outer covering (e.g., the husk of maize, the shell of a walnut, etc.) *bot*

shuttle vector: a vector that can replicate in more than one organism *biot*

sib: short for sibling; one of two or more offspring with both parents in common but deriving from different gametes *tax*

sib mating: a form of inbreeding in which progeny of the same parents (siblings) are crossed *meth*

sibbed: mated individuals having the same parentage *gene*

sibing: the transfer of pollen between different plants of the same variety *meth*

sibling chromatid >>> sister chromatid

sibling species: morphologically similar or identical populations that are reproductively isolated *gene eco*

sibling technique: a method for the isolation of bacteria containing a specific cloned gene (e.g., in shotgun cloning experiments) whose phenotype cannot be easily recognized on individual colonies but can be detected with great sensitivity in a large population where only a small minority of the cells contains the gene wanted *biot*

siblings >>> full sibs

side grafting >>> inarching >>> graft

sieve cell: long, slender, tapering cells that form part of the sieve tube; they lack nuclei but retain cytoplasm; each sieve cell ends in a sieve plate *bot*

sigmoid (curve): S-shaped (e.g., the growth curve of a plant) *stat*

sign: a visible manifestation of a causal agent of plant disease *phyt*

sign test: designed to test a hypothesis about the location of a population distribution; it is most often used to test the hypothesis about a population median; the sign test does not require the assumption that the population is normally distributed; in many applications, this test is used instead of the one sample t-test when the normality assumption is questionable; it is a less powerful alternative to the WILCOXON signed ranks test, but does not assume that the population probability distribution is symmetric; this test can also be applied when the observations in a sample of data are ranks (i.e., ordinal data rather than direct measurements) *stat*

signal peptide: a short segment of about 15-30 amino acids, which is found on the N-terminal of secreted proteins; during processing the protein of the cell metabolites recognizes the signal sequence and allows the secretion or penetration through the membranes of organelles; later in mature proteins, the sequence disappears; it is removed by a protease *phys*

signal transduction: the biochemical events that conduct the signal of a hormone or growth factor from the cell exterior, through the cell membrane, and into the cytoplasm; this involves a number of molecules, including receptors, proteins, and messengers *phys*

significance level: the significance level of a statistical test is the preselected probability of (incorrectly) rejecting the null hypothesis when it is in fact true; usually a small value, such as 0.05 is chosen; if the P value (probability value) calculated for a statistical test is smaller than the significance level, the null hypothesis is rejected *stat*

significance test: statistical test designed to distinguish differences due to sampling error from differences due to discrepancy between observation and hypothesis *stat*

signpost >>> landmark

silage: a type of foodstuff for livestock, prepared from green crops and by fermentation; it contains about 65 percent moisture *agr*

silage crops: crops grown for harvest and storage while in the green and high-moisture condition; maize, grass and sorghum are used for this purpose and cured while in storage *agr*

silencer sequence: a DNA sequence, usually located distant from the core promoter, that may inhibit gene transcription *biot*

silent mutation: a mutation in a gene that causes no detectable change in the biological characteristics and/or gene product *gene*

siliceous: composed of or abounding in silicle *bot*

silicious >>> siliceous

silicle: a short, broad silique *bot*

silicula >>> silicle

siliqua: a dry, dehiscent fruit that has a central partition; it is elongated; it is produced by many members of *Brassicaceae bot*

silique >>> siliqua

silk: in maize, the stigma and style of the female flower, through which the pollen tube grows to reach the embryo sac *bot*

silking: the female flowering in maize (i.e., the moment when silks emerge from the husk) *agr*

silky: covered with close-pressed soft and straight pubescence *bot*

silver staining: using silver nitrate ($AgN0_3$) and an appropriate cell pretreatment interphase nuclei can be made visible without phase contrast microscope; they appear yellow to light brown and nucleoli brown; at metaphase, silver nitrate stains nucleolus organizer regions which have been active in the preceding interphase; although the amount of silver nitrate is in some relation to the amount of activity or number of active rDNA copies, it cannot be used for an accurate quantitative study *cyto*

silviculture: the theory and practice of controlling the establishment, composition, growth, and quality of forest stands to achieve the objectives of its management *fore*

simple fruit: derives from a single pistil *bot*

simple sequence repeat (SSR) DNA marker technique: a genetic mapping technique that utilizes the fact that microsatellite sequences repeat (appear repeatedly in sequence within the DNA molecule) in a manner enabling them to be used as markers *biot*

simplex type >>> autotetraploid >>> nulliplex type

simultaneous selection: when in a selection process several traits are considered in the same generation *meth*

single copy: a gene or DNA sequence are only present once in a haploid genome; most of the structural genes (coding for proteins) represent single copy genes *gene*

single cross: a cross between two parental plants and/or lines *meth* >>> Figures 22, 31

single germ >>> monogerm(ous)

single hybrid >>> double cross

single replication: an experimental plot design in which each variant is present in one replication either randomized or nonrandomized *stat*

single-copy probe: a radioactive-labeled DNA or RNA sequence with highly specific activity; it is used to identify complementary regions applying the Northern and Southern hybridization techniques or colony hybridization *meth biot*

single-copy region: a unique or nonrepeated DNA region occurring only once in the haploid genome; most of the structural genes (protein-coding) are single-copy genes *gene*

single-cross parent: the F1 offspring of two inbred parents, which in turn is used as a parent, usually with another single-cross parent to produce a double-cross hybrid (e.g., in maize) *meth* >>> Figures 22, 31

single-nucleotide polymorphisms (SNPs): variations (in individual nucleotides) that occur within DNA at the rate of approximately one in every 1,300 bp in most organisms; SNPs usually occur in the same genomic location in different individuals *biot*

single-plant plot: used for the replicated evaluation of experimental lines or cultivars by the, for example, honeycomb design; the number of plants evaluated is equal to the number of replications in the experiment; the plots are organized in a systematic manner to permit comparison of a plant of one line with adjacent plants of other lines *meth* >>> honeycomb design >>> Figure 32

single-row plot: an experimental field design in which a single plant row of different length represents one plot *meth*

single-seed descent (SSD): derivation of plants by a selection procedure in which F2 plants and their progeny are advanced by single seeds until genetic purity is achieved; single-seed descent methods (single-seed, single-hill, multiple-seed) are easy ways to maintain populations during inbreeding; natural selection cannot influence the population, unless genotypes differ in their ability to produce viable seeds; artificial selection is based on the phenotype of individual plants, not on the progeny performance; natural selection cannot influence the population in a positive manner, unless undesirable genotypes do not germinate or set seeds *meth* >>> Figure 16

singling peg >>> pricking-out peg

sinigrin >>> glucoside

sister chromatid: derives from replication of one chromosome, as opposed to nonsister chromatids, which derive from the other homologous or nonhomologous chromosomes *cyto*

sister chromatid exchange (SCE): an event, similar to crossing over, that can occur between sister chromatids at mitosis and meiosis; it may be detected in "harlequin" chromosomes (sister chromatids that stain differentially so that one appears dark [usually red-violet] and the other light) *cyto*

site selection: variation in the production of the soil is commonly referred to as soil heterogeneity, caused by variation of the soil type, availability of nutrients, and moisture; the variation cannot be completely eliminated for breeding experiments, but it can be minimized by careful selection of the area in a field where plots will be placed *agr stat*

site-directed mutagenesis (SDM): the process of introducing specific base pair mutations into a gene; a technique that can be used to make a protein that differs slightly in its structure from the protein that is normally produced; single mutation is caused by hybridizing the region in a codon to be mutated with a short, synthetic oligonucleotide; this causes the codon to code for a different specific amino acid in the protein gene product; site-directed mutagenesis holds the potential to create modified (engineered) proteins that have desirable properties not currently available in the proteins produced by the plant *meth gene biot*

skiophilous: shade-loving *bot*

skiophyte >>> skiophilous

slip: a cutting from a mother plant *hort meth*

slit planting: prying open a cut made by a spade, mattock, or planting bar; inserting a young tree; closing the cut on the latter by pressure *fore hort*

slow rusting (genotype): a genotype in which rust develops slowly but never reaches a high degree of severity; it belongs to the type of partial or incomplete resistance; it is quantitatively inherited *phyt*

small grains: small-grain cereals that are monocotyledonous plants belonging to the order of Poales and to the family of Gramineae; they are characterized by specific morphological traits such as stubble, spikelets, scutellum, etc., and by grains that are rich in carbohydrates; there are six main groups cultivated—wheat, barley, rye, triticale, oats, and rice *agr*

smear: direct spreading of cells in a semifluid tissue over the surface of a microscopic slide with a flat-honed scalpel or other tools, and the immediate inversion of the slide over a dish of fixative *prep cyto*

smothering crop: it refers to a crop that suppresses the growth of weeds by its heavy growth and shadowing effect *agr meth*

smudging >>> fumigation

smut: a plant disease caused by a fungus of the order Ustilaginales; the symptoms include the formation of masses of black, sootlike spores; infected plants often show some degree of distortion *phyt* >>> bunt

snow line: the line, as on mountains, above which there is perpetual snow *eco*

SNPs >>> single-nucleotide polymorphisms

sobole: a shoot, stolon or sucker *bot*

sod: the upper stratum of grassland, containing the roots of grass and any other herbs, or a piece of this grassy layer pared or pulled off (turf, divot, fail) *agr*

sodium (Na): a soft, silver-white, chemically active metallic element that occurs naturally only in combination *chem*

sodium chlorate: a colorless water-soluble solid, $NaClO_3$, used as a mordant, oxidizing, or bleaching agent *chem* >>> chemical dessication

sodium dodecyl sulfate-polyacrylamide (SDS-PAA): an ionic detergent, irritant *chem biot*

soft endosperm >>> chalky

soft rot: a decomposition of plant tissue (fruits, roots, stem, etc.) by fungi or bacteria resulting in a tissue becoming soft (e.g., caused by *Erwinia carotovora* in potato) *phyt*

soil: the natural space-time continuum occurring at the surface of the earth and supporting plant life *agr* >>> crop rotation

soil aggregation: the process whereby primary soil particles (sand, silt, clay) are bound together, usually by biological activity and substances derived from root exudates and microbes *agr*

soil heterogeneity: variation in the production of the soil is commonly referred to as soil heterogeneity, caused by variation of the soil type, availability of nutrients, and moisture *agr* >>> crop rotation

soil pasteurization: a sterilization of soil that destroys unacceptable organisms without chemically altering the soil *meth hort* >>> solarization

solarization: the process of sterilizing the soil and killing soil pests by covering the moist planting area with clear plastic *meth hort*

solid medium: as opposed to liquid medium, any in vitro culture medium solidified with agar or other jellying agents; it is widely used in vitro propagation *prep biot*

solum: the upper part of a soil profile; roots, plants, and animal-life characteristics of the soil are mainly confined to the solum *agr*

solvent: a substance that dissolves another to form a solution *chem*

somaclonal variation: variation found in somatic cells dividing mitotically in culture; it can be genetically caused or can be a sort of habituation *biot*

somaclone: a plant regenerated from a tissue culture of somatic cells *biot*

somatic: body cells *bot* >>> somatic cell

somatic apospory >>> apospory

somatic cell: a diploid, body cell, or other cells than the germ cells *bot*

somatic cell fusion: the fusion of somatic cells (usually protoplasts) by different means in vitro *biot* >>> protoplast

somatic crossing-over: crossing-over during mitosis of somatic cells such that parent cells heterozygous for a given allele, instead of giving rise to two identical heterozygous daughter cells, give rise to daughter cells—one of which is homozygous for one of these alleles, the other being homozygous for the other allele *cyto gene*

somatic embryo: an organized embryonic structure morphologically similar to a zygotic embryo but initiated from somatic cells; somatic embryos develop into plantlets in vitro through developmental processes that are similar to those of zygotic embryos *biot*

somatic embryogenesis >>> somatic embryo

somatic fusion >>> somatic cell fusion

somatic hybridization: the fusion of genetically different somatic cells by different means, usually to overcome natural crossing (incompatibility) barriers *biot*

somatic mutation: a mutation occurring in a somatic cell; if the mutated cell continues to divide, the individual will develop a patch of tissue with a genotype different from the cells of the rest of the body *gene*

somatic regeneration >>> regeneration

somatoplastic sterility: the collapse of zygotes during embryonic stages due to disturbances in embryo-endosperm relationships *bot* >>> cytoplasmic male sterility

sorting: a process utilized to sort and/or separate different cells, particles, chromosomes, etc.; some automated means of cell sorting include (1) biochips, utilizing controlled electrical fields to collect specific cell types onto electrodes in the biochip, (2) fluorescence-activated cell (FAC) sorter machines, or (3) magnetic particles (e.g., attached to antibodies) are in use for different applications; the ploidy level can also be determined by cell sorting equipment *biot meth*

source community: a community from which a local variety or a seed lot originated *seed*

source plant: a donor plant from which an explant is taken in order to initiate an in vitro or in vivo culture *meth*

source-identified seed: commonly, a class of true seed defined as seed from natural stands or plantations with known geographic source and elevation according to specific standards *seed*

SOUTHERN transfer: a method in biochemistry named after E. SOUTHERN; this technique is also called a SOUTHERN blot; as the result of a restriction analysis, an agarose gel will contain fragments of DNA; the DNA fragments in the gel are transferred to a cellulose nitrate filter (or other suitable medium) during the SOUTHERN transfer; the moistened receiving material is placed on top of the gel; absorbent materials are placed on tops of the filter; DNA is eluted onto the filter because solvent is drawn up through the bed by capillary action; the filter is used for complementary hybridization tests, for example, if the filter is placed in a solution of suitable radioactively labeled RNA or denatured DNA probe, strands bound to the filter will, in turn, bind labeled probe that is complementary; after washing to remove unbound probe, autoradiography may be used for detection of positive DNA fragments; the method is used to map restriction sites for a single gene within a complex genome *meth biot*

sowing brick: a prepared block or ball of loam, peat, plastic, or foam into which one or more seeds are pressed; when planting out, the emergent seedling can have a better start in an unfavorable environment *fore*

sowing time: the optimum dates for seeding; temperature and moisture are the major determinants of sowing time *agr*

sp. (spp. *pl*): abbreviation of "species"; the expression follows the name of a genus when the single species indicated is unknown or for any other reason not specified *tax*

spacer: noncoding and/or nongenic DNA sequences that may occur between genes *gene* >>> spacer sequence

spacer sequence: a portion of DNA sequences that does not code any RNA; it is rich in adenine and thymine, and found between the transcriptional DNA segments *gene*

span length: the distance spanned by a specified percentage of the fibers in a fibrograph test beard used for quality determination in cotton *meth*

spatial order: chromosomes or genomes may occupy certain domains within the nucleus and/or dividing cells *cyto*

spawn: a common term applied to a mixture of fungal mycelium or a nutritive organic material for the artificial propagation of mushrooms *hort*

SPEARMAN rank correlation coefficient: usually calculated on occasions when it is not convenient, economical, or even possible to give actual values to variables, but only to assign a rank order to instances of each variable; it may also be a better indicator that a relationship exists between two variables when the relationship is nonlinear *stat*

speciation: the splitting of an originally uniform species into daughter species that coexist in time *evol*

species: the individuals of one or more populations, which can interbreed, but which in nature cannot exchange genes with members belonging to other species *tax* >>> Table 12 >>> *cf* Important List of Crop Plants

species hybridization: the mating of two different species; based on species hybridization, several wild species are represented in the ancestry of the cultivated crop plants (e.g., of wheat); induced species hybridization is also used for introgressing of desirable genes into the cultivated crop plants *meth* >>> synthetic amphiploid >>> triticale >>> Figure 2

specific combining ability >>> combining ability

specific resistance >>> vertical resistance

spectral karyotype: a visualization of all chromosomes of a complement, each labeled with a different fluorescence color; this technique is useful for identifying structural chromosome abnormalities or alien chromosomes *cyto* >>> FISH

spectrometry: the measurement of the absorption or emission of light by a substance at a specific wavelength *phy meth*

spectrophotometer: an optical system used in biology to compare the intensity of a beam of light of specified wave length before and after it passes through a light-absorbing medium *meth*

spelt forms >>> speltoid

speltoid: a mutation that arises spontaneously (e.g., in wheat); it resembles spelt wheat in certain features *bot* >>> Table 1

sperm: a male gamete *bot*

sperm nucleus (sperm nuclei *pl*): in angiosperms, the two male gametes that are formed by division of the generative cell; they migrate down the pollen tube behind the vegetative nucleus; when the pollen tube enters the embryo sac the tip of the tube breaks down to release the generative nuclei; one fuses with the egg nucleus to form the zygote and the other usually fuses with the polar nuclei or definitive nucleus to form the primary endosperm nucleus *bot* >>> Figure 25

spermatogenesis: the process of differentiation of a mature sperm cell from an undifferentiated germ-line cell, including the process of meiosis *bot*

sperma(to)phyta: seed-bearing plants *bot*

sphaeroplast: a cell with a partially degraded cell wall *biot*

***Sphagnum* moss:** a bog moss belonging to the genus *Sphagnum;* it was frequently used as a rooting medium for plants *bot hort meth*

spherical aberration: inaccurate focusing of light due to curved surface of lens *micr*

spherosome: oil storage bodies of plant cells *bot*

spica >>> spike

spice: the fragrant vegetable condiments used for the seasoning of food; a wide range of plants, such as pepper, allspice, nutmeg, ginger, cinnamon, cloves, marjoram, oregano, etc. are bred for spice production *hort*

spiciform: spike-shaped *bot*

spicula: a small or secondary spike *bot*

spicule(s): minute teeth occurring on the nerves and other parts of the lemma and glume of, for example, wheat, barley, oats, grasses, etc. *bot*

spike: a "flower" head on which the spikelets are borne without a stalk as, for example, in wheat, barley, rye or other grasses *bot* >>> Figure 34

spike density: the number of spikelets per unit of spike length; it is a distinctive character of certain cereal varieties *bot agr* >>> Figure 34

spikelet: the unit of inflorescence in the cereals and in some grasses consisting of a pair of glumes and one or more florets *bot*

spikelet glume: a chafflike bract; specifically one of the two empty chaffy bracts at the base of the spikelet in the grasses *bot*

spindle: the set of microtubular fibers that appear to move the chromosomes of eukaryotes during mitotic and meiotic cell division *cyto*

spindle attachment >>> centromere

spindle poison: any poison affecting the correct formation or function of the spindle of the dividing cell *cyto*

spindle pole: the regions to which the spindle fibers pull the chromosomes or chromatids during anaphase *cyto*

spine: a sharp woody or rigid outgrowth from a stem, leaf or other plant part *bot*

spinous: having narrow sharply pointed processes (spines) *bot*

spiny >>> spinous

spiral separator: a type of seed separator with no moving parts; the seeds enter at the top and slide or roll down an inclined spiral runway; the speed of seed movement and centrifugal force allows separation of the heavier, round, fast-moving seeds from those that move slower *seed*

splicing: the removing of the intron (noncoding sequence) from an mRNA gene sequence and the fusion of the exons (coding sequence) during the mRNA processing *biot*

splitting >>> fission

spontaneous haploids: the spontaneous occurrence of haploids, usually by asexual processes *bot* >>> apomixis >>> parthenogensis >>> polyembryony >>> Figures 17, 26 >>> Table 7

spontaneous mutation: a naturally occurring mutation, as opposed to one artificially induced by chemicals or irradiation; usually such mutations are due to errors in the normal functioning of cellular enzymes *gene*

spore: a minute reproductive unit in fungi and lower plant forms *bot*

sporophyte: the diploid generation of plants; it is the conspicuous generation in the higher plants *bot* >>> Figure 28

sporophytic self-incompatibility: self-incompatibility is based on the genotypic and phenotypic relationship between the female and male reproductive system; alleles in cells of the pistil determine its receptivity to pollen; the phenotype of the pollen, expressed as its inability to effect fertilization, may be determined by the maternal plant, referred to as sporophytic incompatibility *gene* >>> self-incompatibility

sport: a sudden deviation from type; a somatic mutation *gene* >>> bud sport

sporulation: the period of active spore production *phyt*

spreader: a substance added to fungicide or bactericide preparations to improve contact between the spray and the sprayed surface; in resistance testing, the variety, line, or genotype infected with the pathogen; from the spreader the pathogen is distributed by wind, rain, or insects to the tester genotypes; the spreader and the testers can be arranged in different experimental and field designs *phyt* >>> Figure 30

spreader bed >>> spreader

spreader strip >>> spreader

sprig: a small part of a plant, such as stolons used for propagations, twigs, bearing flowers, etc. *meth hort*

sprigging >>> sprig

sprout: a shoot of a plant, as from a germinating seed, a rootstock, tuber, runner, etc., or from the root (a sucker), stump, or trunk of a tree *bot hort* >>> shoot

sprouting: germination of seeds on mature spikes, usually before harvest in wet years or when harvest is delayed; it can cause severe loss of quality in rye, wheat, barley, or oats *bot agr*

spur: a short or stunted branch or shoot, as of a tree *hort*

spur pruning: a method of pruning (fruit) trees, by which one or two eyes of the previous year's wood are left and the rest cut off, so as to leave spurs or short rods *hort*

SSD >>> single-seed descent

ssDNA: single-stranded DNA *biot*

ssp.: subspecies >>> sp.

SSR >>> simple sequence repeat

stabilizing selection: a type of selection that removes individuals from a population (i.e., from both ends of a phenotypic distribution divided by deviation of a sample of means) *meth stat*

stacked genes >>> gene stacking

staddle: a foundation of trunk and main branches for grafting, either of rootstock or stem builder *meth hort*

stage: the small platform of a microscope on which the object to be examined is placed *micr*

stage of maturity >>> maturation

stalk: a stemlike supporting structure such as a peduncle or pedicel *bot*

stalk diameter: the diameter of a stalk, usually at a designated node or internode *prep agr*

stalk tunneling: the longitudinal tunnels in plant stalks produced by different insects *phyt*

stamen (stamina *pl*): the part of the flower that produces the pollen *bot*

staminal (staminate) flower: a flower bearing stamens but not functional pistils (i.e., a male flower) *bot*

stand: term used for an established field crop or tree plantation *agr*

stand canopy >>> canopy

standard: a criterion for evaluating performance and results; it may be a quantity or quality of output to be produced, a rule of conduct to be observed, a model of operation to be adhered to, or a degree of progress toward a goal; in general, agreeing upon rules for the specification, design, and development of entities within a given class *stat*

standard deviation: a measure of the variability in a population of items *stat*

standard error (SE): a measure of variation of a population of means *stat*

standing corn >>> standing crop

standing crop: the mass of the individuals of a crop in a field *agr*

standing power: a trait of a crop, usually of cereals, that describes the stalk, which resists lodging; it is frequently scored on a scale of 1-9 *agr*

starch: a homopolysaccharide, consisting of glucose molecules, which is the major storage carbohydrate of plants and the major nutritionally important carbohydrate in the human diet; because of its unique physical properties, it is a valuable raw material in many food and nonfood industries; it is synthesized in large amounts in leaves during the day, and degraded during the subsequent night; its utilization depends on the granular structure; starch granules are supramolecular organized structures, made up of ordered and disordered, or amorphous regions; the ordered parts consisting of double helices formed from short branches of amylopectin molecules; most of the double helices are further organized into crystalline lamellae, making the granule into a so-called "semicrystalline" structure; two crystalline forms are found in starch molecules, A and B, the helices in the A-form being more densely packed than in the B-form *chem phys*

starch gel electrophoresis: a method for high-resolution separation of soluble proteins using starch as a matrix; it was introduced by O. SMITHIES in 1955; M. POULIK (a collaborator of SMITHIES) introduced discontinuous buffer systems and enzyme-histochemical staining of electrophoretically separated enzymes, which were later called "zymograms"; the problem of the method is the fact that, due to the presence of various anionic residues, such as glucuronic acid in the purest starch fractions, the measured mobilities are always biased by electroosmotic flow of water in the gels; since the gels have net negative charges, the water moves relative to the stationary gels to compensate; in addition, the concentration of glucuronic acid residues varies in starch preparations from batch to batch; therefore, reproducibility is often severely compromised *chem*

starch plants >>> nutrient enhanced varieties >>> potato

start condon: the condon AUG, in RNA, which represents the start of a new protein in DNA *biot gene*

starter >>> start codon >>> primer

stationary phase: the plateau of the growth curve after log growth, during which cell number remains constant; new cells are produced at the same rate as older cells die *biot*

statistic: a quantity that is calculated from a sample of data in order to give information about unknown values in a corresponding population *stat*

statistical analysis: analyzing collected data for the purposes of summarizing information to make it more usable and/or making generalizations about a population based on a sample drawn from that population *stat*

statistical genetics >>> quantitative genetics

statistical significance: the degree to which a value is greater or smaller than would be expected by chance; typically, a relationship is considered statistically significant when the probability of obtaining that result by chance is less than 5 percent if there were, in fact, no relationship in the population *stat*

statistical test: a type of statistical procedure that is applied to data to determine whether the results are statistically significant *stat*

statistics: the scientific discipline concerned with the collection, analysis, and presentation of data *stat*

statocyte: a cell that is present in pulvinus (thickened leaf sheath base) parenchyma; it contains starch grains and/or calcium oxalate crystals, which are thought to act as statoliths (i.e., balancing stones); it may be gravity-sensing; in cereals, if a culm lodges, pulvinus cells on the lower side will elongate, but not divide; this helps to straighten up the culm *bot*

statolith: any of the granules of lime, sand, calcium oxalate, etc. contained within a statocyte *bot* >>> statocyte

stearic acid: a colorless, waxlike, sparingly water-soluble fatty acid, $C_{18}H_{36}O_2$, occurring in some vegetable oils *chem phys*

steckling: the plantlet in the first year of the biennial sugarbeet stored over winter and planted for the production of seeds; it is grown from unthinned plantlets *agr seed*

steeping: the soaking of barley grains in the production of malt prior to germination *meth*

stem: in vascular plants, the part of the plant that bears buds, leaves and flowers; it forms the central axis of the plant and often provides mechanical support *bot*

stem apex: the top or tip of a stem *bot*

stem cell: any of reproductive and/or generative cells in an organism, as opposed to somatic cells *bot;* in biotechnology, immature cells capable of developing into various (specialized) cell types *biot*

stem cutting: a cutting taken from a portion of stem *hort meth*

stem tip >>> stem apex

stereomicroscope: a microscope having a set of optics for each eye so arranged that the sets view the object from slightly different directions and make it appear in the three dimensions *micr*

sterile: unable to produce reproductive structures (i.e., unable to reproduce) *gene*

sterile spikelet: the nonfertile lateral spikelet in two-row barley *bot*

sterility: failure to produce functional gametes or viable zygotes *bot gene*

sterility gene: a gene that causes sterility *gene*

stickiness: the genetically or artificially induced agglutination of chromosomes; usually it leads to different sorts of chromosome aberrations *cyto*

sticky ends: single-stranded ends of DNA fragments, usually produced by restriction enzymes of the type II *gene biot* >>> blunt ends

stigma: the part of the female reproductive organs on which pollen grains germinate; it is the receptor of the pollen *bot* >>> Figure 35

stigmatic hairs: small tiny hairs and/or hair-like structures on a stigma *bot*

stimulant: a chemical or other substance that excites an organ or tissue to a specific activity (e.g., the application of a plant regulator to a stem to induce root formation) *phys*

sting: a sharp, hollow, glandular hair that secretes an irritating or poisonous fluid (e.g., in nettle) *bot*

stinging hair >>> sting

stipule: a leafy or linear appendage, found, usually in pairs, at or near the base of the petiole of a leaf (e.g., in pea) *bot*

stirps: a race or permanent variety of plants *tax gene*

stochastic: a process with an indeterminate or random element as opposed to a deterministic process that has no random element *stat*

stock: that part of a plant, usually consisting of the root system together with part of the stem, onto which is grafted a scion *bot*; in genetics, an artificial and/or experimental mating group *gene* >>> graft

stock collection >>> stock seeds

stock plant: a mother plant kept for cuttings to reproduce the plant and/or variety *hort fore meth*

stock seeds: the supply of seeds, tubers, or other propagules reserved for planting, multiplication, or used as a source of germplasm for maintaining and increasing seed of crop varieties or genetic tester lines *seed meth*

stolon: a stem that grows horizontally, which roots at its tip to produce a new plant, as opposed to a runner *bot*

stoma (stomata *pl*): a small opening, many of which are found in the epidermal layers of plants, allowing access for carbon dioxide and regress for water; they are surrounded by guard cells, which control the pore size *bot* >>> carbon dioxide

stone fruit >>> drupe

stoner: a modification of the gravity separator, especially constructed to separate stones from crop seeds; the machine separates seed on the same principle as conventional specific gravity machines, however it discharges only at each end; the desirable seeds flow to the lower end and are discharged, while stones and heavier concreted earthen material drop down to the upper end of the deck and are discharged; it useful for conditioning seed of field beans *seed* >>> Table 11

stool: the root or stump of a tree, bush, grass, or cane that produces shoots each year, or the mother plant from which young plants are propagated by the process of layering *hort agr bot*

stop codon: a codon for which there is no corresponding tRNA molecule to insert an amino acid into the polypeptide chain; the protein synthesis is hence terminated and the completed polypeptide released from the ribosome; there are three stop codons (UAA, UAG, UGA) *gene*

stopping >>> disbud

storage organ >>> storage protein(s)

storage protein(s): proteins that are found in specialized storage organs, such as endosperm tissue in cereals or others; for example, the endosperm of wheat contains a great number of nonenzymatic storage proteins that are the component of gluten, one of the most intricate naturally occurring protein complexes; gluten may be subdivided, on the basis of differential solubility, into gliadin and glutenin; in wheat breeding, these proteins may provide markers for specific quality characteristics *phys* >>> Table 15

storage protein genes: structural genes coding the synthesis of storage proteins; for example, in wheat, each homoeologous chromosome group contains structural genes having similar positions within each genome; the differences in the strength and elasticity of dough is under control of several storage protein loci (the high- and low-molecular-weight (HMW, LMW) glutenins (*Glu*-loci) and the gliadins (*Gli*-loci); the HMW glutenins appear responsible for dough strength, the LMW glutenins and gliadins for dough extensibility and elasticity; by using electrophoretic methods it has been possible to identify five loci controlling these proteins; they are located on the homoeologous chromosomes 1 and 6; alleles at these loci are highly variable and correlate with differences between good and poor bread-making or biscuit-making quality *gene prep* >>> Table 15

storage tissue >>> storage proteins

strain: a term sometimes used to designate an improved selection within a variety; a group of individuals from a common origin; generally, a more nar rowly defined group than a variety *tax* >>> variety

strain building: improvement of cross-fertilizing plants by any one of a number of methods of selection *meth*

stratification: the placing of seeds between layers of moist peat or sand and exposing them to low temperature in order to encourage germination and/or breaking dormancy; the method can also be realized in a refrigerator between +4-10°C *seed* >>> prechilling

stratified mass selection: mass selection in which the population is split into subpopulations that are grown under different environmental conditions (i.e., in different fields or in different parts of a field); plants for next generation seed are selected from the different subpopulations *meth*

straw walker: the part of a combine that moves the straw to the rear of the machine *agr* >>> coarse shaker

strawbreaker of cereals >>> eyespot disease

streptavidin: a protein produced by *Streptomyces avidinii;* it shows affinity to the vitamin "biotin;" it is used for detection of biotinylated hybrid DNA *meth micr prep* >>> FISH >>> GISH

streptomycin: an antibiotic; it inhibits the elongation of protein biosynthesis; it is produced by some strains of *Streptomyces;* in genetic experiments, streptomycin resistance is used as a selection marker *meth*

stress: a specific response by the plant to a stimulus that disturbs or interferes with the normal physiological equilibrium *bio*

stress protein(s): these are proteins made by plant cells (and other organisms) when those cells are stressed by environmental conditions (chemicals, pathogens, heat) (e.g., when maize is stressed during its growing season by high nighttime temperatures), the plant switches from its normal production of (immune system defense) chitinase to the production of heat-shock (i.e., stress) proteins *biot phys*

stress tolerance >>> biological control

striate: displaying narrow parallel streaks or bands *bot*

strike: to take root (e.g., of a slip of a plant) *hort meth*

stringency: reaction conditions (e.g., temperature, salt, and pH) that dictate the annealing of single-stranded DNA/DNA, DNA/RNA, or RNA/RNA hybrids; at high stringency, duplexes form only between strands with perfect one-to-one complementarity; lower stringency allows annealing between strands with some degree of mismatch between bases *biot meth*

strip cropping: the practice of growing crops in strips or bands along the contour in an attempt to reduce run-off, thereby preventing erosion or conserving moisture *agr*

strip tillage: planting and tillage operations that are limited to a strip not to exceed one-third of the distance between rows; the area between is left untilled with a protective cover of crop residue on the surface for erosion control *agr*

"strong" flour: strong flour contains strong gluten with good elastic properties (e.g., in wheat); suitable for bread-making, as opposed to "weak" flour, which contains weak gluten; it is less elastic but more extensible than that contained in strong flour; suitable for biscuit-making *prep*

structural aberration: structural changes of the entire chromosome (e.g., deletions, duplications, translocations, inversions, telocentrics or isosomics) *cyto* >>> Figure 37

structural gene: a gene that codes for the amino acid sequence of a protein *gene*

structural heterozygosity: heterozygosity for chromosome mutations *cyto* >>> translocation >>> Figure 37

structural homozygosity: homozygosity for chromosome mutations *cyto*

struggle for existence: the phrase used by C. DARWIN to describe the competition between organisms for environmental resources such as food or a place to live, hide, or breed *evol*

STS >>> sequence-tagged-site

stubble: the bases of cereal stems that remain after harvest *agr*

stubble crop: a crop plant from which stubble remains after combine harvesting (e.g., canola, wheat, barley, and sugarcane) *agr*

STUDENT's test: a statistical method used to determine the significance of the difference between means of two samples *stat*

stunt: to retard or stop plant growth (e.g., through exposure to harsh weather or lack of water or nutrients) *agr hort*

style: an extension of the carpel, which supports the stigma *bot*

subcloning: transplantation of a piece of DNA from one vector to another *biot*

subculture: the aseptic transfer of a part of a stock culture to a fresh medium *biot*

suber >>> cork

suberiferication: the process by which the cut surface of a stem forms a protective, corky layer, especially in conditions of high temperature and high humidity *bot*

suberin: a complex fatty substance found especially in the cell walls of cork *bot*

sublining: dividing a breeding population into several to many smaller populations; all controlled crosses for forward selection are made within a subline, leading to inbreeding within sublines *meth fore*

submedian: not quite in the middle *cyto* >>> Figure 11

submedian centromere >>> submedian

submicroscopic: objects that are too small to be resolved and made visible by the ordinary light microscope *micr*

subpopulation: either an identifiable fraction or subdivision of a population or specific biotype within a natural population *tax* >>> Table 12

subsexual reproduction: a partially asyndetic meiosis that leads to a restitution nucleus with limited crossing over, which leads to genetic variability *bot*

subsoil: the soil that lies beneath which is cultivated; it is usually hard and infertile *agr*

subspecies: used of a species forming a population or group of populations that are distinguishable from other members of the species and often partially reproductively isolated from them *tax* >>> race >>> Table 12 >>> *cf* List of Important Crop Plants

substitution line: a line in which one or more chromosomes are replaced by one or more chromosomes of a donor variety or species; it is used for genetic analysis and gene transfer *cyto*

substrate: a substance that is acted upon, as by an enzyme; also used for a culture medium *prep meth biot*

succession: the gradual supplanting of one community of plants by another, the sequence of communities being termed a "sere" and each stage "seral" *agr hort fore*

succession cropping: the method of growing new plants in the space left by the harvested ones *agr*

succession of crops >>> crop rotation

succulence: having watery or juicy tissue *bot*

succulent plant >>> succulence

sucker: a vegetative shoot of subterranean origin *bot* >>> root sucker

sucrose >>> saccharose

sucrose gradient centrifugation: used for separation of DNA, RNA, or proteins according to size and conformation *meth biot*

suffrutex (*syn* subshrub): low-growing woody shrub or perennial with woody base *bot*

sugar: any sweet, soluble, crystalline, lower-molecular-weight carbohydrate *phys chem* >>> saccharose

sugarbeet leaf curl: caused by the sugarbeet leaf curl virus *phyt*

sui generis: literally, "of its own kind"; an intellectual property right system on plant varieties *seed agr*

sulfonamide: a chemical having a bacteriostatic effect *chem phyt*

sulfur (S): a yellow mineral; an element that is needed by the plant; it is found covalently bound, especially in proteins, where it stabilizes their structures; deficiency leads to chlorosis and etiolation *chem phys;* it is also used as a dust or wet table powder for control of various diseases and insects, including powdery mildew, rust, and mites *phyt*

sulfuration: a treatment with sulfur, with fumes of burning sulfur, sulfur dioxide, or with sulfites as in fumigation, bleaching, or preserving; it is used in greenhouses in order to reduce and/or prevent leaf diseases (e.g., in cereals) *phyt meth*

supercoiled: natural conformation of DNA molecules *biot*

superdominance: interaction of the dominant and recessive alleles of a single gene locus *gene* >>> overdominance

super-elite: very high-grade seed used primarily for further seed production *seed*

superfemale: a female individual that carries higher doses of female determiners *gene*

supergene: a group of neighboring genes on a chromosome that tend to be inherited together and sometimes are functionally related; this chromosome segment is protected from crossing-over and so is transmitted intact from generation to generation, like a recon *gene*

superior: an ovary when the other organs of the flower are inserted below it *bot*

superior pelea >>> upper palea >>> pale

supermale: a male individual that carries higher doses of male determiners *gene*

supernumerary chromosomes: chromosomes present, often in varying numbers, in addition to the characteristic invariable complement of chromosomes *cyto*

supernumerary spikelet: a rudimentary or partially developed spikelet in wheat usually borne immediately below a normal spikelet at the same rachis node *bot*

suppressive soil: a soil in which certain diseases fail to develop because of the presence in the soil of microorganisms antagonistic to the pathogen *phyt*

suppressor mutation: a second mutation that masks the phenotypic effects of an earlier mutation; it occurs in a different site in the genome *gene*

suppressor-sensitive mutation: a mutation whose phenotype is suppressed in a genotype that also carries an intergenic suppressor of that mutation, e.g., amber, ochre, or opal *gene*

surface tension: tension exerted by a liquid surface due to molecular cohesion and apparent at liquid boundaries *phy*

surfactant >>> spreader

survival index: the degree of effectiveness of a given phenotype in promoting the ability of that organism to contribute offspring to the future population *stat*

survival of fittest: the corollary of DARWIN's theory of natural selection, namely that as a result of the elimination by natural selection of those individuals least adapted to the environment, those that ultimately remain are the fittest *evol*

susceptibility: the inability of a host plant to suppress or retard invasion by a pathogen or pest, or to withstand adverse environmental influences *phyt*

susceptible: being subject to infection or injury by a pathogen *phyt*

suspension culture: cells and groups of cells dispersed in an aerated, usually agitated, liquid culture medium *biot*

suspensor: the group or chain of cells produced from the zygote that pushes the developing proembryo toward the center of the ovule in contact with the nutrient supply *bot*

sustainability: the capability of entities to survive without human intervention *meth*

sustainable agriculture: a systematic approach to agriculture that focuses on ensuring the long-term productivity of human and natural resources for meeting food and industrial needs; it is an integrated system of plant and animal production practices having a site-specific application that will, over the long term (1) satisfy human food or industrial needs; (2) enhance environmental quality and the natural resource base upon which the agricultural economy depends; (3) make the most efficient use of nonrenewable resources and on-farm resources and integrate, where appropriate, natural biological cycles and controls; (4) sustain the economic viability of farm operations; and (5) enhance the quality of life for farmers and society as a whole *agr*

sustainable crops >>> sustainable agriculture

sustainable forestry: management of forested area in order to provide wood products in perpetuity, soil and watershed integrity, persistence of most native species, and maintenance of highly sensitive species or suitable conditions for continued evolution of species *fore* >>> sustainable agriculture

sustained yield: an output of renewable resources that does not impair the productivity of the resource; it implies a balance between harvesting and incremental growth or replenishment *agr*

sward: a stand of forage grasses or legumes *agr*

swath: a wind-row; a row of cut or pulled crop usually waiting for drying or curing before further harvesting *agr*

sweet potato whitefly: an insect pest of cotton, fruit, vegetable, and greenhouse crops *phyt*

symbiont: an organism living in a state of symbiosis *eco* >>> symbiosis >>> symbiosis

symbiosis: the living together for mutual benefit of two organisms belonging to different species *eco*

sympatric: of populations or species that inhabit, at least in part, the same geographic region, as opposed allopatric *eco* >>> parapatric

sympetalous >>> gamopetalous >>> gametogamy

symplastic coupling: a special type of cell-cell interaction in which two or more plant cells, connected by plasmodesmata, can exchange macromolecules such as introduced dyes, proteins, or RNA; cells interconnected in this way can form a network of cells that behave as a supracellular domain *phys*

symptom: a visible response of a host plant to a pathogenic organism *phyt*

synapsis: the side-by-side pairing of homologous chromosomes during the zygotene stage of meiotic prophase *cyto*

synaptene >>> zygotene

synaptonemal complex (SC): ribbonlike structures observed in electron micrographs of nuclei in the synaptic stages of meiosis; the ribbon represents a system that promotes synapsis of homologous chromosomes; the SC seems to be the essential prerequisite for homologous chromosome pairing and crossing-over *cyto*

syncarp: a structure consisting of several united fruits, usually fleshy *bot* >>> aggregate fruit

syncarpy: plants with flowers having two or more carpels, all fused together *bot*

synchronous culture: an in vitro culture in which a large proportion of the cells are in the same phase of the cell cycle at the same time *biot*

syndesis >>> synapsis

synergid (synergidae *pl*): one of two haploid cells that lie inside the embryo sac, beside the ovum; they nourish the ovum *bot* >>> Figures 25, 35

syngamy: sexual fusion of the sperm and egg cell *bot* >>> fertilization

syngraft >>> isograft

synkaryon: a nucleus resulting from the fusion of two genetically different nuclei, sometimes from two different species *bot biot*

synteny: chromosomal association of genes (linkage) established in somatic-cell culture *phys gene;* in biotechnology, the conservation of the gene order on the chromosomes; it allows understanding of the evolutionary relationships between species *biot evol*

synthetic (variety): a variety produced by crossing *inter se* a number of genotypes selected for good combining ability in all possible hybrid combinations, with subsequent maintenance of the variety by open pollination; usually the first generation of a synthetic variety is obtained by a polycross involving a certain number of components with a good general combining ability; the components are maintained by identical reproduction, either by vegetative propagation (clones) or by continued sib mating (inbred populations) *seed meth* >>> Table 5

synthetic amphiploid: an artificially produced amphiploid *cyto*

systematics: the science of describing, naming, and systematically classifying organisms *tax*

systemic: absorbed into the sap stream and passed to other parts of the plant *phys* >>> systemic pesticide

systemic pesticide: an agent that is distributed throughout the plant; it protects the entire host against pests for a certain time *phyt* >>> *disease protection*

T

T chromosome: a chromosome in which a terminal (T) region shows neocentric activity *cyto* >>> neocentric activity

T1 (generation): a term used in plant genetics; it refers to the progeny resulting from self-pollination of the primary transformant regenerated from tissue culture *gene*

T4 phage: a type of a bacteriophage used as a source of commonly used ligase, DNA polymerase, and polynucleotide kinase *biot*

tabular root: the main, downward-growing root of a plant, which grows deeply and produces lateral roots along its length *bot*

tail: single-strand DNA extension added by terminal deoxynucleotidyl transferase *biot*

tailings: partly threshed material that has passed through the coarse shakers or straw walkers and is eliminated at the rear of a threshing machine *seed*

take-all >>> saponin >>> take-all disease

take-all disease: a fungal disease *(Gaeumannomyces graminis, syn Ophiobolus graminis)* that attacks wheat plant roots; it causes dry rot and premature death of the plant; certain strains of *Brassica* plants and *Pseudomonas* bacteria act as natural antifungal agents against the fungus *phyt* >>> biological control

tandem duplication >>> tandem repeat

tandem repeat: a chromosomal mutation in which two identical chromosome segments lie adjacent to each other, with the same gene order; the DNA, which codes for the rRNA, contains many tandem repeats *gene*

tandem selection: in the case of successive multiple selection the selection concerns other traits in the first few generations than in later generations *meth*

tannic acid >>> tannin

tannin: a generic term for complex, nonnitrogenous compounds containing phenols, glycosides, or hydroxy acids, which occur widely in plants (e.g., in the testa of cocoa and beans) *chem phys*

Tapesia yallundae: sexual stage of *Pseudocercosporella herpotrichoides phyt* >>> white leaf spot >>> eyespot disease

tapetal cell >>> tapetum

tapetal layer >>> tapetum

tapetum: a layer of cells, rich in food, which surrounds the spore mother cells *bot*

tapping: driving spouts into the trunks of maple or pine trees to let the sap out for food and industrial use *fore*

taproot: a large, descending, central root *bot* >>> tabular root

Taq (DNA) polymerase: a DNA-dependent RNA polymerase from phage T7, which recognizes a very specific promoter sequence; it is used in many expression vectors *biot*

***Taraxum* type:** diplospory where the spore mother cell enters the meiotic prophase but because of synapsis there is no pairing and the univalents remain scattered over the whole spindle; the first meiosis results in a restitution nucleus; the second meiosis results in an unreduced dyad *bot*

target population: the target population is the entire group of individuals a breeder is interested in; the group about which he or she wishes to draw conclusions by several means *stat*

tassel: something resembling this, as at the top of a stalk of maize *bot*

tasseling (of maize): the time or process when maize emerges tassels *agr* >>> tassel >>> detasseling

TATA box: a canonic DNA sequence; part of a plant promoter; promotes the transcription of DNA *gene*

taxis: a change of direction of locomotion in a motile cell, made in response to certain types of external stimulus, such as temperature, light, nutrients etc. *bot*

taxon (taxa *pl*): a group of organisms of any taxonomic rank (i.e., family, genus, species, etc.) *tax* >>> Tables 12, 17

taxonomy: the scientific classification of organisms *tax bio* >>> Tables 1, 12, 14, 17

t-distribution >>> STUDENT's test

T-DNA >>> Ti plasmid

TE: Tris-EDTA buffer *prep chem*

teleutospore >>> teliospore

teliospore: a thick-walled resting spore produced by rust and smut fungi *bot*

telium (telia *pl*): pustule containing teliospores *bot*

telocentric (chromosome): the chromosomal centromere lies at the end *cyto* >>> Figures 11, 37

telochromomere: a chromomere that is terminally located *cyto* >>> chromomere

telochromosome >>> telosome

telomere: one of the two terminal chromomeres of a chromosome; telomeres certain nucleotide sequences, which promote the replication of the DNA double strand; a telomere also plays a role in the spatial orientation of chromosomes within the interphase nucleus *cyto*

telophase: the fourth and final phase of mitosis and the two divisions of meiosis, during which (1) the spindle disappears, (2) nucleoli reappear, (3) the nuclear membrane start to develop around the two groups of daughter chromosomes and/or chromatids, and (4) the chromosomes return to their extended state, in which they are no longer visible; the nuclei then enter a resting stage as they were before division occurred *cyto*

telosome: shorthand of telochromosome; a chromosome with a terminally located centromere *cyto* >>> Figures 11, 37

telosomic >>> telosome

telotrisomic: in allopolyploids, such as hexaploid wheat, a cell or individual with one missing chromosome but having a telocentric and an isochromosome for the same arm of the missing one *cyto* >>> Figures 11, 37

template: the DNA single strand, complementary to a nascent RNA or DNA strand, that serves to specify the nucleotide sequence of the nascent strand *gene*

T-end: a chromosome showing a terminal centromere *cyto*

tendril: part of a stem, leaf, or petiole that is modified as a delicate, commonly twisted, thread-like appendage; it is an aid to climbing (e.g., in pea) *bot*

tepal: one of the perianth members in those flowers where there is no distinction between calyx and corolla *bot*

teratology: the science of malformation *bio*

terminal association: in meiosis, the achiasmatic association of homologous chromosomes just by unspecific end-to-end attachments *cyto*

terminal centromere >>> telosome

terminalization: a progressive shift of chiasmata from their sites of origin to more terminal positions *cyto*

terminase: an enzyme of phage lamda, which generates the staggered cuts at the "cos" sites during packaging *biot* >>> lamda phage

termination: the incorporation of the final amino acid into a polypeptide chain and the release of the complete chain from the ribosomes during protein biosynthesis *gene*

termination codon >>> stop codon

terminator: a nucleotide sequence that acts as a signal for the termination of transcription *biot*

terminator seeds: a descriptive term used for seeds that have been genetically engineered to produce a crop whose first generation produces sterile seeds, thus preventing a second generation from being grown from seeds saved from the first; it might be a way to build patent protection directly into high-value, genetically engineered crop variety and thus recoup high research investment costs *biot*

terpene: a hydrocarbon that is composed of two or more isoprene units; they may be linear or cyclic molecules or combinations of both, and include important biological compounds, such as vitamins A, E, and K *chem phys* >>> growth inhibitor

tertiary gene-pool: gene transfer from a species of the tertiary gene pool to the cultivated species, since the primary gene pool usually requires special crossing and embryo rescue techniques in order to get viable hybrids *evol* >>> primary gene pool >>> secondary gene pool

tertiary tiller >>> secondary tiller

tertiary trisome: a chromosome present in addition to the normal diploid complement but as a result of a reciprocal interchange between two standard chromosomes; frequently it occurs in the progeny of a translocation heterozygote *cyto* >>> balanced tertiary trisomic >>> Figure 14

test cross: a cross between a heterozygote of unknown genotype and an individual homozygous for the recessive genes in question; in general, each cross that contributes to the solution of an experimental question by using more or less defined crossing partners *meth*

test mating >>> test cross

test of significance >>> significance test

test tube: a hollow cylinder of thin glass with one end closed; used to hold chemicals or specimens in laboratory experimentation and analysis *prep*

testa: the seed coat; it derives from secondary outgrowths of the nucellus, and later the ovule; they are called collars or integuments; the inner and outer integuments become the testa of the mature ovule; it is commonly composed of cuticle, palisade layer, sandwich cells, and parenchym cells *bot* >>> caryopsis >>> seed coat

tester (plant): plants of like kind and similar physiological condition used in experiments to measure performance or quality characters *meth* >>> Figure 30

tetraallelic: in tetraploids, when multiple alleles loci all have different (e.g., four alleles, *A1A2A3A4*) *gene*

tetracycline: an antibacterial antibiotic from *Streptomyces* spp. *chem phys* >>> achromycin

tetrad: four homologous chromatids in a bundle during the first meiotic prophase and metaphase *cyto;* in meiosis, the four haploid cells resulting from a single diploid cell during gametogenesis *bot*

tetrad analysis: the use of tetrads to study the behavior of chromosomes and genes in crossing-over during meiosis; particularly used in studies of fungi *gene*

tetraploid: having four sets of chromosomes in the nucleus *gene cyto*

tetraploidization: the mitotic or meiotic procedure in order produce tetraploids *meth cyto*

tetraploidy >>> tetraploid

tetrasome: a chromosome present four times *cyto* >>> tetrasomic

tetrasomic: having one or more chromosomes of a complement represented four times in each nucleus *cyto* >>> Figure 37

tetrasomy: the state of having one or more chromosomes as four copies *cyto* >>> Figure 37

tetrazolium test: a quick test to determine seed viability; tetrazolium is a class of chemicals that have the ability to accept hydrogen atoms from dehydrogenase enzymes during the respiration process in viable seeds; it is the basis of the test during which the tetrazolium chemical undergoes a color change, usually from colorless to red (formazan); the method was developed in Germany in the early 1940s by G. LAKON; the test is used throughout the world as a highly regarded method of estimating seed viability; it can be completed in only a few hours *seed*

TGW >>> thousand-grain weight

THC >>> hemp

theca: usually referring to the pollen sac in flowering plants or the capsule in bryophytes *bot* >>> Figure 35

theobromine: a mutagenically active purine analogue *chem phys*

thermoperiodism: in some plants (e.g., *Chrysanthemum,* tomato), the floral induction is accomplished by repeated exposure to low night temperatures, separated by periods of higher temperature *bot*

thermophil(ic) (plants): plants preferring moderate temperature and/or those that cannot cope with low temperatures or frosts *phys*

thermotaxis >>> taxis

thiamine: contributes to the formation of the important coenzyme thiamine pyrophosphate, which is involved in the oxidative decarboxylation of alpha-keto acids and transketolase reactions *phys*

thin: in a cultivated crop, to remove some plants in order to increase the area available to others *agr*

thinning >>> thin

thiol: any of a class of odiferous, sulfur-containing compounds *chem phys*

thorn: a hard, sharp outgrowth on a plant (e.g., a sharp-pointed aborted branch) *bot*

thousand-grain weight (TGW): equals 1,000-grain weight given in grams; it refers to a measure for seed weight, and thus indirectly for seed size; from a seed lot 1,000 seeds are randomly taken and get balanced in grams *meth*

3' end: describes the different, complementary ends of a DNA single strand; ends with an OH-group *biot gene*

three-parent cross >>> three-way cross

three-point cross: a series of crosses designed to determine the order of three, nonallelic, linked genes upon a single chromosome on the basis of their crossing-over behavior *gene* >>> three-point test cross

three-point test cross: cross involving one parent with three heterozygous gene pairs and another (tester) with three homozygous recessive gene pairs *gene meth* >>> three-point cross

three-way cross: a first generation hybrid between a single-cross and an inbred line or pure line variety *meth* >>> Figure 31

three-way hybrid: a hybrid between an inbred line and single cross hybrid *seed meth* >>> Figure 31

threonine (Thr): an aliphatic, polar alpha-amino acid *chem phys*

threshing: breaking the seeds free from the seedpods and other fibrous material or the separation of seed from chaff *agr*

threshing machine: a device that breaks the seeds free from the seedpods and other fibrous material or the separation of seed from chaff *agr*

threshold value: a critical value on an underlying scale of liability above which individuals manifest a trait or disease *stat*

thrips: any of several minute insects of the order *Thysanoptera* that have long, narrow wings fringed with hairs and that infest and feed on a wide variety of weeds and crop plants *zoo phyt*

throw-back >>> atavism

thylakoid: one of the membranaceous discs or sacs that form the principal subunit of a granum in chloroplasts *bot*

thymidine: a nucleoside with thymine as its base; it acts as an essential growth factor for microorganisms *chem phys*

thymine (T): the pyrimidine base that occurs in DNA *chem gene*

TI >>> transcript imaging

Ti plasmid: Ti = tumor inducing; established tumors in plants contain only a part of the total plasmid (about 10 percent); this part is called T-DNA; this part of the pathogen is integrated into the chromosomes of the host cell *gene*

tigella: a short stem *bot*

tillage: the preparation of soil for seeding; it includes manuring, ploughing, harrowing, and rolling land, or whatever is done to bring it to a proper state *agr*

tiller(s): shoots, some of which will eventually bear spikes, which arise from the base of the stem in the grasses *bot* >>> Figure 28

tillering >>> tiller

tillering node: a node on the base of the stem in grasses from which shoots arise *bot*

tilth: the condition of the soil after preparation for seeding *agr*

timberline *US* >>> tree limit

tissue: a group of cells with similar origin and structurally organized into a functional unit; the organs of multicellular organisms are made up of combinations of tissues, of one or more types of cells *bot*

tissue culture: the maintenance or growth of tissue in vitro in a way that may allow further differentiation, preservation, or regeneration *biot*

titer: the amount of a standard reagent necessary to produce a certain result in a titration *chem*

titrate >>> titration

titration: a method of determining the amount of some substance present in a solution by measuring the amount of a reagent, which must be added to cause a defined chemical change *chem*

tocopherol: any of several oils that constitute vitamin E *chem phys*

tolerance: the ability of a plant to endure attack by a pathogen without severe loss of yield *phyt*

tolerant: the ability of a host plant to develop and reproduce fairly efficiently while sustaining disease *phyt*

tonoplast: a membrane that borders the vacuole of a cell *bot*

topcross: a cross between a selection, line, clone, etc. and a common pollen parent, which may be a variety, inbred line, single cross, etc.; the common pollen parent is called the top cross or tester parent; in maize, a top cross is commonly called an inbred-variety cross; usually, it is used in order to test the "general combining ability" *meth* >>> Figure 19

topcross progeny: progeny from outcrossed seed of selections, clones, or lines to a common pollen parent *meth* >>> Figure 19 >>> topcross

topcross test >>> topcross

top-dressing: fertilization when crop plants are already developed, usually before flowering (i.e., an additional fertilization to the basic dressing before sowing or close behind); in horticulture, using material, such as compost or manure, that is applied to the surface around the plant to aid in drainage, decrease erosion, prevent moisture loss, and keep weeds down *agr hort*

topsoil: the fertile, upper part of the soil *agr*

torpedo stage: the stage of somatic and zygotic embryogenesis in which the embryo or mass of cells are torpedo-shaped *biot*

torus: the receptacle of a flower *bot*

total mean square >>> variance

totipotency: the potential ability of a cell to express all its genetic information under appropriate conditions and to proceed through all the stages of development to produce a fully differentiated adult *gene*

totipotent(ial) >>> totipotency

tough rachis: a nonbrittle rachis of a spike; for example, in wheat, the nonbrittleness is determined by two recessive genes on the short arms of chromosomes 3A and 3B; the consequence of threshing is spikelets, rather than grains *gene bot agr*

toxicity: the quality, relative degree, or specific degree of being toxic or poisonous *meth prep*

toxigenic >>> toxicity

trabant >>> satellite

trace element: an element required only in minute amounts by an organism for its normal growth *phys* >>> micronutrient

trachea: wood vessel *bot*

training: the operation of forming (young) tree plants to a wall or espalier, or causing them to grow in a desired shape *hort meth*

trait: a recognizable quality or attribute resulting from the interaction of a gene or group of genes with the environment *meth gene* >>> character

transaminase (aminotransferase): an enzyme that catalyzes a transamination reaction *chem phys*

transamination: the transfer of an amino group from an amino acid to a keto acid in a reaction catalyzed by a transaminase *chem*

transcript: the RNA product of a gene *gene*

transcript imaging (TI): a technique based on AFLP analysis applied to cDNA; it provides a quantified view of all the transcripts in a sample on an electronic image where approximately 20,000 AFLP bands, each representing a transcript, can be detected and quantitated; differences between strains, samples, or treatment can be detected and the differential bands can be sequenced; the sequence of the band leads to the identification of the transcript *meth biot*

transcription: the polymerization of ribonucleotides into a strand of RNA in a sequence complementary to that of a single strand of DNA; by this means the genetic information contained in the latter is faithfully matched in the former; the process is mediated by a DNA-dependent RNA polymerase *gene*

transcriptome: DNA sequences of the expressed genome *biot*

transduce >>> transduction

transduction: the transfer of bacterial genetic material from one bacterium to another via phage *gene*

transfectant: a genetic transformation of a cell by free (naked) DNA *gene*

transfection: introduction of pure (naked) phage DNA into a cell *biot*

transfer RNA (tRNA): a generic term for a group of small RNA molecules, each composed of 70-80 nucleotides arranged in a clover-leaf pattern stabilized by hydrogen bonding; they are responsible for binding amino acids and transferring these to the ribosomes during the synthesis of a polypeptide *gene*

transferase: an enzyme that catalyzes the transfer of a functional group from one substance to another *chem phys*

transformant: the cell or individual that was transformed during a transformation procedure *biot*

transformation: in general, the heritable modification of the properties of a plant; in biotechnology, the transfer of genetic information to a recipient strain of bacteria by DNA extracted from a donor strain, and recombination of that DNA with the DNA of the recipient *biot*

transformation efficiency: the number of bacterial and/or plant cells that uptake and express plasmid DNA and/or a foreign gene, respectively, divided by the mass of plasmid used *biot*

transgene: a gene introduced into a host genome by transfection or other similar means *biot*

transgenic >>> transgenic plants

transgenic genotype >>> transgenic plants

transgenic plants: a plant that contains an alien or modified DNA (gene) introduced by biotechnological means, and which is more or less stable inherited *biot*

transgression: segregants in a segregating population that fall outside the variation limit of parental lines *gene*

transgressiv(e) >>> transgenic plants

transgressive segregation: the segregation of individuals in the F2 or a later generation of a cross that shows a more extreme development of a character than either parent *gene*

transient expression: the temporary expression of a gene or genes shortly after the transformation of a host cell (i.e., cells in which the transgene has not been physically incorporated into the genome [stable transformation], but is carried as an episome that can be lost) *biot*

transition: a type of mutation that involves the replacement in DNA or RNA of one purine with another or of one pyrimidine with another *gene*

translation: the polymerization of amino acids into a polypeptide chain whose structure is determined genetically *gene*

translocation: a change in the arrangement of genetic material, altering the location of a chromosome segment; the most common forms of translocation are reciprocal, involving the exchange of chromosome segments between two nonhomologous chromosomes *cyto*

translocation tester set: a series of more or less defined homozygously reciprocal translocations; they may be utilized in test crosses to identify accessory or unknown chromosomes by homologous pairing in MI of meiosis (e.g., this sort of tester set was produced in diploid rye and barley) *meth cyto*

transmission: the spread of a disease agent among individual hosts *phyt*

transpiration: the loss of water vapor from a plant to the outside atmosphere; it takes place mainly through the stomata of leaves and the lenticels of stems *phys*

transplant: to relocate or remove to a new growing place *meth;* in biotechnology, the cultured tissue or explant, relocated or transferred to a new site in vitro *biot*

transplanting board: a simple device having regularly spaced slots for the individual plants in order to ensure proper spacing and lining out in the new bed *fore hort*

transposable element: a chromosomal locus that may be transposed from one spot to another within and among the chromosomes of the genome; it happens through breakage on either side of these loci and their subsequent insertion into a new position either on the same or a different chromosome *gene*

transposition: in molecular biology, the process of moving a transposon or other inserts from one position to another within a genome *cyto* >>> translocation

transposon: chromosomal loci capable of being transposed from one spot to another within and among the chromosomes of a complement *gene* >>> transposable element >>> transposition

transposon tagging: the blocking activity of functional genes by insertion of foreign DNA *biot*

transverison: a mutation in which a purine is replaced with a pyrimidine or vice versa *biot*

trap nursery: sets of plant genotypes are assembled that carry specific resistance to the pathogen in question; it is grown in different geographic locations; it provides information on pathogen populations and may provide resistant genotypes for local breeding *phyt*

trap plants >>> trap nursery

tree: a woody plant, which may grow >10 m tall *bot*

tree limit: the altitude above sea level at which timber ceases to grow *eco*

tree line >>> tree limit

tree ring >>> annual ring

trench planting: setting out young trees in a shallow trench or a continuous slit *meth fore hort*

triazine: any of a group of three compounds containing three nitrogen and three carbon atoms arranged in a six-membered ring and having the formula $C_3H_3N_3$; some of these compounds are used as herbicides *chem phys phyt*

tribe: a rank between family and genus, comprising genera whose shared features serve to distinguish them from other genera within the family *tax* >>> Table 12

trichome: a hairy outgrowth on a plant's surface, as a prickle *bot*

trigeneric hybrid: a spontaneous or experimental hybrid consisting of three genomes of different genera *bot cyto gene*

trihybrid: progeny resulting from a cross of parents differing in three genes *gene*

triisosomic: in allopolyploids, such as hexaploid wheat, when a cell or individual lacks one chromosome pair while three homologous isosomes for the same arm are present *cyto* >>> Figure 37

trioecious: species having male, female, and hermaphroditic flowers on different individuals *bot*

triplet: a unit of three successive bases in DNA and RNA, which code for a specific amino acid *gene*

triplex (type) >>> autotetraploid >>> nulliplex type

triploid: applied to a cell or individual with three sets of chromosomes in its nucleus *cyto*

triploidy: a state in which three chromosome sets are present *cyto*

tripping mechanism: a pollen dispersal mechanism of some legumes, in which the staminal column is sprung free of the keel and exposed; it can also be initiated by hand in order to imitate insect activity and to stimulate self-pollination, for example, in broad bean *(Vicia faba) bot*

triradial (chromosome configuration): a chromosomal pairing configuration of three in which the homologous chromosomes are arranged like star *cyto* >>> Figure 15

trisome >>> trisomic

trisomic: a genome that is diploid but that contains an extra chromosome, homologous with one of the existing pairs, so that one kind of chromosome is present in triplicate *cyto* >>> Figures 14, 15, 37

trisomic analysis: a method for mapping gene loci on individual chromosomes by comparing disomic and trisomic segregation patterns of a series of individuals *gene* >>> Figure 37 >>> Table 4

trisomic series: a complete set of trisomics of a given plant species, in which all different chromosomes of the complement are available as trisomics in appropriate individuals (e.g., in barley, rye, tomato, maize, etc.) *cyto* >>> Figures 14, 15, 37

tristyly: three lengths of the style relative to the anthers (short, medium, long) *bot*

trivalent: an association of three homologous chromosomes in meiosis *cyto* >>> Figures 14, 15

tRNA >>> transfer RNA

trophic >>> tropism

tropism: a directional response by a plant to a stimulus *bot*

truck crop: *US* vegetables and fruits grown in large quantities *hort*

true breeding: a situation in which a group of identical individuals always produce offspring of the same phenotype when intercrossed *meth gene*

truncated: appearing as though abruptly cut across toward the apex *bot*

truncation selection: a breeding method in which individuals in whom quantitative expression of a phenotype is above or below a certain value (i.e., truncation point) are selected as parents for the next generation *meth*

trunk >>> stem

trunkless >>> boleless

truthfully labeled seed: seed with label of the producer with information on the seed quality *seed*

tryphine >>> entomophilous

trypsin: an enzyme of the pancreatic juice, capable of converting proteins into peptone *chem phys*

tryptophan(e) (Trp): a heterocyclic, nonpolar, alpha-amino acid *chem phys*

tube nucleus >>> vegetative nucleus

tube planting: setting out young plants in narrow, open-ended cylinders of various materials *meth hort fore*

tuber: a swollen stem or root that functions as an underground storage organ *bot*

tuber blight >>> late blight

tuberous >>> bulbous

tuberous roots: they look like tubers, but are actually swollen, nutrient-storing root tissue (e.g., in dahlias); during the growing season, they put out fibrous roots to take up moisture and nutrients; new growth buds, or eyes, form at the base of the stem; this area is called the crown *bot*

tubiform floret: a small flower in a flower head or other cluster showing a tubelike shape *bot*

tubular floret >>> tubiform floret

tuft: many stems in a close cluster at ground level; not spreading (e.g., as in some grasses) *bot*

tunic: a loose, outer covering or skin surrounding some corms and bulbs (e.g., in onion and tulip) *bot*

turf: the surface of grassy land, consisting of soil or mold filled with the roots of grass and other plants *agr*

turgescent >>> turgid

turgid: the crisp, fresh condition found when the cells of the plant are amply supplied with water to the extent that they are fully extended, as opposed to wilted *bot*

turgor: the rigidity of a plant and its cells and organs resulting from hydrostatic pressure exerted on the cell walls *phys*

turgor pressure >>> turgor

twig: a shoot or small branch of a tree or shrub *bot*

twin: a pair of individuals produced at one birth *gene*

twin seedling: a common feature of plants; frequently, one of the seedlings is diploid, while the other is haploid via apomictic development (e.g., in asparagus, rye, etc.); in the past, twin seedlings were used for haploid selection *bot meth* >>> haploid

two-rowed >>> barley

tyrosinase: an enzyme that converts tyrosine to dopa and oxidizes this to dopa quinone *chem phys*

tyrosine (Tyr): an aromatic, polar alpha-amino acid *chem phys*

U

UAG stop codon >>> amber codon

ultracentrifugation: centrifugation carried out at high rotor speeds (<100,000 rpm) and therefore under high centrifugal forces (<750,000 g) *meth*

ultracentrifuge: a set-up for high-speed centrifugation between 65,000-100,000 rounds per minute *prep*

umbel: an inflorescence in which all the pedicels arise at the apex of an axis *bot*

umbellate >>> umbelliferous plants

umbelliferous plants: tap-rooted plants with minute flowers aggregated into flat or umbrella-shaped heads (e.g., carrot, parsnip, celery, dill, parsley, etc.) *bot*

unavailable nutrients: plant nutrients that are present in the soil but cannot be taken up by the plant roots because they have not been released from the rock by weathering or from organic matter *agr*

unavailable water: water that is present in the soil but cannot be taken up by the roots because it is strongly adsorbed on to the surface of particles *agr*

unbalanced diallelic: a genotype involving a multiple allelic locus in autotetraploids where two alleles are represented an unequal number of times *gene*

unbalanced experimental design: an experiment or set of data in which all treatments or treatment combinations are not equally represented; a common cause of unbalanced experiments is unequal mortality among entries in a test *meth* >>> randomized-block design

unbalanced translocation: a type of chromosome translocation in which a loss of chromosomal segments results in a deleterious genetic effect *cyto*

uncomplete block design: an experimental design that is preferable where large numbers of cultivars are compared in a single yield trial; the entries in each replication are subdivided into smaller blocks, in a manner designed to reduce the error caused by soil variation; usually, it refers to a lattice design, considering the restriction that (1) the number of testers must be harvested, (2) inferior strains cannot be discarded prior to harvest to reduce harvest expenses, and (3) researchers still must analyze the experiment as a lattice design *meth*

unconscious selection: indirect selection by breeders (sometimes called "parallel selection") when in addition to a target characteristic another trait is taken to the next generation; it can be a morphological, biochemical, or physiological trait genetically linked to the target characteristic *meth*

underplant crop: in horticulture, adding one or more complementary, low-growing plants beneath and around taller plants *hort;* in agriculture, main crops can be underplanted with vegetables; it helps these plants to have some shade in the heat of the summer and one can often get a crop longer than otherwise; wide ranges of combinations between main and underplant crops are possible and are applied for several purposes *agr*

underreplication: certain heterochromatic chromosome regions and ribosomal DNA that show a slower replication as compared to the remaining genomic DNA *gene*

understock: the bottom or supporting part of a graft composed of either root or stem tissue or both *hort* >>> rootstock >>> graft

unequal crossing-over: a crossing-over after improper pairing between chromosome homologues that are not perfectly aligned; the result is, for example, one crossing-over chromatid with one copy of the segment and another with three copies *cyto* >>> Figure 24

unifactorial >>> monogenic

uniflorous: showing one flower only *bot*

uniform crossover: in genetic algorithms, a breeding technique in which it is randomly decided for each element of a breeding pair of individuals whether they should be switched *stat*

uniformity: describes the state of a population or group in which all the individuals are genetically identical; it is a typical feature of clonal varieties (e.g., potato); in general, lack of diversity within and between plant species is apparent in modern cultivars *gene seed*

unilateral: the type of panicle (e.g., in oats) where the branches are all turned to one side like a pennant *bot*

unilateral inheritance: inheritance that is associated with linkage in sex chromosomes *gene*

Union for Protection of New Varieties of Plants: Union pour la Protection des Obtentions Vegetales (UPOV); it is an intergovernmental organization with headquarters in Geneva, Switzerland; it is based on the International Convention for the Protection of New Varieties of Plants, as revised since its signature in Paris on December 2, 1961; on April 16, 1993, the Union consisted of 23 member States; the objective of the convention is the protection of new varieties of plants by an intellectual property right

unipolar (spindle): a cell spindle with only one pole *cyto*

unisexual: a flower that possesses either stamens or carpels but not both (i.e., a plant possessing only male or female flowers) *bot*

United States Department of Agriculture (USDA): the U.S. agency responsible for regulation of biotechnology products in plants and animals; the major laws under which the agency has regulatory powers include the Federal Plant Pest Act (PPA), the Federal Seed Act, and the Plant Variety Act (PVA); in addition, the Science and Education (S and E) division has nonregulatory oversight of research activities that the agency funds *agr*

univalent: a single chromosome observed during meiosis when bivalents are also present; it has no pairing mate *cyto* >>> Figure 15

univalent shift: a spontaneous change in monosomy from one chromosome to another; it is caused by partial asynapsis or desynapsis during meiosis *cyto*

univalent switch >>> univalent shift

universe >>> population *stat*

unloader: a device on a combine to transport the grains from a bunker to a transporter *agr*

unreduced gametes: gametes not resulting from common meiosis, and so showing the number of chromosomes per cell that is characteristic of a sporophyte; they spontaneously arise as a consequence of irregular division in anaphase I of meiosis; they may contribute to spontaneous (meiotic) polyploidization; in rye, they were used for production of tetraploids via valence crosses *cyto*

unsaturated fatty acid: a fatty acid that has a double bond between the carbon atoms at one or more places in the carbon chain; hydrogen can be added at the site of the double bond *chem phys*

unspecific resistance >>> horizontal resistance

upgrading: the reprocessing of a seed lot to remove low quality seeds or other materials; the remaining seeds are of higher quality than the original *seed meth*

UPOV >>> Union for Protection of New Varieties of Plants

upper palea: upper glume >>> pale

upstream: a term used for description of the position of a DNA sequence within a DNA or protein molecule; it means that the position of the sequence lies away from the direction of the synthesis of a DNA or protein molecule *gene*

uracil (U): a pyrimidine base that occurs in RNA *chem gene*

urea: a compound, $CO(NH_2)_2$, occurring in urine and other body fluids as a product of protein metabolism; an important plant fertilizer *chem agr*

uredospore: a sexual spore of the rust fungi *bot*

USDA >>> United States Department of Agriculture

USDA-ARS >>> United States Department of Agriculture, Agricultural Research Service

V

vacilin >>> globulin

vacuole: a transparent vesicle that is usually large and singular in mature cells but small in some meristematic cells; it is filled with a dilute solution that is isotonic with the cytoplasm *bot*

valence cross: crossing of individuals of different ploidy level; a method, for example, used in production of tetraploid rye varieties by crossing a tetraploid genotype as female and a diploid as male, respectively; tetraploid F1 seeds could be selected by green-grained xenia on a pale-grained mother plant, caused by fusion of reduced gametes of the mother plant (carrying a recessive allele for pale seed color) with unreduced gametes of the male parent (carrying the dominant allele for green seed color) *meth*

valine (Val): an aliphatic, nonpolar amino acid *chem phys*

value-added grains >>> nutrient-enhanced varieties

value-enhanced grains >>> nutrient-enhanced varieties

variability: the sum of different genetic or phenotypic characters within different taxa *gene*

variance: when all values in a population are expressed as plus and minus deviations from the population mean, the variance is the mean of the squared deviations; it is a measure of variation of a population; it can be divided into phenotypic variance, genotypic variance, and environmental variance *stat*

variate: a single observation or measurement *stat*

variation: differences in form or function between individuals or other taxa *gene*

variegate >>> variegation

variegation: the phenomenon in some plants in which patches of two or more different colors occur on the leaves or flowers; it may be an inherited characteristic or may be due to virus infection *bot*

varietal hybrid: the product resulting from the mating of two varieties *meth*

varietal protection >>> protected variety

variety: a plant differing from the other member of the species to which it belongs by the possession of some hereditary traits; breeding varieties can be classified according to the manner of propagation, such as clone varieties (maintained by vegetative propagation), line varieties (maintained by self-fertilization), panmictic varieties (propagated by cross-fertilization) or hybrid varieties (produced by directed crosses) *meth*

variety listing: it refers to a list of new varieties recommended for agriculture and horticulture; after official performance tests, only those varieties that meet the standards are accepted for a national or international variety list; important for variety listing are the testing for value and use as well as distinctness, homogeneity, and stability of the candidates *seed agr*

variety mixture: a composite population made up of a random mixture of different varieties; it may exhibit considerable phenotypic variation in one or more characters, but which have one or more desirable agronomic traits in common *agr* >>> blend

vascular: furnished with vessels or ducts *bot*

vascular bundle: a discrete, longitudinal strand that consists principally of vascular tissue *bot*

vector: an organism capable of transmitting inoculum *phyt;* in genetics, a vehicle, such as a plasmid or virus, for carrying recombinant DNA into a living cell *gene*

vegetation: a collection of plants of diverse or the same species *eco*

vegetational analysis: any of various methods of studying small (sample) areas of constituent plants, often counting the numbers of plants of communities to make extrapolations to a larger area (e.g., estimating the yield) *meth agr*

vegetative: applied to a stage or structure that is concerned with feeding and growth rather than with sexual reproduction *bot gene*

vegetative cone: in flowering plants, the vegetative cone consists of three different layers; they are called dermatogen (L1), subdermatogen (L2), and corpus (L3); from the dermatogen the epidermis is formed, from the subdermatogen the mesophyll and the gametes, and from the corpus the vascular bundle, the flesh, the pith, and the adventitious roots; the cell division of the dermatogen and the subdermatogen is usually anticline in other words, vertical to the layer (in this way the layers show surface expansion); the cell division of the corpus is usually periclinal (i.e., the cells divide parallel to the layers) *bot*

vegetative meristem: gives rise to parts, such as stems, leaves, roots, etc. *bot*

vegetative nucleus: the tube-nucleus of a pollen grain in a flowering plant *bot* >>> Figures 25, 35

vegetative propagation: a reproductive process that is asexual and so does not involve a recombination of genetic material (e.g., cloning of potato) *meth* >>> Figure 28

vegetative reproduction >>> vegetative propagation

vegetative stage >>> whorl stage

vein >>> veinure

veinure: threads of fibrovascular tissue in a leaf or other organ, especially those that branch *bot*

venation >>> nervature

venom >>> toxicity

venomous: toxic *phys*

ventral: the inner side, furrowed in the grain of, for example, wheat, barley and in the caryopsis of oats *bot*

ventral furrow: the groove running along the length of the ventral side of the caryopsis *bot*

vermiculite: a porous form of mica, a mineral that makes good rooting media for seed germination because of its capacity to retain moisture and permit aeration *meth*

vernalin: a hypothetical hormonelike substance found in plant meristematic regions, produced by vernalization; this substance is apparently graft transmissible, but has not yet been identified; different cold-requiring species may form different substances during vernalization *phys*

vernalization: the treatment of germinating seeds with low temperatures to induce flowering at a particular preferred time; in the genetic model plant, *Arabidopsis thaliana,* a gene *Flc* suppresses the formation of flowers during cold periods; another gene *Vrn2* triggers that suppression; however, in spring the suppression is raised in a way that the gene "remembers" the previous cold period, because the gene *Vrn2* itself is cold-insensitive; there is a similar gene in *Drosophila melanogaster,* which also serves like "chemical memory" *meth phys*

vernation: the arrangement of bud scales or young leaves in a shoot bud *bot*

vertical gene transfer >>> outcrossing

vertical resistance: the existence of differential levels of resistance to different races of a given pathogen conditioned by one or a few qualitative genes *phyt* >>> resistance

verticilate: whorled *bot*

viability: the probability that a fertilized egg will survive and develop into an adult organism; the term is also often applied to plant germination experiments with comparisons across phenotypic classes under standard specific environmental conditions *phys seed*

viable: capable of germinating, living, growing, or sufficiently developed physically as to be capable of living *phys seed*

vibrator separator: a machine utilizing a vibrating deck for separating seeds on the basis of their shape and differing surface textures *seed*

vicilin: a protein common in broad bean *phys*

video microscopy: microscopy that takes advantage of video as an imaging, image-processing, or controlling device *micr*

vignin >>> globulin

vine: climbing plants with woody or herbaceous stems that climb, twist, adhere, or scramble over other taller objects; they can climb by tendrils, aerial roots, twining stems, twining leafstalks, adhesive disks, or hooks *bot*

vine-growing >>> viticulture

vine-louse: a plant louse *(Phylloxera vitifoliae)* that injures the grapevine *phyt*

viniculture >>> viticulture

virescence: greening of tissue that is normally devoid of chlorophyll (e.g., the abnormal development of flowers in which all organs are green and partly or wholly transformed into structures like small leaves) *phyt*

virion: an individual virus particle *gene*

viroid: a piece of infectious nucleic acid; in plant pathology, any of a class of plant pathogenic agents consisting of an infectious, single-stranded, free RNA molecule *gene*

virulence: the relative ability of a microorganism to overcome the resistance of a host *phyt* >>> aggressiveness

virulent >>> virulence

viruliferous aphids: aphids acting as vector for plant viruses *phyt zoo*

virus: a type of noncellular "organism," that has no metabolism of its own; a nucleoprotein entity that can replicate within living cells; it passes through bacterium-retaining filters *bot gene*

virus-free plant: a plant that shows no sign of viral particles or symptoms *phyt*

viscid: sticky *bot*

vital coloring test >>> vital staining

vital dye >>> vital staining

vital staining: a stain that is capable of entering and staining a living cell without causing an injury *meth cyto seed* >>> fluorescin diacetate (FDA) staining

vitamin: an organic compound produced by plants (often functions as a coenzyme) that is required in relatively small amounts in the diet for the normal growth of animal organisms *phys*

vitamin B_1 >>> thiamine

vitamin H >>> biotin(e)

viticulture: the science, culture, or cultivation of grapes and grapevines *hort*

vitreous grain: characterizing slightly translucent kernels *agr*

viviparity >>> viviparous

viviparous: applied to a plant whose seeds germinate within and obtain nourishment from the fruit; it refers also to a plant that reproduces vegetatively from shoots rather than an inflorescence *bot*

vivipary >>> viviparity

vivisection: the action of cutting into or dissecting a living body *meth*

vivotoxin >>> pathotoxin

volunteer >>> self-sown cereals >>> volunteer plants

volunteer plants: plants that have resulted from natural propagation, as opposed to having been deliberately planted by humans *seed*

v/v: may indicate simple proportion (e.g., 3:1 v/v); may indicate percent volume in volume *meth*

vybrid: the first- and subsequent-generation progenies of crosses of heterozygous facultative apomicts *bot*

W

W chromosome: a sex chromosome that is limited to the female sex *cyto*

water sprout: a shoot arising from a bud located on wood that is not older than one year *hort*

water-absorbing capacity: in breadmaking, a high capacity to absorb water is required; this is associated with hard milling texture, high protein content, and the degree of starch damage during the milling process *prep*

waterlogged: soil saturated with water *agr*

wax coating: a thin layer covering the stem, leaves, flowers, and fruits of most plants; waxes are manufactured as oily droplets in epidermal cells, from which they migrate to the outer surface of the plant via tiny canalculi in cell walls, and crystallize as rods and platelets; their pattern of deposition is sometimes used as a micromorphological character below the genus level; the wax coating reduces the water transpiration of the plant and is involved in water balance and resistance mechanisms against diseases *bot*

waxiness: the phenomenon of whitish, powdery, or waxy covering of plant leaves, stems, or flowers *bot*

waxy hull-less barley (WHB): a barley mutant that is rich in soluble fiber and low in fat content; these characteristics make it a nutritionally valuable ingredient for food products; it has been shown that the soluble fiber, beta-glucan, reduces cholesterol and lowers blood glucose and insulin response following a meal *gene*

waxy maize: maize that produces kernels in which the starch that is contained within those kernels is at least 99 percent amylopectin, versus the average of 72-76 percent amylopectin in common starch *seed gene*

"weak" flour >>> "strong" flour

weathering: all the physical, chemical, and biological processes that cause the disintegration of rocks at or near the surface *agr*

weed: a plant that occurs opportunistically on land that has been disturbed by human activity or under cultivated land where it competes for nutrients, water, sunlight, or other resources with cultivated plants *phyt*

weed killer >>> herbicide

weediness: unwanted effects of a plant *agr*

weeding: remove weeds from the crop stand *agr*

Western blot >>> Western blotting

Western blotting: a technique similar to SOUTHERN blotting but for the analysis of proteins instead of DNA *biot meth*

wet milling: process in which feed material is steeped in water, with or without sulfur dioxide, to soften the grains in order to help separate the kernel's various components *meth agr*

white leaf disease (of oats): caused by copper deficiency, often on peaty soils; can be compensated by application of copper sulfate (2 kg per 400 l water/ha) to leaves *phyt*

white leaf spot (of rape, *Pseudocercosporella capsellae*): characterized by large white spots with dark margins and gray-black centers in older lesions on leaves; pods can be similarly infected *phyt* >>> *Tapesia yallundae*

whitefly: insects whose adults resemble tiny moths but are related to aphids; nymphs that suck sap and damage leaves of several Brassicaceae and other horticultural plants *phyt*

whorl: an arrangement of leaves, etc., in a circle around a stem >>> verticilate

whorl stage: the developmental stage of a grass plant prior to the emergence of the inflorescence *phys*

wide cross >>> wide hybridization

wide hybridization *syn* wide cross: cross combinations between taxonomically remote species or genera *meth*

wide hybrids >>> wide hybridization

wide row planting: the method of wide-spaced sowing seeds in multiple rows (e.g., for better selection of young breeding material); or several rows with wide parallel channels for irrigation *meth agr*

wild type: the most frequently observed phenotype, or the one arbitrarily designated as "normal" *gene* >>> Table 35

wilt: a type of disease in which wilting is a principal symptom *phyt*

wilting point: the percentage of water remaining in the soil when the plants wilt permanently *phys*

wind pollination: pollination by wind-borne pollen *bot* >>> allogamy

wind-row: a loose, continuous row of cut or uprooted plants placed on the surface of the ground for drying to facilitate harvest *agr*

wing(s): the two expanded parts of the glume in, for example, wheat, which lie on each side of the keel; in general, a membranous or thin and dry expansion or appendage of a seed or fruit *bot*

winged fruit >>> wing(s)

winnower: a simple device for seed cleaning from weeds and chaff using an air flow *seed*

winnowing mill >>> winnower

winter annuals: plants from autumn-sown seed that bloom and fruit in the following spring, then die *phys* >>> winter-type

winter killing >>> killing frost

winter spore >>> teleutospore >>> resting spore

winter-and-spring wheat: facultative growth habit *bot*

winter-type (of growth habit): plants germinating in autumn, requiring vernalization during the wintertime for flower induction during the following year *bot* >>> vernalization

witch's broom: massed outgrowth of branches of woody plants caused by fungi (e.g., by rusts) *phyt*

withertip: death of the leaf beginning at the tip, usually in young leaves *phys*

working collection: a collection of germplasm kept under short-term storage conditions, commonly used by breeders or researchers *meth*

world collection of crop plants: a global collection of samples of a species or genera; it is a coordinated activity of several countries and institutions under the IPGRI *meth*

w/v: weight in volume, as the number of grams of constituent in 100 ml solution *meth* >>> v/v

X

x: designates the basic number of chromosome sets *cyto* >>> basic number

X chromosome: a sex chromosome found in a double dose in the homogametic sex and in a single dose in the heterogametic sex *cyto*

X rays: electromagnetic radiation having wavelengths in the range of approximately 0.1-10 nm, between ultraviolet radiation and gamma-rays, and capable of penetrating solids *meth*

xanthophyll: yellowish-brownish (oxygen-containing) carotinoids occurring in the chloroplasts (e.g., the lutein of leaves) *bot*

xenia: a situation in which the genotype of the pollen influences the developing embryo of the maternal tissue (endosperm) of the fruit to produce an observable effect on the seed *bot*

xenogamy (cross-pollination): intercrossing between flowers of different individuals, as opposed to geitonogamy *bot*

xenotransplant: the implantation of an organ or limb from one species to another organism in a different species *hort* >>> heterograft

xeric: dry *bot*

xerograft >>> heterograft

xerophyte: a drought-resistant plant or plants that grow in extremely dry areas *bot*

xerophytic: growing on dry conditions *bot*

Xta >>> chiasma

xylan: a polysaccharide of xylose and a component of hemicellulose *phys*

xylem: a plant tissue consisting of various types of cells that transports water and dissolved substances toward the leaves *bot*

xylene >>> xylol

xylol: a liquid solvent *prep* >>> refraction index

xylose: an aldopentose sugar that is commonly found in plants and especially in woody tissue *chem phys*

Y

Y chromosome: plays a role in sex determination *cyto*

YAC >>> yeast artificial chromosomes

YAC clone >>> yeast artificial chromosomes

yearling: a one-year-old seedling and/or plantlet *agr fore hort*

yeast: a general term for a fungus that can exist in the form of single cells, reproducing by fission or by budding *bot*

yeast artificial chromosomes (YAC): a yeast artificial chromosome is used to clone very large DNA fragments in yeast; although most bacterial vectors cannot carry DNA pieces that are larger than 50 base pairs, YACs can typically carry DNA pieces that are as large as several hundred base pair *gene biot*

yellows: a plant disease characterized by yellowing and stunting of the host plant *phyt*

yield: commonly, the aggregate of products resulting from the growth or cultivation of a crop and usually expressed in quantity per area *agr* >>> Table 33

yield appraisal >>> vegetational analysis

yield monitoring: collecting data on the amount of production at regular intervals and by certain means (e.g., GPS readings); the resulting yield map is basic to decisions about fertilization, pest control, and other adjustments in a system of precision farming *agr*

yield structure: an analysis used to determine the numerous morphological and physiological components of a plant contributing to the final yield (given in different measures) *agr* >>> vegetational analysis >>> Table 33

yield trial: a nursery or experimental design in order to determine the yield capacity of a crop or yield components *meth* >>> Table 33

Z

Z chromosome: a sex chromosome that is limited to the male sex *cyto*

ZADE method >>> long-plot design

zDNA: a left-handed, alternate form of double-stranded DNA in which the backbone phosphates zigzag *gene*

zeatin: a mitogen isolated from maize kernels *chem phys* >>> Table 15

zein: a storage protein of maize found in the endosperm *chem phys* >>> Table 15

ZELENY test: a test to measure the protein quality; the grain is milled to form a white flour and mixed with a suspension agent; the resulting suspension volume is then measured in millimeters; for example, wheats with a ZELENY volume between 20 and 30 are acceptable (<19 low; 25 medium; 35 high; 45 very high, >50 extremely high) *meth*

zinc (Zn): a blue-white metallic element that is a trace element (essential element) that is required by plants; it is found in various enzymes; it functions as the prosthetic group of a number of enzymes; deficiency prevents expansion of leaves and internodes, giving a rosette style of plant *chem phys*

zinc finger (protein): a protein motif involved in the recognition of DNA; the structure contains a complex zinc ion and consists of an antiparallel protein-chain hairpin and loop followed by a helix *gene*

zone of hybridization: a geographical area in which different plant taxa have the same habitat allowing spontaneous intercrossing if some sexual and reproductive prerequisites are given (e.g., in the evolution of hexaploid wheat such zones of hybridization played in important role), for example, *Aegilops* species intercrossed with primitive wheats in Asia Minor *eco*

zoom: to control, by magnifying or reducing, the size of an image, either optically or electronically *micr*

zoospore: temporarily mobile sex organs of some lower plants (e.g., in *Vaucheria sessilis*) *bot*

Z-type: in sugarbeet breeding varieties with high sugar content (Z = Zucker = sugar) but normal yielding capacity *agr*

zygomere: hypothetical pairing sites along the chromosomes; for example, it is suggested that in barley the zygomeres are located proximal, as opposed to rye by which the zygomeres are located terminal *cyto*

zygomorphic: the condition of having only one plan of symmetry (e.g., orchid flowers) *bot*

zygomorphic corolla: dorsiventral carolla; a flower composed by one symmetric axis (e.g., *Antirrhinum majus*) *bot*

zygonema >>> zygotene

zygote: the fertilized ovum formed from the fusion of male and female gametes; a diploid zygote is formed by the union of a sperm nucleus of the pollen with the egg cell of the embryo sac; the zygote divides mitotically to form the embryo of the seed; union of a sperm nucleus with the two polar nuclei of the embryo sac results in a triploid ($3n$) cell that divides mitotically to form the endosperm of the seed *bot* >>> Figure 25

zygotene: the stage of first meiosis when the homologous chromosomes are associating side by side (synapse); for example, a diploid crop species with $2n = 20$ chromosomes would form ten pairs of homologous chromosomes *cyto*

zygotic >>> zygote

zygotic embryo: an embryo that derives from fusion of male and female gametes *bot*

zymogens: the enzymatically inactive precursors of certain proteolytic enzymes; the enzymes are inactive because they contain an extra piece of peptide chain; when this peptide is hydrolyzed by another proteolytic enzyme the zymogen is converted into the normal, active enzyme *phys*

zymogram: diagrammatic presentation of enzymes separated by means of electrophoresis *meth*

LIST OF IMPORTANT CROP PLANTS AND OTHER PLANTS OF THE WORLD

A

abaca: *Musa textilis syn Cannabis gigantea* (Moraceae), $2n = 2x = 20$ *agr*

abaca banana >>> abaca

abaca-fibre >>> abaca

acha >>> hungry rice

acidanthera: *Acidanthera bicolor* var. *murielae* (Iridaceae) *hort*

adlay: *Coix lacryma-jobi* (Gramineae), $2n = 2x = 20$ *hort agr*

adley >>> adlay

Adzuki bean: *Phaseolus angularis syn Vigna aconitifolia* (Leguminosae), $2n = 2x = 22$ *hort* >>> pulse

***Aegilops*:** a genus of grasses including several species of breeding and genetic interest; some of them contributed to the evolution of bread wheat; they are also used as donors of resistances and other genes for the cropped wheat, *Aegilops* spp. (Gramineae) >>> Figure 10

African millet >>> finger millet

African rice: *Oryza glaberrima* (Gramineae), $2n = 2x$, A'A' = 24 *agr*

African walnut: *Tetracarpidium conophorum* (Euphorbiaceae) *hort*

agave: a semiwoody perennial native of the American continents; used for the production pulque, aquamiel, mescal, and tequila, *Agave americana* et spp. (Agavaceae) *agr*

Agropyron: *Agropyron* spp. (Gramineae)

akee (apple): *Blighia sapida* (Sapindaceae) *hort*

Alexandrinian clover: a clover of the multicut group, *Trifolium alexandrinum* (Leguminosae), $2n = 2x = 16$ *agr*

alfalfa: an autotetraploid species; because of its good in vitro culture ability it early became a subject of biotechnological approaches; *Medicago sativa* (Leguminosae), $2n = 2x$, SS, $4x$, SSSS = 16, 32 *agr*

alkanet root >>> alkanna

alkanna: *Alkanna tuberculata syn A. tinctoria* (Boraginaceae), $2n = 2x = 14$ *hort*

alligator pear >>> avocado

almond: *Amygdalus communis syn Prunus amygdalus* (Rosaceae), $2n = 2x = 16$ *hort*

alsike (clover): *Trifolium hybridum* (Leguminosae), $2n = 2x = 16$ *agr*

amaranth, grain: *Amaranthus A. caudatus* or *hypochondriacus syn A. frumentaçeus syn A. leucocarpus* (Amaranthaceae), $2n = 2x = 32$ *agr* >>> dye amaranth

amaranth, salad: *Amaranthus lividus* (Amaranthaceae) *hort*

amarelle: *Cerasus vulgaris* var. *capроniana* (Rosaceae) *hort*

American chestnut: *Castanea dentata* (Fagaceae), $2n = 2x = 24$ *hort*

American hazel: *Corylus americana* (Corylaceae), $2n = 2x = 22$ *hort*

American plum: *Prunus americana* (Rosaceae), $2n = 2x = 16$ *hort*

amorpha >>> bastard indigo

Amur grape: *Vitis amurensis* (Vitaceae) *hort*

anise: *Pimpinella anisum* (Umbelliferae) *hort*

annual meadow grass: *Poa annua* (Gramineae) *agr*

ape >>> giant taro

apple: *Malus domestica syn M. pumila* (Rosaceae), $2n = 2x, 3x = 34, 51$ *hort*

apricot: *Armeniaca manshurica syn Prunus armeniaca* (Rosaceae), $2n = 2x = 16$ *hort*

aralia: *Aralia* spp. (Araliaceae) *hort*

Armenian cherry >>> amarelle

aroids, edible: *Alocasia* spp., *Colocasia* spp., *Cyrtsperma* spp., *Xanthsoma* spp. (Araceae) *hort*

arrowhead: *Sagittaria sagittifolia* (Alismataceae), $2n = 2x = 22$ *hort*

arrowroot (starch): *Maranta arundinacea* (Marantaceae) *hort*

arrowroot canna: *Canna glauca* (Cannaceae) *hort* >>> Queensland arrowroot

artichoke: *Cynara cardunculus syn C. scolymus* (Compositae), $2n = 2x = 34$ *hort*

arvi >>> taro

asp: *Populus tremula* (Salicaceae) *fore*

asparagus: as a dioecious vegetable, it has been used for 3,000 years in Egypt; as a medical plant against cough and urinary sufferings it has been used for more than 3,000 years in China; since the fifteenth century it has grown in Europe; Germany, France, Spain, Greece, Belgium are the European countries with the highest production; Peru became the biggest exporter of asparagus; it is grown from seeds or rhizomes for its young, succulent shoots; breeding is a challenge since this dioecious crop is inevitably outpollinating; useful genotypes can be maintained by vegetative propagation; it is also possible to produce diploid homozygous hermaphrodite asparagus by diploidization of haploids obtained from twin seedlings; selfed seeds from those doubled haploids are highly uniform and available for seed-propagated asparagus, *Asparagus officinalis* (Liliaceae), $2n = 2x = 20$ *hort*

asparagus bean: *Vigna unguiculata* ssp. *unguiculata syn Phaseolus unguiculata* ssp. *sesquipedalis* (Leguminosae), $2n = 2x = 14$ *hort*

asparagus pea: *Tetragonolobus purpureus syn Lotus tetragonolobus* (Leguminosae) *hort*

aspen >>> asp

aster: *Aster* spp. (Compositae)

astragalus: *Astragalus cicer* (Leguminosae) *agr*

aubergine >>> eggplant

aucuba: *Aucuba japonica* (Cornaceae) *hort*

auricula: *Primula* spp. (Primulaceae) *hort*

Australian nut >>> Macadamia nut

Australian rice: *Oryza australiensis* (Gramineae), $2n = 2x$, EE = 24 *agr*

avocado: *Persea americana syn P. gratissima* (Lauraceae), $2n = 2x = 24$ *hort*

awnless brome >>> smooth brome

awusa nut >>> African walnut

azalee: *Rhododendron simsii* (Ericaceae) *hort*

B

Bahia grass: not mentioned as a turf till 1938; it was first found on a sodded sand bank in Florida (USA) and called "pensacola"; this excellent growing plant may have arrived as a stowaway on a fruit boat from Central or South America; as a crop it was released in 1944; in 1946, it was discovered as a diploid plant ($2n = 20$), which reproduces sexually based on variable progenies; native to eastern Argentina, the pensacola genotype has become one of the major forage grasses of the southeastern United States; pensacola is more cold tolerant than tetraploid Bahia grass cultivars; the tetraploid Bahia grass cytotype *P. notatum* var. *latiflorum* is the most common botanical variety in tropical and subtropical America, *Paspalum notatum* (Gramineae), $2n = 2x, 4x = 20, 40$ *agr*

bajori: *Pennisetum echinurus syn P. typhoideum* (Gramineae) *agr*

balsam >>> garden balsam

balsam pear >>> bitter gourd

bambara groundnut: *Vigna subterranea syn Voandzeia subterranea* (Leguminosae), $2n = 2x = 22$ *hort* >>> pulse >>> gynophore

bamboo cane: *Bambusa* spp. (Gramineae) *agr*

banana: *Musa* spp. (Musaceae), $2n = 2x$, AA = 22; cultivated hybrids, acubal, $2n = 3x$, AAA, AAB, ABB = 33 *hort*

barberry: *Berberis vulgaris* (Berberidaceae) *bot hort phyt*

barley: a commonly diploid cereal crop; it is ancient as the origins of agriculture itself; barley grain is used as feed for animals, malt, and human food; barley was a staple food as far back as 18,000 years ago; it was the energy food of the masses; its use as human food was very popular during the Roman Empire and it continued to be the main food cereal of northern Europe until the sixteenth century; barley is still an important staple food in several developing countries; in the highlands of Tibet, Nepal, Ethiopia, in the Andean countries, in some areas of North Africa, Turkey, Iran, Afghanistan, India, and Russia, barley is used as human food either for bread making (usually mixed with bread wheat) or for specific recipes; the largest use of food barley is found in regions where other cereals do not grow well due to altitude, low rainfall, or soil salinity; it evolved from wild forms of *Hordeum spontaneum;* there are types of two or six rows of ears; it is the fourth most commonly grown cereal in the world, with a global production of about 200 million tons; it is the cereal that has the most widespread natural distribution, *Hordeum* spp.; two-rowed barley = *H. distichon,* six-rowed barley = *H. vulgare* (Gramineae), $2n = 2x$, VV = 14, 2C DNA content 13 pg *agr* >>> Tables 15, 16, 30, 32, 35 >>> waxy hull-less barley

barnyard grass: *Echinochloa crus-galli* (Gramineae), $2n = 4x = 36$ *bot agr*

basil: *Ocimum basilicum* (Labiatae) *hort*

basket willow: *Salix viminalis, S. dasyclados, S. amygdalina, S. americana, S. purpurea, S. daphnoides, S. pentandra* (Salicaceae) *fore agr*

bastard indigo: *Amorpha fructicosa* (Leguminosae) *hort agr*

bay-tree >>> laurel(-tree)

beach grass: *Ammophila arenaria* (Gramineae) *bot agr*

bean: *Phaseolus* ssp. (Leguminosae) *hort*

beardgrass: *Andropogon ischaemum* (Gramineae) *bot agr*

beet: *Beta vulgaris* ssp. *cicla* (Chenopodiaceae) *hort*

beet rape: *Brassica rapa* spp. *rapifera* (Cruciferae) *agr* >>> Figure 8

beetroot >>> red beet

begonia: *Begonia* spp. (Begoniaceae) *hort*

Bengal gram >>> chickpea

bent >>> bent grass

bent grass: *Agrostis vulgaris* (Gramineae) *bot agr*

bergamot: *Monarda didyma* (Labiatae) *hort*

Berlandieri grape: *Vitis berlandieri* (Vitaceae) *hort*

bermudagrass: common bermudagrass, *Cynodon dactylon,* and its interspecific hybrids with *C. transvaalensis* are the most popular turfgrasses for golf courses and sports turfs, as well as for lawns and roadsides throughout the southern United States; first recorded release of an improved cultivar was in the early 1940s; most cultivars are vegetatively propagated by plugs, sod, or sprigs; first improved seeded bermudagrass was the "Guyman" cultivar released in 1982; until that time, tetraploid bermudagrass ($2n$=36) was the only seeded bermudagrass being sold in the trade as 'Arizona common' or simply 'common' bermudagrass, *Cynodon dactylon* (Gramineae), $2n = 2x$, AA = 18 *agr*

Berseem clover >>> Egyptian clover

betel nut (palm): *Areca catechu* (Palmae), $2n = 4x = 32$ *hort fore*

bilberry >>> blueberry

bilimbi: *Averrhoa bilimbi* (Oxalidaceae) *hort*

billion dollar grass >>> barnyard grass

birch: *Betula* spp. (Betulaceae) *fore*

bird rape: *Brassica rapa* var. *silvestris f. oleifera* (Brassicaceae) *agr* >>> Figure 8

bird's-foot >>> finger millet

bird's-foot trefoil: *Lotus corniculatus* (Leguminosae), $2n = 2x, 4x = 12, 24$, 2C DNA content 4.0 pg *agr*

bitter apple >>> colocynth

bitter gourd: *Momordica charantia* (Cucurbitaceae), $2n = 2x = 22$ *hort*

black bent: *Agrostis gigantea* (Gramineae) *bot agr*

black currant: *Ribes nigrum* (Grossulariaceae), $2n = 2x = 16$ *hort*

black locust: one of the most important stand-forming tree species; for example, it covers about 20 percent of the forest area and provides about 18 percent of the annual timber output in Hungary, *Robinia pseudoaccacia* (Leguminosae) *fore*

black medick: *Medicago lupulina* (Leguminosae) *agr*

black mustard: *Brassica nigra* (Cruciferae), $2n = 2x$, BB = 16 *agr* >>> Figure 8

black pepper: *Piper nigrum* (Piperaceae), $2n = 4x = 52$ *hort*

black salsify: *Scorzonera hispanica* (Compositae) *hort*

blackberry: *Rubus rosa* (Rosaceae) *hort*

black-eyed pea >>> cow bean

blackgram >>> urd bean

blackwood acacia: *Acacia melanoxylon* (Leguminosae) *fore*

blue lupin: *Lupinus angustifolius* (Leguminosae), $2n = 2x = 48$ *agr*

blue pea >>> green pea

blue sisal: *Agave amaniensis* (Agavaceae), $2n = 2x = 60$ *agr*

blueberry: *Vaccinium myrtillus* (Ericaceae) *hort* >>> highbush blueberry >>> lowbush blueberry

Bombay hemp >>> cantala

borage: *Borago officinalis* (Boraginaceae) *hort*

bottle gourd: *Lagenaria siceraria* ssp. *siceraria* (Cucurbitaceae), $2n = 2x = 40$ *hort*

bow-string hemp: *Sansevieria* spp. (Agavaceae), $2n = 2x = 40$ *hort*

box-tree: *Buxus sempervirens* (Buxaceae) *hort*

bramble >>> blackberry

Brazil nut: *Bertholletia excelsa* (Lecythidaceae) *hort fore*

bread wheat >>> wheat

breadfruit: *Artocarpus altilis syn A. communis* (Moraceae), $2n = 2x = 56$ *hort*

brewing yeast: *Saccharomyces cerevisae* (Saccharomycetaceae) *biot*

broad bean: *Vicia faba* (Leguminosae), $2n = 2x = 12$, 2C DNA content 28.0 pg *agr hort* >>> Tables 16, 35

broad red clover: *Trifolium pratense* (Leguminosae) *agr*

broccoli: *Brassica oleracea* var. *italica* (Cruciferae) *hort* >>> Figure 8

bromegrass: *Bromus carinatus* (Gramineae) *bot agr*

brown mustard: *Brassica juncea* (Cruciferae), $2n = 4x = 36$ *agr* >>> Figure 8

brunching onion >>> Welsh onion

Brussels sprouts: *Brassica oleracea* var. *gemmifera* (Cruciferae) *hort* >>> Figure 8

buckbean >>> marsh trefoil

buckwheat: *Fagopyrum esculentum* (Polygonaceae), $2n = 2x = 16$ *agr* >>> Tatary buckwheat >>> notch-seeded buckwheat

buffalo grass: *Cenchrus cillaris* (Gramineae) *agr*

bulb barley: *Hordeum bulbosum* (Gramineae), $2n = 2x$, $4x$, $H^{bul}H^{bul} = 14$, 28 *bot*

busy lizzie: *Impatiens walleriana* (Balsaminaceae) *hort*

butter bean >>> Lima bean

butter cabbage: *Brassica napus* ssp. *arvensis* (Brassicaceae) *hort* >>> Figure 8

butterhead (lettuce): *Lactuca sativa* (Compositae) *hort*

C

cabbage, brassicas: any of several cultivated varieties of a plant, *Brassica oleracea* ssp. *capitata*, of the mustard family, having a short stem and leaves formed into an edible head, *Brassica oleracea* (Cruciferae), $2n = 2x$, C'C' = 18 *hort* >>> Figure 8

cacao: *Theobroma cacao* (Sterculiaceae), $2n = 2x = 20$ *hort*

cactus fig: edible fruits used for juices, liqueurs, and jams, and as an herb plant, *Opuntia stricta* var. *dilleni* (Cactaceae) *hort agr*

cactus pear >>> cactus fig

calabash (gourd): *Crescentia cujete* (Bignoniaceae) *hort* >>> bottle gourd

calabrese >>> sprouting broccoli

calamus: *Acorus calamus* (Araceae), $2n = 2x$, $4x = 18$, 36 *hort*

calceolaria: *Calceolaria* spp. (Scrophulariaceae) *hort*

California bromegrass >>> bromegrass

camomile >>> chamomile

Canada thistle: *Cirsium arvense* (Asteraceae) *bot agr*

canahua: *Chenopodium pallidicaule* (Chenopodiaceae), $2n = 2x = 18$ *hort*

Canary grass: *Phalaris canariensis* (Gramineae), $2n = 2x$, $4x = 14$, 28 *agr*

candytuft: *Iberis* spp. (Brassicaceae) *hort*

canna: *Canna edulis* (Cannaceae) *hort*

cantala: *Agave cantala* (Agavaceae), $2n = 3x = 90$ *agr*

cantaloup >>> cantaloupe

cantaloupe: *Cucumis melo* (Cucurbitaceae), $2n = 2x = 24$, 2C DNA content 2.85-3.89 pg *hort*

caper (spurge): *Euphorbia lathyris* (Euphorbiaceae) *hort*

caraway (seed): *Carum carvi* (Umbelliferae) *hort*

cardamom: *Amomum maximum syn Elettaria cardamomnum* (Zingiberaceae), $2n = 4x = 48$ *hort*

cardoon: *Cynara cardunculus* (Compositae) *hort*

carnation: *Dianthus* spp. (Caryophyllaceae) *hort*

carpet bent: *Agrostis stolonifera* (Gramineae) *bot agr*

carrot: *Daucus carota* (Umbelliferae), $2n = 2x = 18$ *hort*

cashew: *Anacardium occidentale* (Anacardiaceae), $2n = 2x = 42$ *fore hort*

cassava: a perennial shrub producing a high yield of tuberous roots in 6 months to 3 years after planting; originating in Central and South America, cassava spread rapidly and arrived on the west coast of Africa via the Gulf of Benin and the River Congo at the end of the 16th century and on the east coast via the Reunion Island, Madagascar, and Zanzibar at the end of the eighteenth century; by the beginning of the nineteenth century cassava arrived in India, but controlled breeding did not begin until the 1920s; for many farmers it is the primary staple but also used as a cash crop to produce industrial starches, tapioca, and livestock feeds; world production in 1995, all from developing countries, was about 165.3 million tons from about 16.2 million ha; currently, Nigeria, Brazil, the Democratic Republic of Congo, Thailand, and Indonesia are the world's largest producers, *Manihot esculenta* (Euphorbiaceae), $2n = 2x = 36$ *hort*

castor (bean): *Ricinus communis* (Euphorbiaceae), $2n = 2x = 20$ *hort* >>> Table 16

Caucasian clover >>> Kura clover

Caucasian persimmon: *Diospyros lotus* (Ebenaceae), $2n = 2x = 30$ *hort* >>> persimmon

cauliflower: *Brassica oleracea* var. *botrytis*, (Cruciferae) *hort* >>> Figure 8

celeriac >>> celery

celery: *Apium graveolens* (Umbelliferae), $2n = 2x = 22$ *hort*

chamomile: *Matricaria chamomilla* (Compositae) *hort*

chard >>> beet

charlock: *Sinapis arvensis* (Cruciferae) *bot agr*

chayote: *Sechium edule* (Cucurbitaceae), $2n = 2x = 28$ *hort*

cheat >>> rye brome

cherimoya: *Annona cherimola* (Annonaceae) *hort*

cherry: *Prunus* spp. (Rosaceae) *hort* >>> Japanese cherry >>> sour cherry >>> sweet cherry

cherry plum: *Prunus cerasifera* (Rosaceae), $2n = 2x, 3x, 4x, 6x$, CC = 16, 24, 32, 48 *hort*

chervil: *Anthriscus cerefolium* (Umbelliferae) *bot hort*

chess >>> rye brome

chestnut: *Castanea* spp. (Fagaceae) *fore* >>> American chestnut >>> Chinese chestnut >>> Japanese chestnut >>> sweet chestnut

Ch'iao >>> rakkyo

chickling pea: *Lathyrus sativus* (Leguminosae), $2n = 2x = 14$ *hort*

chickpea: the only common cultivated annual species is *C. arientinum*, although *C. soongaricum* is also cultivated as a food plant in some parts of Afghanistan, eastern Himalayas, and Tibet; the center of diversity is in western Asia, probably in the Caucasus region/Minor Asia, *Cicer arietinum* (Leguminosae), $2n = 2x = 16$ *agr*

chicory: *Cichorium intypus* (Compositae), $2n = 2x = 18$ *hort*

Chilean strawberry: *Fragaria chiloensis* (Rosaceae) *hort*

chili: *Capsicum annuum* (Solanaceae) *hort*

China aster: *Callistephus hortensis syn C. chinensis* (Compositae) *hort*

China-grass >>> ramie

Chinese cabbage: *Brassica olearacea* var. *chinensis (syn* var. *pekinensis)* (Cruciferae), $2n = 2x$, AA = 20 *hort* >>> Figure 8

Chinese chestnut: *Castanea mollissima* (Fagaceae), $2n = 2x = 24$ *hort*

Chinese chive: *Allium tuberosum* (Alliaceae), $2n = 2x, 4x = 16, 32$ *hort*

Chinese garlic: *Allium macrostemon* (Alliaceae), $2n = 2x\text{-}6x = 18\text{-}72$ *hort*

Chinese gooseberry: *Actinidia chinensis* (Actinidiaceae), $2n = 2x = 116$(?)

Chinese jute: *Abutilon avicennae, A: theophrasii* (Malvaceae), $2n = 2x = 42$ *agr*

Chinese leek: *Allium ramosum* (Alliaceae), $2n = 4x = 32$ *hort*

Chinese mustard >>> pak-choi

Chinese sugar cane >>> sweet sorghum

Chinese tallow tree: *Sapium sebiferum* (Euphorbiaceae), $2n = 2x = 36$ *fore*

Chinese wisteria: *Wistaria sinensis* (Leguminosae) *fore hort*

chireweed >>> stitchwort

chive: *Allium schoenoprasum* (Alliaceae), $2n = 2x, 3x, 4x = 16, 24, 32$ *hort*

christophine >>> chayote

chumberas >>> pencas

cibol >>> Welsh onion

cinchona: *Cichona* spp. (Rubiaceae) *hort fore*

cinnamon: *Cinnamomum camphora* (Lauraceae) *hort fore*

citron: *Citrus* spp. (Rutaceae), $2n = 2x = 18$ *hort* >>> hesperidia

citrus >>> citron

clove >>> carnation

clove: *Eugenia caryophyllis* (Myrtaceae) *hort*

clove-pink >>> carnation

club wheat: *Triticum compactum* (Gramineae) >>> wheat

cluster bean >>> guar

clustered clover: *Trifolium glomeratum* (Leguminosae) *agr*

cobnut: *Corylus* spp. (Corylaceae) *hort*

coca: *Erythroxylon coca* (Erythroxylaceae) *hort fore*

cocksfoot: three levels of polyploidy occur: diploid, tetraploid, and hexaploid; tetraploids are the most common and generally distributed forms; it was distributed from Europe to North America, Australia, and New Zealand; *Dactylis glomerata* (Gramineae), $2n = 4x = 28$ *agr*

cockspur: *Echinochloa crus-galli* (Gramineae) *agr*

cocoa: *Theobroma cacao* (Sterculiaceae) *hort fore*

coconut: *Cocos nucifera* (Palmae), $2n = 2x = 32$ *fore hort*

cocozelle: *Cucurbita pepo* (Cucurbitaceae), $2n = 2x = 40$ *hort*

cocyam >>> tanier

coffee (Arabian): *Coffea arabica* (Rubiaceae), $2n = 4x$, AABB = 44 *hort*

colocynth: *Colocynthis citrullus syn Citrullus vulgaris* (Cucurbitaceae) *hort*

columbine: *Aquilegia* spp. (Ranunculaceae) *hort*

common bulb onion >>> onion

common chickling: *Lathyrus sativus* (Leguminosae) *agr*

common chicory >>> chicory

common corncockle: *Agrostemma githago* (Caryophyllaceae) *bot agr*

common grape >>> European grape

common horsetail: *Equisetum arvense* (Equisetaceae) *bot agr*

common milkweed: *Asclepias syriaca* (Asclepiadaceae) *bot agr*

common millet: *Panicum miliaceum* (Gramineae), $2n = 4x = 36$ *agr*

common onion >>> onion

common osier >>> basket willow

common petunia: *Petunia hybrida* (Solanaceae) *hort*

common velvet grass: *Holcus lanatus* (Gramineae) *bot agr*

common vetch: *Vicia sativa* ssp. *sativa* (Leguminosae), $2n = 2x = 12$, 2C DNA content 4 pg *agr*

cone wheat >>> rivet wheat >>> wheat

coracan millet >>> finger millet

coriander: *Coriandrum sativum* (Umbelliferae) *hort*

corktree: *Phellodendron* spp. (Rutaceae) *fore agr*

corn >>> maize

corn bindweed: *Convolvulus tricolor* (Convolvulaceae) *bot agr*

corn flower: *Centaurea cyanus* (Asteraceae) *bot agr*

corn salad: *Valerianella olitoria, syn V. locusta* (Valerianaceae) *hort*

corn thistle >>> Canada thistle >>> creeping thistle >>> cursed thistle

corn toadflax: *Linaria arvensis* (Scrophulariaceae) *bot agr*

cos lettuce: *Lactuca capitata* var. *romana* (Compositae) *hort*

cotton: belongs to a genus with 33 diploid and 6 allotetraploid species native to the tropical and subtropical regions of the world; the cultivars, consisting almost entirely of the allotetraploids, are grown in the warmer regions for their seed and fiber; *Gossypium* spp. (Malvaceae), $2n = 4x$, AADD = 52 *agr* >>> short American staple cotton

cottonwood >>> poplar

couch-grass >>> *Agropyron*

cow bean: *Phaseolus unguiculata* ssp. *sinensis syn Vigna unguiculata* (Leguminosae), $2n = 2x = 22$ *hort* >>> pulse

cowpea >>> cow bean

crambe: *Crambe abyssinica* (Cruciferae) *agr*

creeping thistle >>> Canada thistle

cress: *Lepidium sativum* (Cruciferae), $2n = 2x, 4x = 16, 32$ *hort*

crested dogstail: *Cynosurus cristatus* (Gramineae) *bot agr*

crimson clover: *Trifolium incarnatum* (Leguminosae), $2n = 2x = 14$ *agr*

crisphead (lettuce): *Lactuca sativa* (Compositae) *hort*

crowfoot: *Ranunculus* spp. (Ranunculaceae) *bot agr*

crown imperial: *Fritillaria imperialis* (Liliaceae) *hort*

cucumber: *Cucumis sativus* (Cucurbitaceae), $2n = 2x = 14$, 2C DNA content 1.37-2.48 pg *hort*

cucumber tree >>> bilimbi

cucurbits: *Cucumis* spp., *Citrullus* spp., *Cucurbita* spp., *Lagenaria* spp. (Cucurbitaceae), $2n = 2x = 40$ *hort*

cucuzzi >>> bottle gourd

cumin: *Cuminum cyminum* (Umbelliferae) *hort*

curled lettuce >>> cutting lettuce

curly kale >>> kale

cursed thistle >>> Canada thistle

custard apple: *Annona cherimolia* (Annonaceae), $2n = 2x = 14$ *hort*

cutting lettuce: *Lactuca sativa* var. *crispa* or var. *secalina* (Compositae) *hort*

D

daffodil >>> narcissus

dahlia: *Dahlia pinnata* (Asteraceae) *hort*

daikon: *Raphanus sativus* convar. *sativus syn Raphanus raphanistroides* (Cruciferae) *hort*

daimyo oak: native to Japan, Korea, and northern and western China; *Quercus dentata* (Fagaceae), $2n = 2x, 4x = 24, 48$ *fore*

daisy fleabane: *Erigeron* spp. (Compositae) *bot agr*

Damson plum: *Prunus insititia* (Rosaceae), $2n = 6x = 48$ *hort*

dandelion: *Taraxacum* spp. (Compositae) *hort bot agr*

darnel ryegrass: *Lolium temulentum* (Gramineae) *bot agr*

dasheen: *Colocasia esculenta* var. *globulifera* (Araceae), $2n = 3x = 42$ *hort*

date (palm): *Phoenix dactylifera* (Palmae), $2n = 2x = 36$ *fore hort*

deadnettle: *Lamium* spp. (Lamiaceae) *bot agr*

derris: *Derris elliptica* (Leguminosae), $2n = 2x = 22$ *agr*

dill: *Anethum gravolens* (Umbelliferae) *hort*

Dinkel wheat: *Triticum spelta* (Gramineae) >>> wheat

drooping brome: *Bromus tectorum* (Gramineae) *bot agr*

durian: *Durio zibethinus* (Bombacaceae), $2n = 2x = 14$ *fore*

durum wheat: an ancient wheat grown since Egyptian times; a wheat used to make bread and other bakery products; the hard, flinty kernels are specially ground and refined to obtain semolina, a granular product used in making pasta items such as macaroni and spaghetti; most durum wheats are grown in Mediterranean countries, Russia, North America, and Argentina, *Triticum durum* (Gramineae), $2n = 4x = 28$ *agr* >>> wheat

Dutch turnip >>> stubble turnip

dwarf bean >>> French bean

dwarf French bean: *Phaseolus vulgaris* var. *nanus* (Leguminosae) *hort*

dwarf sisal: *Agave angustifolia* (Agavaceae), $2n = 2x = 60$ *agr*

dye amaranth: *Amaranthus cruentus* (Amaranthaceae), $2n = 2x = 32$ *hort*

dyer's alcanet >>> alkanna

dyer's chamomile: *Anthemis tinctoria* (Compositae), $2n = 2x = 18$ *hort*

E

earth-smoke >>> fumitory

Eastern white pine: *Pinus strobus* (Pinaceae) *fore*

East-India arrowroot: *Curcuma* ssp. (Zingiberaceae) *hort*

eddo >>> taro

eggapple >>> eggplant

eggplant: *Solanum melongena* (Solanaceae), $2n = 2x = 24$ *hort*

Egyptian clover: *Trifolium alexandrinum* (Leguminosae) *agr*

Egyptian tree onion >>> tree onion

einkorn: one-grained wheat, so called because it has a single seed per spikelet; a wheat cultivated since Stone Age times, *T. monococcum, T. boeoticum* (Gramineae), $2n = 2x$, AA = 14, 2C DNA content 14 pg *agr* >>> wheat

elephant garlic >>> Levant garlic

elephant grass: *Pennisetum purpureum* (Gramineae), $2n = 4x = 28$ *agr*

eleusine >>> finger millet

emmer wheat: two-grained wheat; grown since the Neolithic times, *Triticum dicoccum, T. dicoccoides* (Gramineae), $2n = 4x$, AABB = 28, 2C DNA content 25 pg *agr* >>> wheat

endive: *Cichorium endivia* et spp. (Compositae), $2n = 2x$, $3x = 18$, 27 *hort*

English marigold: *Calendula officinalis* (Compositae) *hort bot*

English ryegrass: *Lolium perenne* (Gramineae) *agr*

estragon: *Artemisia dracunculus* (Compositae) *hort*

Ethiopian mustard: *Brassica carinata* (Cruciferae), $2n = 4x = 34$ *agr* >>> Figure 8

European elymus >>> lyme grass

European grape: *Vitis vinifera* ssp. *sativa* (Vitaceae) *hort*

European plum: *Prunus domestica* (Rosaceae) *hort*

F

fall rose >>> China aster

false >>> bastard indigo

false acacia >>> locust tree

false flax: *Camelina sativa* (Cruciferae) *agr*

false goat's beard: *Astilbe* spp. (Saxifragaceae) *hort*

false indigo >>> bastard indigo

false saffron >>> safflower

feijoa: *Feijoa sellowiana* (Myrtacaceae) *hort*

fennel: *Foeniculum vulgare* (Apiaceae) *hort*

fenugreek: *Trigonella foenum-graecum* (Leguminosae) *agr*

field bean >>> broad bean

field brome: *Bromus arvensis* (Gramineae) *bot agr*

field horsetail >>> common horsetail

field pea: *Pisum sativum* ssp. *arvense* (Leguminosae) *agr*

field pink: *Dianthus campestris* (Caryophyllaceae) *bot agr*

field salad >>> corn salad

field scabios: *Knautia arvensis* (Dipsacaceae) *bot agr*

field speedwell: *Veronica agnestis* (Scrophulariaceae) *bot agr*

fig: *Ficus carica* (Moraceae), $2n = 2x = 26$ *hort*

filbert: *Corylus maxima* (Corylaceae), $2n = 2x = 22$ *hort*

finger millet: *Eleusine coracan* ssp. *coracan* (Gramineae), $2n = 4x = 36$ *agr*

flax (linseed): an annual dicot plant, 40 to 80 cm in height ; the fruit is a capsule containing less than 10 seeds whose oil content varies from 35 to 45 percent; it is cultivated either as a textile plant, for the fibres contained in the stem, or for its oleo-protinaceous seeds; winter flax varieties, with their procumbent growth at the beginning of their development, are differentiated from spring flax varieties, which grow erect and are sensitive to cold; textile flax has been cultivated in Europe since the Middle Ages, but has declined since the appearance of cotton and synthetic fibers; the long stem is slightly branched at the top and is rich in fibres; planting occurs in spring; harvesting occurs by uprooting when the capsules are yellow-green; retting permits decomposition of cements that bind the fibers; flax seeds produce an oil used for industrial purposes and are also used in animal feed; sown in March, the oil-yielding flax is harvested when the seeds are mature, drying may be necessary; this crop is found in France, Germany, and Great Britain, *Linum usitatissimum* (Linaceae), $2n = 2x = 30$, 2C DNA content 1.4 pg *agr*

fleur-de-lys >>> iris

fodder beet: *Beta vulgaris* var. *crassa* (Chenopodiaceae) *agr*

fodder radish: *Raphanus sativus* var. *oleiformis* (Cruciferae) *agr*

fonio >>> hungry rice

foxglove: *Digitalis* spp. (Scrophulariaceae) *hort*

foxtail lily: *Eremurus* spp. (Liliaceae) *hort bot*

foxtail millet: *Setaria italica* (Gramineae), $2n = 2x$, AA = 18 *bot agr*

freesia: *Freesia* spp. (Iridaceae) *hort*

French bean: *Phaseolus vulgaris* (Leguminosae), $2n = 2x = 22$ *hort* >>> pulse

fuchsia: *Fuchsia* spp. (Onagraceae) *hort*

fumitory: *Fumaria* spp. (Papaveraceae) *bot agr*

G

gambier: *Uncaria gambir* (Rubiaceae) *hort*

garden aster >>> China aster

garden balsam: *Impatiens balsamina* (Balsaminaceae), $2n = 2x = 14$ *hort*

garden beet >>> red beet

garden cress >>> cress

garden hydrangea: *Hydrangea hortensis* (Saxifragaceae) *hort*

garden leek: *Allium porrum* (Alliaceae), $2n = 4x = 32$ *hort*

garden lettuce >>> butterhead

garden orach >>> mountain spinach

garden pansy: *Viola* spp. (Violaceae) *hort*

garden pea: despite its importance as a traditional crop plant, the pea has a history as an organism used for genetic studies going as far back as MENDEL (1866); pea also later became a subject of intensive genetic studies, and thus one of the best genetically investigated plants; it has led to the identification and symbolization of more than 600 classical genes, in addition to about 2,500 genes identified and preserved in collections; particularly, pea mutants were used in breeding; *Pisum sativum* ssp. *hortense* (Leguminosae), $2n = 2x = 14$ *hort* >>> fasciata type of pea >>> Table 16

garden thyme: *Thymus vulgaris* (Labiatae) *hort*

gardenia: *Gardenia* spp. (Rubiaceae) *hort*

garlic: *Allium sativum* (Alliaceae), $2n = 2x = 16$ *hort*

gerbera: *Gerbera jamesonii* (Compositae) *hort*

German chamomile: *Matricaria chamomilla* (Compositae) *bot hort agr*

germander speedwell: *Veronica chamaedrys* (Scrophulariaceae) *bot agr*

gherkin >>> cucumber

giant swamp taro: *Cyrtosperma chamissonis syn C. edule, C. merkusii* (Araceae), $2n = 2x = 26$ *hort*

giant taro: *Alocasia macrorhiza syn A. indica* (Araceae), $2n = 2x = 28$ *hort*

gillyflower: *Matthiola* spp. (Cruciferae) *hort*

ginger: *Zingiber officinale* (Zingiberaceae), $2n = 2x = 22$ *hort*

globe artichoke >>> artichoke

Goa bean: *Psophocarpus* spp. (Leguminosae) *hort*

goat grass >>> *Aegilops*

golden chamomile >>> dyer's chamomile

golden gram >>> mungbean

golden oatgrass: *Trisetum flavescens* (Gramineae) *bot agr*

gombo >>> okra

good king Henry: *Chenopodium bonus-henricus* (Chenopodiaceae) *bot agr*

gooseberry: *Ribes uva-crispa* (Grossulariaceae), $2n = 2x = 16$ *hort*

goosy grass >>> drooping brome

grape: *Vitis* spp. ($2n = 2x = 38$), *Muscadinia* spp. ($2n = 2x = 40$) (Vitaceae), *hort* >>> Amur grape >>> Berlandieri grape >>> European grape

grape hyacinth: *Muscari* spp. (Liliaceae) *hort*

grapefruit: *Citrus paradisi* (Rutaceae), $2n = 2x = 18$ *hort* >>> hesperidia

grapevine >>> European grape

grass: several species of the family *Gramineae bot agr* >>> Table 35

grasscloth >>> ramie

greater bird's-foot trefoil: *Lotus uliginosis* (Leguminosae) *bot agr*

great-headed garlic >>> Levant garlic

green pea: *Pisum sativum* convar. *sativum* or convar. *vulgare* (Leguminosae) *hort*

green pepper >>> sweet pepper

greengram >>> mungbean

groundnut >>> peanut

guar: *Cyamopsis tetragonoloba* (Leguminosae), $2n = 2x = 14$ *hort*

guava: *Psidium guayava* (Myrtaceae), $2n = 2x, 3x = 22, 33$ *hort*

Guinea grass: *Panicum maximum* (Gramineae), $2n = 4x = 32$ *agr*

gumbo >>> musk okra

H

hairy vetch: *Vicia villosa* (Leguminosae) *bot agr*

hairy vetchling: *Lathyrus hirsutus* (Leguminosae) *bot agr*

haricot >>> French bean

harlequin flower: *Sparaxis tricolor* (Iridaceae) *hort*

hawkbit >>> dandelion

hazel (nut): *Corylus* spp. (Corylaceae), $2n = 2x = 22$ *hort* >>> Siberian hazel >>> American hazel >>> Turkish cobnut >>> cobnut >>> filbert

head cabbage: *Brassica oleracea* convar. *capitata* (Brassicaceae) *hort* >>> Figure 8

head lettuce: *Lactuca sativa* var. *capitata* (Compositae) *hort*

hedge hyssop: *Gratiola officinalis* (Scrophulariaceae) *bot agr*

hemp: cultivation and use of hemp for fiber can be traced back to 2,800 B.C. in China; for many centuries hemp has been cultivated as a source of strong stem fibers, seed oil, and psychoactive drugs in its leaves and flowers; environmental concerns and recent shortages of wood fiber have renewed interest in hemp as a raw material for a wide range of industrial products; hemp is a herbaceous annual that develops a rigid woody stem ranging in height from 1 - 5 m; the stalks have a woody core surrounded by a bark layer containing long fibers that extend nearly the entire length of the stem; breeding has developed hemp varieties with increased stem fiber content and very low levels of delta-9-tetrahydro-cannabinol (THC), the psychoactive ingredient of marijuana; although hemp is well adapted to the temperate climatic zone and will.grow under varied environmental conditions, it grows best with warm growing conditions, an extended frost-free season, highly productive agricultural soils, and abundant moisture throughout the growing season; hemp yields range from 2.5 to 8.7 tons of dry stems per acre; hemp is dioecious plant having both staminate (male) and pistillate (female) plants, each with distinctive growth characteristics; staminate plants are tall and slender with few leaves surrounding the flowers, while pistillate plants are short and stocky with many leaves at each terminal inflorescence; staminate plants senesce and die soon after their pollen is shed, while pistillate plants remain alive until the seeds mature; quite stable monoecious varieties have been developed, *Cannabis sativa* (Moraceae); $2n = 2x = 20$, XY male, XX female *agr*

henbane: *Hyoscyamus niger* (Solanaceae) *bot agr*

Henequen agave: on the Canary Islands used as a fiber plant, *Agave fourcroydes* (Agavaceae), $2n = 5x = 40$ (?) *agr*

hevea (rubber) >>> rubber

hickory: *Hicoria* spp. (Juglandaceae) *fore hort*

highbush blueberry: *Vaccinium corymbosum* (Ericaceae), $2n = 4x = 48$ *hort*

holcos: *Holcos lanatus* (Gramineae) *bot agr*

hollyhock: *Alcea rosea* ssp. *plena* (Malvaceae) *hort*

hop clover: *Trifolium campestre* (Leguminosae) *agr*

hops: a crop in central Europe for more than 1,000 years; despite the early origin of cultivation, hops never developed into a major crop because they are only used by the brewery industry to flavor fermented malt beverages, primarily beer and ale; it is a dioecious species; male plants are only used for breeding or yield stimulation; vegetatively propagated female plants are grown for commercial production in about 30 countries worldwide; in diploids, a 1:1 sex ratio of seedling progenies is expected for an XX (female) and XY (male) sex mechanism, however males are less frequent; *Humulus lupulus* (Moraceae), $2n = 2x = 20$, XY male *agr*

horseradish: *Armoracia rusticana* (Cruciferae), $2n = 4x = 32$ *hort*

hot paprika >>> hot pepper

hot pepper: *Capsicum annum* var. *accuminatum* (Solanaceae) *hort*

huauzontle: *Chenopodium nuttalliae* (Chenopodiaceae), $2n = 4x = 36$ *hort*

huckleberry >>> blueberry

Hungarian vetch: *Vicia pannonica* (Leguminosae), $2n = 2x = 12$ *agr*

hungry rice: *Digitaria exilis* (Gramineae) *bot agr*

hyacinth: *Hyacinthus orientalis* (Liliaceae) *hort*

hyacinth bean >>> lablab

hyssop: *Hyssopus officinalis* (Labiatae) *hort*

I

ice plant: *Mesembrianthemum crystallinum* (Aizoaceae) *bot agr*

Indian dwarf wheat: *Triticum sphaerococum* (Gramineae) >>> wheat

Indian hemp: a special type of *C. sativa* that is cultivated in India as a source of narcotics, *Cannabis indica* (Moraceae) *agr*

Indian lettuce: *Lactuca indica* (Compositae), $2n = 2x = 18$ *hort*

Indian rape: *Brassica campestris* ssp. *dichotoma* (Cruciferae), $2n = 2x$, AA $= 20$ *hort*

indigo: *Indigofera* spp. (Leguminosae), $2n = 2x = 16$ *hort agr*

indigo woad: *Isatis tinctoria* (Brassicaceae) *hort agr*

intermediate wheatgrass: *Agropyron intermedium* (Gramineae) *bot agr*

iris: *Iris* spp. (Iridaceae) *hort*

isanu: *Tropaeolum tuberosum* (Tropacolaceae) *hort*

Isfahan wheat: *Triticum turgidum* (Gramineae), $2n = 4x$, AABB $= 28$, 2C DNA content 25 pg *agr*

Italian ryegrass: *Lolium multiflorum* (Gramineae), $2n = 2x = 14$ *agr*

ivory nut: *Phytelephas macrospora* (Palmae) *hort bot*

J

jack bean: *Canavalia ensiformis* (Leguminosae) *hort* >>> pulse

jackfruit: *Artocarpus heterophyllus* (Moraceae), $2n = 2x = 56$ *hort*

Japan laurel >>> aucuba

Japanese aucuba >>> aucuba

Japanese barnyard millet: *Echinochloa frumentacea* (Gramineae) *bot agr*

Japanese bunching onion >>> Welsh onion

Japanese cherry: *Prunus serrulata rosea* (Rosaceae) *hort*

Japanese chestnut: *Castanea crenata* (Fagaceae), $2n = 2x = 24$ *hort*

Japanese laurel >>> aucuba

Japanese mint: *Mentha arvensis* (Labiatae), $2n = 8x$, $R^aR^aSSJJAA = 96$ *hort*

Japanese persimmon >>> persimmon

Japanese plum: *Prunus salicina* (Rosaceae), $2n = 2x, 4x = 16, 32$ *hort*

Japanese privet: *Ligistrum japonicum* (Oleaceae), $2n = 2x = 44$ *hort*

Japanese radish: *Raphanus sativus* ssp. *niger* (Cruciferae) *hort*

Japanese snake gourd: *Trichosanthes cucumerina* (Cucurbitaceae), $2n = 2x, 4x = 22, 44$ *hort*

Java cantala >>> cantala

Java cardamom >>> cardamom

Jerusalem artichoke: *Helianthus tuberosus* (Compositae), $2n = 6x = 102$ *hort* >>> sunflower

jicama: *Pachyrrhizus erosus* (Leguminosae), $2n = 2x = 22$ *hort* >>> pulse

Job's tears >>> adlay

jointed charlock >>> runch

jute: *Corchorus* spp. (Tiliaceae), $2n = 2x = 14$ *agr*

K

kaki >>> persimmon

kale: *Brassica oleracea* ssp. *acephala* var. *sabellica* (Brassicaceae) *agr hort* >>> Figure 8

kamut wheat: "kamut" derives from the ancient Egyptian word for wheat; it is marketed as a new cereal, however it is an ancient relative of modern durum wheat *(Triticum durum);* it is thought to have evolved contemporary with the free-threshing tetraploid wheats; it is also claimed that it is related to *Triticum turgidum,* which also includes the closely related durum wheat; the correct subspecies is in dispute; it was originally identified as *Triticum polonicum;* some other taxonomists believe it is *Triticum turanicum,* commonly called Khorasan wheat; although its true history and taxonomy is not yet clear, its great taste, texture, and nutritional qualities as well as its hypoallergenic properties are unequivocal; it shows two to three times the size of common wheat with 20-40 percent more protein, higher in lipids, amino acids, vitamins and minerals, and a "sweet" alternative for all products that now use common wheat >>> wheat >>> Khorasan wheat

kapok: *Ceiba pentandra* (Bombacaceae) *hort*

Katjang bean: *Phaseolus uniculata* ssp. *cylindrica* (Leguminosae) *hort*

keladi >>> taro

kenaf: a 4,000-year-old crop with roots in ancient Africa; as a member of the hibiscus family, it is related to cotton and okra; it grows quickly, rising to heights of 12-14 feet in as little as 4-5 months; it may yield of 6-10 tons of dry fiber per acre; while the flowering can last 3-4 weeks, each individual flower blooms for only one day; after blooming the flower drops off, leaving a seed pod behind; the stalk consists of two distinct fiber types; the outer fiber is called "bast" and comprises about 40 percent of the stalk's dry weight; the refined bast fibers measure 2.6 mm and are similar to the best softwood fibers used to make paper; the whiter, inner fiber is called "core," and comprises 60 percent of the dry weight; these refined fibers measure 0.6 mm and are comparable to hardwood tree fibers, which are used in a wide range of paper products; upon harvest, the whole plant is processed in a mechanical fiber separator, similar to a cotton gin; the separation of the two fibers allows independent processing and provides raw materials for a growing number of products, *Hibiscus cannabinus* (Malvaceae), $2n = 2x = 36$ *agr*

Kentucky bluegrass: *Poa pratensis* (Gramineae) *agr*

Kersting's groundnut: *Macrotyloma geocarpum* (Leguminosae) *hort* >>> pulse >>> gynophore

Khorasan wheat: *Triticum turanicum, T. orientale* (Gramineae) *agr* >>> wheat >>> kamut wheat

kidney bean: *Phaseolus coccineus, P. multiflorus* (Leguminosae), $2n = 2x = 22$ *hort* >>> pulse

kidney vetch: *Anthyllis vulneraria* (Leguminosae) *bot agr*

Kikuyu grass: *Pennisetum clandestinum* (Gramineae), $2n = 4x = 36$ *agr*

kiwi: *Actinidia chinensis* (Actinidiaceae) *hort*

kodo millet: *Paspalum scrobiculatum* (Gramineae) *agr*

kohlrabi: *Brassica oleracea* var. *gongylodes* (Cruciferae) *hort* >>> Figure 8

kola (nut): *Cola nitida* (Sterculiaceae), $2n = 4x = 40$ *hort*

kolomikta: *Actinidia kolomicta* (Actinidiaceae), $2n = 2x = 112$ (?)

koracan >>> coracan >>> finger millet

kudzu: *Pueraria thunbergiana* (Leguminosae) *hort*

kui ts'ai >>> Chinese chive

kummerovia >>> lespedeza

Kura clover: *Trifolium ambiguum* (Leguminosae) *agr*

kurrat: *Allium kurrat* (Alliaceae), $2n = 4x = 32$ *hort*

L

lablab: *Dolichos lablab syn Lablab purpureus* (Leguminosae), $2n = 2x = 22$ *hort* >>> pulse

lady's fingers >>> okra

lamb's lettuce >>> corn salad

lamb's-quarter: *Chenopodium album* (Chenopodiaceae) *bot agr*

larch: *Larix* spp. (Pinaceae) *fore*

large-seeded false flax: *Camelina sativa* (Cruciferae) *agr*

larkspur: *Delphinium consolida* (Ranunculaceae) *bot agr*

laurel (-tree): *Laurus nobilis* (Lauraceae), $2n = 2x = 48$ *hort*

lavender: *Lavendula angustifolia* (Solanaceae) *hort*

leaf celery: *Apium graveolens* ssp. *secalinum* (Umbelliferae) *hort*

leaf lettuce >>> cutting lettuce

leek >>> garden leek

lemon: *Citrus limon* (Rutaceae), $2n = 2x = 18$ *hort* >>> hesperidia

lemon balm: *Melissa officinalis* (Labiatae) *hort*

lemon for candied peel: *Citrus medica* (Rutaceae), $2n = 2x = 18$ *hort* >>> hesperidia

lentil: the lentil ranks among the oldest and the most appreciated grain legumes of the old world (back to 8,000-7,000 BC originated in the Near East); it is cultivated from the Atlantic coast of Spain and Morocco in the west, to India in the east; the place of origin of the cultivated lentil is not known with certainty; the greatest variability in the cultigen is found in the Himalaya-Hindu-Kush junction area between India, Afghanistan, and Turkestan; presently, the major lentil producing regions are Asia (58 percent of the area) and the West Asia-North Africa region (37 percent of the acreage of developing countries); it is the most important pulse in Bangladesh and Nepal, where it significantly contributes to the diet; farmers also grow lentils in India, Iran, and Turkey; other significant producers in the developing world include Argentina, China, Ethiopia, Morocco, Pakistan, and Syria; global lentil production is growing rapidly; it has risen by 112 percent from 1.3 million tons in the period 1979-1981 to 2.9 million tons in the period 1993-1995, resulting from a 54 percent increase in area to 3.42 million hectares and an increase in productivity of 38 percent from 600 kg/ha to 825 kg/ha; ICARDA has a mandate to improve this crop, *Lens culinaris* (Leguminosae), $2n = 2x = 14$ *agr* >>> ICARDA

lespedeza: *Lespedeza stipulacea* (Leguminosae), $2n = 2x = 20$ *agr*

lesser bindweed >>> corn bindweed

lesser broomrape: *Orobanche minor* (Orobanchaceae) *bot agr*

lettuce: *Lactuca sativa* (Compositae), $2n = 2x = 18$ *hort*

Levant garlic: *Allium ampeloprasum* (Allicaceae), $2n = 4x, 6x$ = AAA'A" 32, 48 *hort*

licorice: *Glycyrhiza glabra* (Leguminosae), $2n = 2x = 16$ *hort fore*

lilac: *Syringa* spp. (Oleaceae) *hort*

lily: *Lilium* spp. (Liliaceae), $2n = 2x = 24$ *hort*

Lima bean: *Phaseolus lunatus* (Leguminosae), $2n = 2x = 22$ *hort* >>> pulse

lime: *Citrus* spp. (Rutaceae), $2n = 2x = 18$ *hort* >>> hesperidia

lime-tree: *Tilia* spp. (Tiliaceae) *fore*

linden >>> lime-tree

linseed >>> flax

litchi: *Litchi chinensis* (Sapindaceae), $2n = 2x = 30$ *hort*

little millet: *Panicum miliare* (Gramineae) *agr*

locust: *Ceratonia siliqua* (Leguminosae), $2n = 2x = 24$ *hort*

locust tree: *Robinia pseudoacacia* (Leguminosae) *fore*

loganberry: *Rubus loganobaccus* (Rosaceae) *hort*

Lolium x hybridum: an artificial grass hybrid between *Lolium perenne* x *Lolium multiflorum;* it is used in agriculture (Gramineae) *agr*

loose-leaved lettuce >>> cutting lettuce

lop grass >>> drooping brome

loquat: *Eriobotrya japonica* (Rosaceae) *hort*

lovage: *Levisticum officinale* (Umbelliferae) *hort*

lovegrass: *Eragrostis* spp. (Gramineae) *bot agr*

love-in-a-mist: *Nigella damascena* (Ranunculaceae) *bot agr*

lowbush blueberry: *Vaccinium angustifolium* (Ericaceae), $2n = 2x = 24$ *hort*

lucerne >>> alfalfa

luffa: *Luffa cylindrica, L. acutangula* (Cucurbitaceae), $2n = 2x = 26$ *hort*

lulos: *Solanum quitoense* (Solanaceae), $2n = 2x = 24$ *hort*

lupin: *Lupinus* spp. (Leguminosae) *bot agr hort* >>> sweet yellow lupin >>> white lupin >>> blue lupin

lychee >>> litchi

lyme grass: *Elymus europaeus* (Gramineae) *bot agr*

M

Macadamia nut: *Macadamia* spp. (Proteaceae), $2n = 2x = 28$ *fore hort*

macaroni wheat >>> durum wheat >>> wheat

Macha wheat: *Triticum macha* (Gramineae) *bot agr* >>> wheat

Madagascar bean >>> Lima bean

magua: *Tropaeolum tuberosum* (Tropaeolaceae) *hort*

maize *U.S. syn* **corn:** maize is the world's fourth most important crop, behind only wheat, rice, and potatoes; there are more than 327 million acres of maize planted each year, worldwide; the United States produces over 526 million U.S. tons per year; other countries that produce a large amount of maize include: Africa, Argentina, Brazil, China, France, India, Mexico, Romania, Russia, and South Africa; the best place to grow maize is in well-aerated, deep, warm soil with a lot of organic matter, nitrogen (N), phosphorus (P), and potassium (K); semi-high summer temperatures, warm nights, and adequate, well distributed rainfall helps it even more during the growing season; the growing season and day length also are a factor in growth; maize evolved in Mexico or Central America about 6,000 years ago; it developed from a small, wild plant with a pod-pop of cob; modern maize has its cob enclosed in the one sheath, which prevents dissemination of seed; it has imperfect flowers; it is monoecious and cross-pollinating; almost all maize grown in the world are hybrid maizes; there are seven types of maize: (1) *flint;* flint maize kernels are hard and smooth and have little soft starch; flint was probably the first maize Europeans ever laid eyes on; is not grown in the United States as much as it is in Asia, Central America, Europe, and South America; in temperate zones, flint maize matures earlier, has better germination, and the plant vigor is earlier than in dent, (2) *flour;* flour maize contains a lot of soft starch, and has almost no dent; though it is not used much anymore, it is grown in the drier sections of the United States and in the Andean region of South America; it is an older type of maize, and was found in a lot of graves of the Aztecs and Incas; since the kernel is so soft, the American Indians could make it into flour, (3) *pop;* it is an extreme form of flint; it has a very small proportion of soft starch; it is a very minor crop, and is grown mostly for humans to eat; the reason it "pops" so well, is because of the horny endosperm, which is a tough, stretchy material that can resist the pressure of steam, which is generated in the hot kernel until it has enough force to explode or "pop," (4) *sweet;* this type of maize has an almost clear, horny kernel when it is still young; the kernels become wrinkled when dry; the ears can be eaten fresh, or can be stored in cans; the only difference between sweet and dent maize is that sweet genotype has a gene that prevents some sugar from being converted into starch; it is grown a lot as a winter crop in the southern United States, (5) *dent;* getting its name from the dent in the crown of the seed, it is grown more than any other type of maize; millions of tons of grain are produced from dent corn; it is used for human and industrial use, and for livestock feed; the starch reaches the summit of the seed, and the sides are also starchy; the denting is caused by the drying and shrinking of the starch; the dent corn grown in the Corn Belt of the United States came from a mix of New England flints and gourseed (an old variety of corn grown by the Indians in southeastern North America, (6) *waxy;* seeds appear waxy; chemically, it has a different type of starch than normal starch; it was developed in China, and some waxy mutations have occurred in American dent strains; very little is grown, and that which is, is used for producing a starch similar to tapioca starch, (7) *pod;* pod maize is not grown commercially, but it is used a great deal in studying the phylogenesis of maize; it resembles varieties of the primitive forms; every seed is enclosed in a pod and the whole ear is also enclosed in a husk, *Zea mays* (Gramineae), $2n = 2x = 20$, 2C DNA content 11.0 pg *agr* >>> Tables 15, 16, 32, 35

Malabar nightshade: *Basella alba, B. rubra* (Chenopodiaceae) *hort*

mallow: *Malva* spp. (Malvaceae) *hort*

mandarin: *Citrus reticulata* (Rutaceae), $2n = 2x = 18$ *hort* >>> hesperidia

mangel: *Beta vulgaris* (Chenopodiaceae)

mangel-wurzel >>> beet

mango: *Mangifera indica* (Anacardiaceae), $2n = 2x = 40$ *hort*

mangold: *Beta vulgaris* var. *cicla* (Chenopodiaceae) *agr hort*

mangosteen: *Garcinia mangostana* (Guttoferae) *hort*

Manila aloe >>> cantala

Manila hemp >>> abaca

manna: *Cassia fistula* (Leguminosae) *hort*

marigold: *Tagetes* spp. (Compositae) *hort*

marjoram: *Origanum majorana* (Labiatae) *hort*

marrow fat pea >>> green pea

marrow-stem kale: *Brassica oleracea* convar. *acephala* var. *medullosa* (Brassicaceae) *agr hort* >>> Figure 8

marsh trefoil: *Menyanthes trifoliata* (Menyanthaceae) *bot agr*

marvel of Peru: *Mirabilis jalapa* (Nyctaginaceae) *hort*

mash >>> urd bean

masterwort: *Astrantia* spp. (Umbelliferae) *bot agr*

maté: *Ilex paraguariensis* (Aquifoliaceae), $2n = 4x = 40$ *hort*

matgrass: *Nardus* spp. (Gramineae) *bot agr*

Mauritius hemp: *Furcraea gigantea* var. *willemettiana* (Agavaceae), $2n = 2x = 60$ *agr*

May beet: *Brassica napus* var. *rapa* (Cruciferae) *agr*

meadow fescue: *Festuca pratensis* (Gramineae), $2n = 2x = 14$ *agr*

meadow foxtail: *Alopecurus pratensis* (Gramineae) *bot agr*

meadow saffron: *Colchicum autumnale* (Liliaceae) *bot agr hort*

meadow soft grass >>> common velvet grass

meadowfoam: *Limnanthes* spp. (Limnanthaceae) *bot agr*

medlar: *Mespilus germanica* (Rosaceae) *bot hort*

melon >>> cantaloupe

Michaelmas daisy >>> aster

mignonette: *Reseda* spp. (Resedaceae) *hort*

milfoil >>> yarrow

milkvetch >>> astragalus

millet: a general name for a variety of species that is grown in similar regions as sorghum; millet is more drought resistant; different millets may have evolved in different parts of the world, including Africa and Asia; it has been grown in China for about 5,000 years; five types are described: (1) common, *Panicum miliaceum,* (2) finger, *Eleusine coracana,* (3) foxtail, *Setaria italica,* (4) pearl, *Pennisetum americanum,* (5) Japanese barnyard, *Echinochloa frumentacea,* (Gramineae) *agr hort* >>> Table 35

moneywort: *Lysimachia nummularia* (Primulaceae) *bot agr*

Mongolian oak: *Quercus mongolica* (Fagaceae), $2n = 2x = 24$ *fore*

monkeyflower: *Mimulus* spp. (Scrophulariaceae) *hort*

moong >>> mung bean

morello: *Cerasus vulgaris* var. *austera* (Rosaceae) *hort*

morello cherry >>> morello

Moricandia >>> wild crucifer

moth bean: *Vigna aconitifolia* (Leguminosae) *hort* >>> pulse

mountain clover: *Trifolium montanum* (Leguminosae) *bot agr*

mountain savory: *Satureja montana* (Labiatae) *hort*

mountain spinach: *Atriplex hortensis* ssp. *viridis* (Chenopodiaceae) *hort*

mugwort: *Artemisia vulgaris* (Asteraceae) *hort*

mulberry: *Morus alba* (Moraceae), $2n = 2x = 28$ *hort*

multiplier (onion): *Allium cepa* var. *aggregatum* (Alliaceae), $2n = 2x = 16$ *hort*

mungbean: most widely cultivated throughout the southern half of Asia including India, Pakistan, Bangladesh, Sri Lanka, Laos, Cambodia, Vietnam, eastern parts of Java, eastern Malaysia, South China, and Central Asia, *Phaseolus radiatus syn Vigna radiata* (Leguminosae), $2n = 2x = 22$, 2C DNA content 0.53 pg *hort* >>> pulse

mungo bean >>> urd bean

mushroom: the cultivated mushroom *(A. bisporus)* or any edible fungus similar to it in appearance, *Agaricus bisporus hort*

musk melon >>> cantaloupe

musk okra: *Hibiscus esculentus* (Malvaceae) *hort*

mustards: *Brassica* spp. and *Sinapis alba* (Cruciferae) *agr* >>> Figure 8

myrtle: any plant of the genus *Myrtus,* for example, *M. communis,* having evergreen leaves, fragrant white flowers,and aromatic berries (Myrtaceae) *bot hort*

N

narbon vetch: *Vicia narbonensis* (Leguminosae), $2n = 2x = 14$ *hort*

narcissus: *Narcissus pseudo-narcissus, N. poiticus* (Iridaceae) *hort*

narrow-leafed lupin >>> blue lupin

nasturzium: *Tropaeolum majus* (Tropaeolacae) *hort*

navet petit de Berlin: *Brassica campestris* var. *rapa* (Cruciferae) *hort* >>> Figure 8

navy >>> French bean

New Zealand spinach: *Tetragonia tetragonoides* (Tetragoniaceae) *hort*

niger seed: *Guizotia abyssinica* (Compositae), $2n = 2x = 20$ *hort agr*

nira >>> Chinese chive

noble cane >>> sugarcane

nopal >>> pencas

notch-seeded buckwheat: *Fagopyrum emarginatum* (Polygonaceae), $2n = 2x = 16$ *agr*

nutmeg: *Myristica fragrans* (Myristicaceae), $2n = 6x = 42$ *hort*

O

oak: *Quercus* spp. (Fagaceae) *fore* >>> Mongolian oak >>> daimyo oak

oats: this group of cereal species shows a wide range of ecological adaptability; usually it is a crop for temperate areas with seaboard climates; it originates from Asia Minor; it seems that it was introduced into Europe and Asia as weeds among the earliest domesticated cereals, *Avena* spp. (Gramineae), $2n = 6x$, AACCDD = 42, 2C DNA content 43 pg *bot agr* >>> Tables 14, 15, 16, 30, 32, 35 >>> wild oat

oca: *Oxalis tuberosa* (Oxalidaceae), $2n = 6x = 66$ *hort*

oil palm: *Elaeis guineensis* (Palmae), $2n = 2x = 32$ *hort* >>> Table 16

oil radish: *Raphanus sativus* var. *oleiformis* (Cruciferae) *agr*

oilseed rape >>> rapeseed

okra: *Abelmoschus esculentus* (Malvaceae), $2n = 4x$, T'T'YY = 130 *hort*

old cocoym >>> taro

olive: *Olea europaea* (Oleaceae), $2n = 2x = 46$ *hort*

onion: plants belonging to the genus *Allium* are recognized by the pungent smell or taste of "onion," which they produce when their tissues are crushed or tasted; they are biennial or perennial bulbous herbs; the bulbs are formed by the swollen leaf bases attached to the base of the underground part of the stem; in a few species there are very long sheathing leaf bases, which are much less swollen, and other have rhizomes or storage roots; onions are widely spread throughout the temperate Northern Hemisphere of the world; more than 500 species are common in the old world; there are more than 80 species found in the new world; chives is the only species found in both the old and the new worlds; onions have been used by humans for several centuries (e.g., during ancient civilizations of Egypt, Rome, Greece, and China); *Allium cepa* (Alliaceae), $2n = 2x = 16$, 2C DNA content = 33.5 pg *hort* >>> chive

opuntia: fruits are eaten, leaves are used as fodder for animals; within the subtribe Opuntioideae there are several species used as crop and horticultural plants; there is special use for production of the stain "carmine red" by the ecto-parasite Cochenille *(Dactylopius coccus) agr hort*

orache: *Atriplex hortensis* (Chenopodiaceae) *hort*

orange: *Citrus sinensis* (Rutaceae), $2n = 2x = 18$ *hort* >>> hesperidia

orchard grass: *Dactylis glomerata* (Gramineae) *agr*

oregano: *Origanum vulgare* (Labiatae) *hort*

oryzopsis: *Oryzopsis miliacea* (Gramineae) *bot hort*

osier >>> basket willow

ox-eye chamomile: *Anthemis tinctoria* (Asteraceae) *hort*

oyster cap fungus: *Pleurotus ostreatus hort*

P

pak-choi: *Brassica campestris* ssp. *chinensis* (Brassicaceae), $2n = 2x$, AA = 20 >>> Figure 8

pansy: *Viola wittrockiana* ssp. *hiemalis* (Violaceae) *hort*

papaya: *Carica papaya* (Caricaceae), $2n = 2x = 28$ *hort*

paprika: *Capsicum annuum* (Solanaceae) *hort*

Para nut >>> Brazil nut

park red fescue: *Festuca rubra genuina* (Gramineae) *agr*

parsley: *Petroselinum crispum* (Umbelliferae), $2n = 2x = 22$ *hort*

parsnip: *Pastinaca sativa* (Umbelliferae), $2n = 2x = 22$ *hort*

passion fruit: *Passiflora edulis* (Passifloraceae), $2n = 2x = 18$ *hort*

pattypan squash: *Cucurbita pepo* convar. *pattisonina* (Cucurbitaceae), $2n = 2x = 40$ *hort*

pea >>> garden pea

peach: *Persica davidiana syn Prunus persica* (Rosaceae), $2n = 2x = 16$ *hort*

peanut: also called groundnut; a four-foliate legume with yellow sessile flowers and subterranean fruits; it is native to South America, and it originated between southern Bolivia and northern Argentina, from where it spread throughout the New World as Spanish explorers discovered its versatility; at present, farmers in Asia and Africa also cultivate it under a wide range of environmental conditions in areas between 40 degrees south and 40 degrees north of the equator; the largest producers of groundnut are China and India, followed by sub-Saharan African countries and Central and South America, most of the crop is produced where average rainfall is 600 to 1,200 mm and mean daily temperatures are more than 20°C; it became a major oilseed crop of the tropics and subtropics; the seeds are rich in protein and oil; the genus *Arachis* is widely distributed in the north and central regions of South America; a special variety is the Spanish peanut, its kernels are small to medium size with smooth skin, and the kernel color ranges form a pale pinkish buff to a light brown during storage; this type of peanut is used predominantly in peanut candy, although significant quantities also are used for salted peanuts and peanut butter; they have a higher oil content than other types of peanuts; the so-called runner peanut is the most widely used peanut for making peanut butter, peanut candies, baked goods, and snack nuts, *Arachis hypgaea* (Leguminosae), $2n = 2x = 20$ *agr* >>> Table 16 >>> Kersting's groundnut >>> gynophore

pear: *Pyrus communis* (Rosaceae), $2n = 2x$, $3x = 34$, 51 *hort*

pearl millet: *Pennisetum americanum* (Gramineae) >>> millet

pecan (nut): *Carya laciniosa, syn C. pecan* (Juglandaceae), $2n = 2x = 32$ *hort*

pencas: edible fleshy twigs and fruits; cropping areas about 50,000 ha (Mexico), ~60,000 ha (Tunisia), ~100,000 ha (Italy), *Opuntia maxima syn Ficus indica* (Cactaceae) *hort* >>> opuntia

pensacola >>> Bahia grass

peony: *Paeonia* spp. (Paeoniaceae) *hort*

pepper >>> sweet pepper

peppermint: *Mentha piperita* (Labiatae) *hort*

perennial ryegrass: *Lolium perenne* (Gramineae), $2n = 2x = 14$, 2C DNA content 4.16 pg *agr*

perennial sweet leek >>> garden leek

Persian clover: *Trifolium resupinatum* (Leguminosae) *agr*

Persian wheat: *Triticum carthlicum, T. persicum* (Gramineae) *bot agr* >>> wheat

persimmon: *Diospyros kaki* (Ebenaceae), $2n = 6x = 90$ *hort*

pe-tsai >>> Chinese cabbage

Philippine sisal >>> cantala

picotee >>> carnation

pigeonpea: the generic name *Cajanus* derived from the word "Katjang" or "Catjang" of Malay language meaning pod or bean; Africa is regarded as the place of origin, *Cajanus cajan* (Leguminosae), $2n = 2x = 22$ *hort* >>> pulse

pimento: *Pimenta dioica* (Myrtaceae), $2n = 2x = 22$ *hort*

pimpernel: *Pimpinella* spp. (Umbelliferae) *hort*

pincushion flower >>> scabious

pine: *Pinus* spp. (Pinaceae) *fore* >>> Ponderosa pine >>> red pine >>> Eastern white pine

pineapple: *Ananas comosus* (Bromeliaceae), $2n = 2x = 50$ *hort*

pink >>> carnation

pistachio (nut): *Pistacia vera, P. mutica* (Anacardiaceae), $2n = 2x = 30$ *hort*

pita >>> Henequen agave

pita savila >>> savila

pitera de gogo >>> savila

plane-tree: *Platanus hybrida* (Platanaceae) *fore hort*

plantain: *Plantago* spp. (Plantaginaceae) *bot agr* >>> plane-tree

plum: *Prunus* spp. (Rosaceae), $2n = 6x$, CCSSSS = 48 *hort* >>> Damson plum >>> European plum

Polish wheat: *Triticum polonicum* (Gramineae) *bot agr* >>> wheat

pomegranate: *Punica granatum* (Punicaceae), $2n = 2x = 16$ *hort*

Ponderosa pine: *Pinus ponderosa* (Pinaceae) *fore*

poplar: *Populus* spp. (Salicaceae) *fore*

poppy: *Papaver somniferum* (Papaveraceae), $2n = 2x$, $4x = 22$, 44 *hort*

portulac: *Portulaca grandiflora* (Portulacaceae) *hort*

potato: domesticated more than 6,000 years ago in the high Andes of South America; in the 16th century, Spanish conquistadors brought the potato from Peru to Europe, where it took two centuries before potatoes were introduced into the European diet; at present, potato is the fourth most important crop in developing countries after rice, wheat, and maize; more than 3 billion people consume potatoes; potato production is expanding at an unprecedented rate; approximately 30 percent of the world's potato crop is currently produced in developing countries, mainly by small-scale farmers; China is the largest producer; the center of origin is South America; the potato derives from the andigena subspecies, which is the progenitor of the tuberosum subspecies; the latter originated from different related diploid populations in different locations through sexual polyploidization; recent molecular studies even demonstrate the close relationship to tomato; particularly in potato haploids became important for breeding and genetics, *Solanum tuberosum* (Solanaceae), $2n = 4x$, AAA'A' = 48 *agr* >>> Tables 17, 35

potato onion >>> multiplier (onion)

prickly burr >>> American chestnut

primrose: *Primula* spp. (Primulaceae) *hort*

pumpkin: *Cucurbita moschata* (Cucurbitaceae), $2n = 2x = 40$ *hort* >>> vegetable marrow

purple vetch: *Vicia benghalensis* (Leguminosae) *bot hort*

pyrethrum: *Chrysanthemum* spp. (Compositae), $2n = 2x = 18$ *hort*

Q

quackgrass >>> *Agropyron*

quassia: *Quassia amara* (Simaroubaceae), $2n = 4x = 36$ *hort fore*

Queensland arrowroot: *Canna edulis* (Cannaceae), $2n = 2x$, $3x = 18, 27$ *hort*

quick-grass >>> *Agropyron*

quince: *Cydonia oblonga* (Rosaceae), $2n = 2x = 34$ *hort*

quinine: *Cinchona officinalis, Q. pubescens syn C. succirubra* (Rosaceae), $2n = 2x = 34$ *fore*

quinoa: *Chenopodium quinoa* (Chenopodiaceae), $2n = 4x = 36$ *hort*

R

radish: *Raphanus sativus* var. *radicula* (Cruciferae), $2n = 2x$, RR = 18 *hort*

ragi >>> finger millet

ragweed: *Abrosia eliator* (Asteraceae) *bot agr*

rakkyo: *Allium cepa syn A. chinense syn A. bakeri* (Alliaceae), $2n = 2x, 3x, 4x = 16, 24, 32$ *hort*

rambutan: *Nephelium lappaceum* (Sapindaceae), $2n = 2x = 22$ *hort*

ramie: one of the oldest vegetable fibers and has been used for thousands of years; the fibers are found in the bark of the stalk; the fiber is very fine and silk-like, naturally white in color and has a high lus; it was used for Chinese burial shrouds over 2,000 years ago, long before cotton was introduced in the Far East; it is classified chemically as a cellulose fiber, just as cotton, linen, and rayon; leading producers of ramie are China, Taiwan, Korea, the Philippines, and Brazil, *Boehmeria nivea* (Urticaceae), $2n = 2x = 14$ *agr*

rampion: *Campanula rapunculus* (Campanulaceae) *hort*

rape: *Brassica* spp. (Cruciferae) *agr* >>> Figure 8

rapeseed: *Brassica napus* ssp. *napus* (Cruciferae), $2n = 4x$, AACC = 38 *agr* >>> Figure 8 >>> Tables 16, 35 >>> Canola

raspberry: *Rubus* spp. (Rosaceae), $2n = 2x = 14$ *hort*

ratabaga >>> rutabaga

red bean >>> Adzuki bean

red beet: *Beta vulgaris* ssp. *vulgaris* convar. *crassa* var. *conditiva* (Chenopodiaceae)

red cabbage: *Brassica oleracea* convar. *capitata* var. *rubra* (Brassicaceae) *hort* >>> Figure 8

red clover: *Trifolium pratense* (Leguminosae), $2n = 2x = 14$ *agr* >>> Table 35

red currant: *Ribes sativum* (Grossulariaceae), $2n = 2x = 16$ *hort*

red fescue: *Festuca rubra* (Gramineae) *agr*

red pepper >>> sweet pepper

red pine: *Pinus resinosa* (Pinaceae) *fore*

red plum >>> American plum

red rice >>> African rice

red top >>> black bent

red-head cabbage >>> red cabbage

reed canary grass: *Phalaris arundinacea* (Gramineae) *agr*

rhea >>> ramie

rhubarb: *Rheum rhaponticum* (Polygonaceae), $2n = 4x = 44$ *hort*

ribbon grass >>> reed canary grass

rice: second to wheat in terms of area and amount of grain produced; it probably originated from India or Southwest Asia, where several wild species are found; rice culture spread to China about 5,000 years ago and to Europe about 2,500 years ago; it is an annual grass, 80-150 cm height; the inflorescence is a loose panicle containing about 100 single flowered spikelets; it is normally self-pollinating; the mature kernels are enclosed in the pelea and lemma and their color varies from white to brown; it grows with its roots in water; rice transports oxygen to its roots from the leaves; categories of rice are based on length of grain: short (5 mm), medium (6 mm), long (7 mm); short-grain types of the *japonica* type have short straw, whereas long-grain types of the *indica* type usually have taller and weaker stems; rice is a short-day plant; rice growing was the key for the development of Asian civilizations and certain African cultures; Asiatic species of rice appear to have diverse origins and to be derived from a complex in which *Oryza rufipogon* and *O. nivara* play a major role; African species are believed to have been domesticated from *O. barthii;* rice remains, along with wheat, the main diet staple of humans, especially in high-density areas in hot and wet tropical and subtropical areas; the annual production is about 500 million tons, *Oryza sativa* (Gramineae), $2n = 2x$, AA = 24, 2C DNA content 2 pg *agr* >>> Tables 15, 32

rice bean: *Vigna umbellata* (Leguminosae) >>> pulse

ricegrass: *Oryzopsis miliacea* (Gramineae) *agr*

rice-paper plant: *Tetrapanax papyriferum* (Araliaceae), $2n = 2x = 24$ *hort*

rivet wheat: *Triticum turgidum* (Gramineae) *agr* >>> wheat

rock maple >>> sugar maple

rocket >>> rocket salad

rocket salad: *Eruca sativa syn E. vesicara* (Cruciferae) *agr hort*

rocoto: *Capsicum pubescens* (Solanaceae), $2n = 2x = 24$ *hort*

roquette >>> rocket salad

rose: *Rosa* spp. (Rosaceae) *hort*

roselle: *Hibiscus sabdariffa* (Malvaceae) *hort*

rosemary: *Rosmarinus officinalis* (Lamiaceae) *hort*

rubber (tree): *Hevea brasiliensis* (Euphorbiaceae), $2n = 2x = 36$ *fore*

runch: *Raphanus raphanistrum* (Cruciferae) *bot agr*

runner bean >>> kidney bean

Russian dandelion: *Scorzonera tau-saghyz* (Compositae) *hort*

rutabaga >>> Swede >>> Figure 8

rye: a cereal that played a major role in the feeding of European populations throughout the Middle Ages, owing to its considerable winter hardiness; the cultivated rye resulted from crossbreeding between *Secale vavilovii* and the perennial species, *S. anatolicum* and *S. montanum;* it was domesticated rather late, having evolved primitively as weeds among the cereals cultivated earlier; the world production amounts about 30 million tons; it is mainly used in baking for black bread, in confectionary for gingerbread, blinis, etc. or for the production of rye whiskey; hybrid rye is widely grown in Germany, Poland, and other European countries, *Secale cereale* (Gramineae), $2n = 2x$, $R^{cer}R^{cer} = 14$, 2C DNA content 18.9 pg *agr* >>> ergot >>> trisomic >>> translocation tester set >>> pentosan >>> Tables 15, 32, 35

rye brome: *Bromus secalinus* (Gramineae) *bot agr*

ryegrass >>> perennial ryegrass

S

safflower: originally grown for the dye extracted from its florets and as minor oil crop; *Carthamus tinctorius* (Compositae), $2n = 2x$, $BB = 24$ *hort*

saffron: *Crocus sativus* (Iridaceae) *hort*

sage: *Salvia officinalis* (Labiatae) *hotr*

sago palm: *Metroxylon sagu* (Palmae) *hort agr fore*

sainfoin: *Onobrychis viciifolia* (Leguminosae), $2n = 2x, 4x = 14, 28$ *agr*

salsify: *Tragopogon porrifolius* (Compositae) *hort*

saltbush: *Atriplex* spp. (Chenopodiaceae) *bot agr*

san chi: *Gynura pinnatifida* (Compositae), $2n = 2x = 20$ *hort*

sanwa millet >>> barnyard grass

sapodilla: *Manilkara zapota* (Sapotaceae), $2n = 2x = 26$ *hort*

sarson >>> yellow-seeded sarson

savila: used in many countries of Africa, the Canary Islands, and America as an herb plant, already described 2,500 years B.C., *Aloe vera* (Aloaceae) *hort agr*

savory: *Satureja hortensis* (Labiatae) *hort*

Savoy cabbage: *Brassica oleracea* var. *sabauda* (Cruciferae) *hort* >>> Figure 8

scabious: *Scabiosa* spp. (Dipsacaceae) *bot agr*

scarlet runner bean >>> kidney beans

scorzonera >>> black salsify

scurvy grass: *Cochlearia officinalis* (Cruciferae) *bot agr*

sea lavender: *Limonium sinuatum* (Plumbaginaceae) *hort*

sea sand-reed >>> beach grass

seakale: *Crambe maritima* (Cruciferae) *hort*

seradella: *Ornithopus sativus* (Leguminosae), $2n = 2x = 14$ *agr*

serpent root >>> black salsify

sesame: the cultivated sesame (edible sesame) has been differentiated into about 3,000 varieties and strains, distributed extensively from the tropical to temperate zones in the world; *Sesamum indicum* (Pedaliaceae), $2n = 2x = 26$ *agr*

shaddok: *Citrus maxima* (Rutaceae) *hort* >>> hesperidia

shallot: *Allium cepa* var. *ascalonicum* (Alliaceae), $2n = 2x = 16$ *hort*

sheep('s) fescue: *Festuca ovina* (Gramineae) *agr*

short American staple cotton: *Gossypium herbaceum* (Malvaceae), $2n = 2x$, $A_1 A_1 = 26$, 2C DNA content 2.1 pg *agr*

shot wheat >>> Indian dwarf wheat

Siberian hazel: *Corylus heterophylla* (Corylaceae), $2n = 2x = 28$ *hort*

Sieva bean >>> Lima bean

silver vine: *Actinidia polygama* (Actinidiaceae), $2n = 2x$, $4x = 58$, 116 *hort*

sisal: used for fiber production, *Agave sisalana* (Agavaceae), $2n = 5x = 138$-149 (?) *agr* >>> Henequen agave

skirret: *Sisum sisarum* (Umbelliferae) *hort*

slender foxtail: *Alopecurus agrestis* (Gramineae) *bot agr*

small melilot: *Melilotus indica* (Leguminosae) *agr*

small radish >>> radish

small timothy: *Phleum bertolonii* (Gramineae) *bot agr*

small-leafed sweet basil >>> basil

smooth brome: *Bromus inermis* (Gramineae) *bot agr*

smooth-stalked meadow grass >>> Kentucky bluegrass

snap bean >>> French bean

snapdragon: *Antirrhinum* spp. (Scrophulariaceae) *hort*

soft brome: *Bromus mollis* (Gramineae) *bot agr*

sorghum: originated from Africa about 6,000 years ago; it then spread to India, China, Europe, and America; it is an annual grass with a panicle containing two types of spikelets (pedicelled and sessile); the sessile spikelets contain perfect flowers, the pedicelled spikelets contain flowers that are either male only or sterile; self-pollination is usual; sorghum exhibits the C4 photosynthetic pathway, which is more efficient than the C3 pathway; it is a short-day plant; the sorghum grain is more or less rounded, about 6 mm in diameter, often with colored lines; the color varies from white to brown or black; there are four types of *Sorghum bicolor* (1) milo—drought resistant, many tillers, early maturing, (2) kafir—thick stalks, large leaves, used for forage and grain, (3) sweet—sweet juice in the stalk, grows up to 3 m height, for animal fodder, (4) broomcorn—has branches, used for making brooms, *Sorghum bicolor* (Gramineae), $2n = 2x = 20$ *agr* >>> millet >>> Tables 15, 16

sorrel: *Rumex patienta, R. acetosa* (Polygonaceae) *hort*

sour cherry: *Prunus cerasus* (Rosaceae), $2n = 4x = 32$ *hort*

soursop: *Annona muricata* (Annonaceae), $2n = 2x = 14$ *hort*

soya >>> soybean

soybean: derives from the wild annual progenitor, *G. soja;* most genetic diversity is found among 12 wild perennial species, which are indigenous to Australia, South Pacific Islands, Taiwan, and southern China; *Glycine max* (Leguminosae), $2n = 2x$, GG = 40 *agr* >>> Table 16

speedwell: *Veronica* spp. (Scrophulariaceae) *bot agr*

spikenard >>> aralia

spinach: *Spinacia oleracea* (Chenopodiaceae), $2n = 2x = 12$ *hort*

spotted medick: *Medicago arabica* (Leguminosae) *agr*

spring oilseed rape >>> rapeseed

spring onion >>> Welsh onion

sprouting broccoli: *Brassica olearacea* convar. *Botrytis* var. *cymosa* (Brassicaceae) *hort* >>> Figure 8

squash: *Cucurbita maxima* (Cucurbitaceae), $2n = 2x = 40$ *hort*

star grass: *Cynodon* spp. (Gramineae) *bot agr*

starwort >>> aster

statice >>> sea lavender

stinging nettle: *Urtica* spp. (Urticaceae) *bot agr*

stitchwort: *Stellaria* spp. (Caryophyllaceae) *bot agr*

stone onion >>> Welsh onion

straw flower: *Helichrysum bracteatum* (Asteraceae) *hort*

strawberry: *Fragaria ananassa* (Rosaceae), $2n = 8x = 56$ *hort*

strawberry clover: *Trifolium fragiferum* (Leguminosae) *bot agr*

strawberry peach >>> Chinese gooseberry

strawberry spinach: *Chenopodium capitatum* (Chenopodiaceae) *hort*

stubble turnip: *Brassica rapa* var. *rapifera* (Brassicaceae), $2n = 2x$, AA = 20 *agr* >>> Figure 8

subterranean clover: *Trifolium subterraneum* (Leguminosae) *hort*

subterranean vetch: *Vicia sativa* ssp. *amphicarpa* (Leguminosae) *bot agr*

suckling clover: *Trifolium dubium* (Leguminosae) *bot agr*

Sudan grass: *Sorghum sudanese* (Gramineae) *agr*

sugar apple >>> sweetsop

sugar maple: *Acer saccharum* (Aceraceae), $2n = 2x = 26$ *fore hort*

sugar palm: *Arenga saccharifera* (Palmae), $2n = 2x = 32$ *fore hort*

sugar pea: *Pisum sativum* var. *axiphium* (Leguminosae) *hort*

sugarbeet: a crop with major economic importance, providing about 45 percent of the world's sugar production; wild relatives of sugarbeet are found in Europe, Asia Minor, and North Africa; polyploid breeding has long been a major concern; at present, the majority of commercial hybrids are triploids produced on diploid seed parents pollinated by tetraploids; *Beta vulgaris* ssp. *vulgaris* convar. *crassa* var. *altissima* (Chenopodiaceae), $2n = 2x$, VV = 18 *agr* >>> Table 35

sugarcane: it is a tropical perennial grass, thriving in humid regions; it is vegetatively propagated by planting a "seed-piece," a piece of cane stalk with at least one bud; it resprouts annually from underground buds on basal portions of old stalks; depending on variety and growing conditions, a 2-4 pound stalk with 15 percent sugar will be produced in about 12 months from an original planting or 9-11 months from regrowth; types of sugarcane can be placed into one of three categories according to its physical and chemical characteristics: (1) chewing cane contains fibers that stick together when chewed, making it easier to spit out the pulp once the sugar has been consumed, (2) cane for crystals must contain a high percentages of sucrose, since this is the sugar type that easily forms into crystals when concentrated, (3) syrup canes contain less sucrose and more of other sugars, allowing the juice to be concentrated into syrup and still not form crystals; several old-named varieties are still available; sugarcane cultivars are hybrid products of at least three or four *Saccharum* spp.; they are high polyploids and are genetically complex; the cultivated forms base on *S. sinense* and *S. barberi* species from India and China; the center of diversity is Indonesia; *Saccharum officinarum* (Gramineae), $2n = 2x = 100\text{-}130$ *agr*

sun hemp: *Crotalaria juncea* (Leguminosae), $2n = 2x = 16$ *agr*

sun plant >>> portulac

sunflower: the genus of cultivated sunflower *(H. annuus)* includes more than 50 species; it contains one other economically important species, *H. tuberosus,* plus several ornamentals; wild species played an important role in genetic improvement of sunflower; *Helianthus* spp. (Compositae), $2n = 2x = 34$, 2C DNA content 6.6-9.9 pg *agr hort* >>> Tables 31, 35 >>> Jerusalem artichoke

swamp meadow grass: *Poa palustris* (Gramineae) *bot agr*

Swede >>> rapeseed

Swede's rape >>> rapeseed

Swedish turnip >>> Swede

sweet basil >>> basil

sweet cherry: *Prunus avium* (Rosaceae), $2n = 2x = 16$ *hort*

sweet chestnut: *Castanea sativa* (Fagaceae) *hort fore*

sweet cicily: *Myrrhis odorata* (Umbelliverae) *hort*

sweet clover: belongs to genus with 19 species native to Eurasia from Central Europe to Tibet; used for forage and for soil enrichment through nitrogen fixation; *Melilotus albus* (Leguminosae), $2n = 2x = 16$ *agr*

sweet corn >>> maize

sweet flag >>> calamus

sweet orange: *Citrus sinensis syn C. aurantium* (Rutaceae) *hort* >>> hesperidia

sweet pea: *Lathyrus odoratus* (Leguminosae) *hort*

sweet pepper: species of *Capsicum* are grown throughout the tropics and are valuable crops under protected cultivation in many temperate countries; peppers with pungent fruits are used as a spice either fresh, dry, or as extracted oleoresin; those with nonpungent fruits are used as a vegetable; the genus is native to the Americas, where the fruits have been used by humans for over 5,000 years; the tabasco pepper is a large-fruited form of domesticated *C. frutescens,* while small-fruited forms of *C. frutescens* are cultivated for oleoresin extraction; most species are self-compatible and are facultative inbreeders; *Capsicum annuum* (Solanaceae), $2n = 2x = 24$ *hort* >>> tabasco

sweet potato: domesticated in Central America more than 5000 years ago; the crop was reportedly introduced into China in the late 16th century; because of its hardy nature and broad adaptability, and because its planting material can be rapidly multiplied from very few roots, sweet potato spread through Asia, Africa, and Latin America during the 17th and 18th centuries; sweet potato has secondary centers of genetic diversity; in Papua New Guinea and in other parts of Asia, many types of sweet potato can be found that are genetically distinct from those found in their area of origin; today widely grown in tropical and temperate regions of the world due to its high yield, high nutritive values, and adaptability to a wide range of soils and drought; *Ipomoea batatas* (Convolvulaceae), $2n = 6x$, BBBBBB = 90 *hort agr*

sweet root >>> calamus

sweet sorghum: *Miscanthus sacchariflorus* (Gramineae), $2n = 4x = 76$ *agr*

sweet williams: *Dianthus barbatus* (Caryophyllaceae) *hort*

sweet yellow lupin: *Lupinus luteus* (Leguminosae) *agr*

sweetsop: *Annona squamosa* (Annonaceae), $2n = 2x = 14$ *hort*

Swiss chard: *Beta vulgaris* var. *cicla* (Chenopodiaceae) *hort*

sword bean: *Canavalia gladiata* (Leguminosae) *hort* >>> pulse

T

ta'amu >>> giant taro

tabasco: *Capsicum frutescens* (Solanaceae) *hort* >>> sweet pepper

table watermelon: *Citrullus vulgaris* var. *edulis* (Cucurbitaceae) *hort*

Tahiti arrowroot: *Tacca leontopetaloides* (Taccaceae) *hort*

tall fescue: *Festuca arundinacia* (Gramineae), $2n = 6x = 42$ *agr*

tall oatgrass: *Arrhenatherum elatius* (Gramineae) *bot agr*

tall wheatgrass: *Agropyron elongatum* (Gramineae) *bot agr*

tamarind: *Tamarindus indica* (Leguminosae), $2n = 2x = 24$ *fore hort*

tancy phacelia: *Phacelia tanacetifolia* (Hydrophyllaceae) *agr*

tangerine >>> mandarin

tania: *Xanthosoma sagittifolium* or *X. violaceum* (Araceae) *hort* >>> tanier

tanier: *Xanthosoma atrovirens* (Araceae), $2n = 2x = 26$ *hort* >>> tania

tara vine: *Actinidia arguta* (Actinidiaceae), $2n = 2x = 116$ (?)

taro: one of the oldest known vegetables and has been grown in some regions of the world for more than 2,000 years, *Colocasia esculenta* (Araceae), $2n = 2x = 28$ *hort*

tarragon: *Artemisia dracunculus* (Compositae) *hort*

Tatary buckwheat: *Fagopyrum tataricum* (Polygonaceae), $2n = 2x = 16$ *agr*

tau-sahyz >>> Russian dandelion

tea: *Camellia sinensis* (Camelliaceae), $2n = 2x = 30$ *hort*

tears of Job >>> adlay

tef(f): *Eragrostis tef* (Gramineae), $2n = 4x = 40$ *agr*

teonochtil >>> pencas

teosinte: *Euchlaena mexicana* (Gramineae) *bot agr*

tepary bean: *Phaseolus acutifolius* var. *latifolius* (Leguminosae), $2n = 2x = 22$ *hort* >>> pulse

thale cress: a species of flowering of the family *Cruciferae;* this plant became a main subject of molecular genome analysis because it has a small and simple genome ($2n = 2x = 10$ chromosomes), over half of which codes for protein and it can be easily cultured, having a life cycle of only 6-8 weeks, *Arabidopsis thaliana* (Cruciferae) *bot*

thistle: *Cirsium* spp. (Asteraceae) *bot agr*

thyme: *Thymus vulgaris* (Lamiaceae) *hort*

tien chi >>> san chi

timothy: *Phleum pratense* (Gramineae), $2n = 6x = 42$ *agr*

tobacco: a tall, herbaceous plant; the leaves of which are harvested, cured, and rolled into cigars, shredded for use in cigarettes and pipes, and processed for chewing or snuff; the main source of commercial tobacco is *Nicotiana tabacum,* although *Nicotiana rustica* is also grown and is used in oriental tobaccos; breeders have developed a wide range of morphologically different types, from the small-leaved aromatic tobaccos to the large, broad-leaved cigar tobaccos; it is the most widely grown nonfood crop in the world; it is not only of agronomic interest but also for its utilization in genetic, physiologic, biochemical, and biotechnological research; *Nicotiana tabacum, Nicotiana rustica* (Solanaceae), $2n = 4x = 48$ *agr*

tomato: as potato tomato originates from South America; as garden pea it became a subject of intensive genetic research and is still one of the best-investigated crop plants; alien gene transfer and genetic engineering were successfully demonstrated on tomato; *Lycopersicon esculentum* (Solanaceae), $2n = 2x = 24$ *hort agr* >>> potato

toria >>> Indian rape

t'ou >>> rakkyo

touch-me-not: *Impatiens balsamina* (Balsaminaceae) *hort*

tree of heaven: *Ailanthus glandulosa* (Simaroubaceae) *bot hort*

tree onion: *Allium cepa* var. *viviparum* (Alliaceae), $2n = 2x = 16$ *hort*

tree sorrel >>> bilimbi

tree tomato: *Cyphomandra betacea* (Solanaceae), $2n = 2x = 24$ *hort*

triticale: a amphiploid hybrid between wheat *(Triticum)* and rye species *(Secale)* in which wheat is the donor of the cytoplasm; it is a cereal crop created by humans from wheat and rye based on work starting at the end of the 19th century (WILSON 1876, UK; RIMPAU 1888, 1891, Germany); the aim was to combine the quality advantages of wheat with the stress insensitivity of rye; growth habit is similar to wheat; it differs from wheat by a greater vigor and larger size of spikes and grains; among three basic ploidy levels so far developed for triticale, the hexaploid type became the most important for breeding and agriculture; there is a taxonomic proposal for classification: genus *Triticum,* section *Triticale,* notospecies *Triticale krolowii* ($2n = 4x = 28$), *Triticale turgidocereale* ($2n = 6x = 42$), and *Triticale rimpaui* ($2n = 8x = 56$) (Gramineae) *agr* >>> Tables 1, 15, 16

tritipyrum: an amphiploid hybrid between wheat *(Triticum)* and *Thinopyrum* species *(Thinopyrum)* in which wheat is the donor of the cytoplasm (Gramineae) *agr*

tritordeum: an amphiploid hybrid between wheat *(Triticum)* and barley species *(Hordeum)* in which wheat is the donor of the cytoplasm (Gramineae) *agr*

true turnip >>> stubble turnip

truffle: an edible fruit-body of Tuber or other Tuberales, usually growing subterranean *bot hort*

tuba root >>> derris

tulip: *Tulipa* spp. (Liliaceae), $2n = 2x, 3x, 4x = 24, 36, 48$ *hort*

tunas >>> pencas

tung (oil tree): *Aleurites fordii, A. montana* (Euphorpiaceae), $2n = 2x = 22$ *fore*

Turkish cobnut: *Corylus colurna* (Corylaceae), $2n = 2x = 22$ *hort* >>> hazel

turkterebinth nut >>> pistachio (nut)

turmeric: *Curcuma longa* (Zingiberaceae) *hort*

turnip: *Brassica campestris* (Cruciferae), $2n = 2x$, AA = 20 *hort*

turnip cabbage >>> kohlrabi

turnip rape: *Brassica campestris* ssp. *oleifera* (Cruciferae), $2n = 2x$, AA = 20 *agr*

turnip-rooted chevil: *Chaerophyllum bulbosum* (Umbelliferae) *hort*

U

ulluco: *Ullucus tuberosus* (Basellaceae), $2n = 2x = 24$ *hort*

underground onion >>> multiplier (onion)

urd bean: the maximum diversity for *V. mungo* exists in upper Western Ghats and the Deccan Hills and a second center in Bihar (i.e., the center of origin of two crops *V. mungo* and *V. radiata* lies in India); *Phaseolus mungo syn Vigna mungo* (Leguminosae), $2n = 2x = 22$ *hort* >>> pulse >>> mungbean

V

valerian: *Valeriana officinalis* (Valerianaceae) *hort*

vanilla: *Vanilla planifolia* (Orchidaceae), $2n = 2x = 32$ *hort*

varnish tree: *Rhus venicifera syn R. verniciflua* (Anacardiaceae), $2n = 2x = 30$ *fore*

vegetable marrow: *Cucurbita pepo* (Cucurbitaceae), $2n = 2x = 40$ *hort*

vetch: *Vicia* spp., *Lathyrus* spp. (Leguminosae) *agr bot* >>> wooly-pod vetch >>> common vetch >>> hairy vetch >>> hairy vetchling >>> Hungarian vetch >> kidney vetch >>> narbon vetch >>> purple vetch >>> subterranean vetch

viper's grass >>> black salsify

W

wallflower: *Cheiranthus cheiri* (Brassicaceae) *hort*

walnut: *Juglans* spp. (Juglandaceae), $2n = 2x = 32$ *hort* >>> African walnut

watercress: *Rorippa nasturtium-aquaticum* (Cruciferae), $2n = 2x$, R'R' = 32 *hort*

watermelon: *Citrullus lanatus, C. vulgaris* (Cucurbitaceae), $2n = 2x = 22$ *hort*

wattle: *Acacia senegal* (Leguminosae), $2n = 2x = 26$ *hort*

wavy hairgrass: *Deschampsia flexuosa* (Gramineae) *agr bot*

wax gourd: *Benincasa hispida* (Cucurbitaceae), $2n = 2x = 24$ *hort*

wax tree: *Rhus succedanea* (Anacardiaceae), $2n = 2x = 30$ *fore*

Welsh onion: *Allium fistulosum* (Alliaceae), $2n = 2x = 16$ *hort*

wheat: a family of related small grains that are descended from the natural crossing of three Middle East grasses (*Aegilops* spp.) centuries ago; it is the most-grown cereal crop in the world; it comprises 14 species; the inflorescence is a spike, containing about 20-30 spikelets; each with about 4-6 florets; one seed is set per floret, although the smaller florets may not bear seeds; it is normally self-pollinating; four main commercial market classes are described: (1) hard red spring, (2) hard red winter, (3) soft red winter, (4) white wheat; wheat is the leading human food resource; a world production >550 million tons; the cultivation of wheat has developed for more than 10,000 years; the areas dedicated to wheat throughout the world exceed the area for all other crops, such as rice, maize, and potato; the various wheats include a group of diploid species characterized by eight genomes and group of tetraploid and hexaploid with nine genomes; two main species are produced: (a) soft wheat (hexaploid), *Triticum aestivum* ($2n = 6x$, AABBDD = 42, 2C DNA content 36 pg), and (b) hard wheat (tetraploid), *Triticum durum, Triticum* spp. (Gramineae) *agr* >>> Figure 10 >>> Tables 1, 15, 16, 30, 32, 35

wheat-grass >>> *Agropyron*

white (head) cabbage: *Brassica oleracea* convar. *capitata* var. *alba* (Brassicaceae) *hort* >>> Figure 8

white campion: one of the few members of the plant kingdom carrying sex chromosomes, *Silene alba* (Caryophyllaceae) *bot biot*

white clover: *Trifolium repens* (Leguminosae) *agr*

white lupin: *Lupinus albus* (Leguminosae), $2n = 2x = 48$ *agr*

white mustard: *Sinapis alba* (Cruciferae) *agr*

white sweet clover: *Melilotus alba* (Leguminosae) *agr*

whortleberry >>> blueberry

wild chamomile >>> German chamomile

wild crucifer: a potential source for alien gene transfer; of special interest is its C3-C4 intermediate photosynthetic and/or photorespiratory mechanism, *Moricandia arvensis* (Cruciferae) *bot biot*

wild emmer wheat: *Triticum dicoccoides* (Gramineae) *agr* >>> emmer wheat

wild oat: *Avena fatua* (Gramineae) *bot agr*

wild mustard >>> charlock

wild radish >>> runch

wild rye >>> lyme grass

willow: *Salix* spp. (Salicaceae) *agr fore bot* >>> basket willow

winged bean: *Psophocarpus tetragonolobus* (Leguminosae), $2n = 2x = 26$ *hort* >>> pulse

winter grape >>> Berlandieri grape

winter (oilseed) rape >>> rapeseed

wooly-pod vetch: *Vicia villosa* ssp. *dasycarpa* (Leguminosae) *bot agr* >>> vetch

wormwood: *Artemisia absinthium* (Compositae) *hort*

wrinkled pea: *Pisum sativum* convar. *medullare* (Leguminosae) *hort*

Y

yam bean: *Sphenostylis stenocarpa syn Pachyrrhizus tuberosus* (Leguminosae) *hort* >>> pulse

yam(s) bean >>> jicama

yams (Asia): *Dioscorea alata* and *D. esculenta* (Dioscoreaceae), $2n = 3x - 10x = 30 - 100$ *hort agr*

yams (Africa): *Dioscorea rotundata* (Dioscoreaceae), $2n = 4x = 40$ *hort agr*

yarrow: *Achillea ssp.* (Asteraceae) *hort*

yautia >>> tanier

year bean: *Phaseolus polyanthus* (Leguminosae) *agr* >>> pulse

yellow gram >>> chickpea

yellow lucerne: *Medicago falcata* (Leguminosae), $2n = 2x, 4x = 16, 32$ *agr*

yellow lupin >>> sweet yellow lupin

yellow sucking: *Trifolium dubium* (Leguminosae) *agr*

yellow sweet clover: *Melilotus officinalis* (Leguminosae) *agr*

yellow trefoil >>> black medick

yellow-seeded sarson: *Brassica campestris* ssp. *tricularis* (Cruciferae), $2n = 2x$, AA = 20 *hort* >>> Figure 8

yew: *Taxus baccata* (Taxaceae) *hort fore*

yielding maguey >>> cantala

Z

zucchini: *Cucurbita pepo* convar. *giromontiia* (Cucurbitaceae), $2n = 2x = 40$ *hort*

TABLES

TABLE 1. Classification of wheats (*Triticum* spp.)

Genome level	Species	Chromosome number (2n =)	Genome formula
Diploid	*urartu*	14	AA
	boeoticum	14	AA
	monococcum	14	AA
	sinskajae	14	AA
Tetraploid	*dicoccoides*	28	AABB
	dicoccum	28	AABB
	paleocolchicum	28	AABB
	carthlicum	28	AABB
	turgidum	28	AABB
	polonicum	28	AABB
	durum	28	AABB
	turanicum	28	AABB
	araraticum	28	AAGG
	timopheevi	28	AAGG
	militinae	28	AAGG
Hexaploid	*spelta*	42	AABBDD
	vavilovii	42	AABBDD
	macha	42	AABBDD
	sphaerococcum	42	AABBDD
	compactum	42	AABBDD
	aestivum	42	AABBDD
	zhukovskyi	42	AAAAGG

TABLE 2. Number of gametes, genotypes, and phenotypes considering one and multifactorial heterozygosity in F1 and F2 generations

Number of heterozygous factors in F1 generation	Number of different gametes in F1 Number of homozygous genotypes in F2 Number of phenotypes in F2, considering complete dominance of all characters	Number of genotypes in F2	Number of combinations (among these numbers all genotypes may be realized in F2)
1	2	3	4
2	4	9	16
3	8	27	64
4	16	81	256
5	32	243	1024
6	64	729	4096
7	128	2187	16384
8	256	6561	65536
9	512	19683	262144
10	1024	59049	1048576
n	**2^n**	**3^n**	**$2^n \times 2^n = 4^n$**

TABLE 3. Segregation of recessive nulliplex genotypes from triplex, duplex, and simplex genotypes considering selfing, random chromosome distribution, and complete dominance

	Triplex	Duplex	Simplex
Genotypes	AAAa	AAaa	Aaaa
Gametes	AA	AA	Aa
	AA	Aa	Aa
	AA	Aa	Aa
	Aa	Aa	aa
	Aa	Aa	aa
	Aa	aa	aa
	3AA:3Aa	1AA:4Aa:1aa	3Aa:3aa
Progeny	9AAAA	1AAAA	9AAaa
	18AAAa	8AAAa	18Aaaa
	9AAaa	18AAaa	9aaaa
		8Aaaa	
		1aaaa	
	No nulliplex aaaa type	**Segregating nulliplex aaaa type of ratio 1:35**	**Segregating nulliplex aaaa type of ratio 9:27**

TABLE 4. Expected F2 segregations of trisomic F1 plants from a critical cross of trisomic by disomic, excluding any selection, male transmission of n + 1 gametes and abnormal chromosome segregation

F1 genotype	**Selfing or backcrossing of disomic recessive**	**Expected segregation depending on the transmission of n + 1 gametes through female**	
		50%	**30%**
Aa	Selfing	3:1	3.0:1.0
Aa	Backcrossing	1:1	1.0:1.0
Aaa	Selfing	17:1	11.9:1.0
AAa	Backcrossing	5:1	3.3:1
Aaa	Selfing	2:1	1.6:1.0
Aaa	Backcrossing	1:1	1.0:1.3

TABLE 5. Basic methods of plant breeding

Selection breeding	Mass selection	Positive mass selection
		Negative mass selection
	Pedigree selection	Autogamous plants
		Allogamous plants
		Vegetatively grown plants
Combination breeding		Autogamous plants
		Allogamous plants
Hybrid breeding		Autogamous plants
		Allogamous plants
Synthetics		

TABLE 6. Genotypic and phenotypic segregation in F2 populations considering two genes and interacting in different manners

Genotypes	Phenotypes* a	b	c	d	e
1 AABB					
2 AABb					
2 AaBB					
4 AaBb					
1 AAbb					
2 Aabb					
1 aaBB					
2 aaBb					
1 aabb					
Segregation	**9:3:3:1**	**12:3:1**	**15:1**	**1:4:6:4:1**	**9:7**

*Phenotype determination:
a = factor *A* is dominant over *a*, factor *B* is dominant over *b*, and there is no gene interaction.
b = factor *A* is epistatic of *Bb* and *bb*.
c = factor *A* is equidirectional substituted by *B*, and factor *B* is equidirectional substituted by *A*.
d = the dominant alleles *A* and *B* show additive interaction in the same direction, i.e. *AABb* = *AaBB*, *AaBb* = *AAbb*, *etc.*
e = factor *A* can be phenotypically expressed only when *A* and *B* are present due to complementary gene action.

TABLE 7. Frequencies and ratios of plants in F2 progeny of doubled haploids, diploids, and tetraploids showing complete recessiveness

Number of alleles	**Frequency of recessive plants as determined by one of the plants given below (1/x)** Doubled haploid	**Diploid**	**Tetraploid**
1	2	4	36
2	4	13	1296
3	8	64	46656
4	16	256	1679616
5	32	1024	60466176
n	**2^n**	**2^{2n}**	**6^{2n}**

TABLE 8. Genome relationships between embryo, endosperm, and ovary after crossing parents with different ploidy levels

Female × Male	Embryo	Endosperm	Ovary
2x × 2x	2x	3x	2x
2x × 3x	3x	4x	2x
4x × 2x	3x	5x	4x
2x × 6x	4x	5x	2x
6x × 2x	4x	7x	6x

TABLE 9. Genetic segregation patterns depending on the number of genes involved

No. of different gene pairs	No. of different gametes	No. of different F1 × F1 gametes	No. of F2 genotypes	No. of homozygous F2 genotypes	No. of heterozygous F2 genotypes	No. of F2 phenotypes*
1	2	4	3	2	1	2
2	4	16	9	4	5	4
3	8	64	27	8	19	16
4	16	256	81	16	65	32
5	32	1024	243	32	211	64
n	**2n**	**4^n**	**3^n**	**2^n**	**$3^n - 2^n$**	**2^n**

*When complete dominance

TABLE 10. Frequencies of homozygotes and heterozygotes in a progeny of a heterozygous individual after subsequent self-pollinations

Number of generation	Genotypes			Relative amount of heterozygotes per population
	AA	Aa	aa	
0	0	1	0	1
1	1/4	2/4	1/4	1/2
2	3/8	2/8	3/8	1/4
3	7/16	2/16	7/16	1/8
4	15/32	2/32	15/32	1/16
5	31/64	2/64	31/64	1/32
10	1023/2048	2/2048	1023/2048	1/1024
n				**$1/2^n$**

TABLE 11. Examples of seed conditioning in some crop plants

Small-seeded grasses	Field beans	Small grains	Small-seed legumes
Scalper	Air screen cleaner	Aspirator	Scalper
Debearder	Gravity separator	Debearder	Huller, scarifier
Air screen cleaner	Stoner	Air screen cleaner	Air screen cleaner
Disc and/or cylinder separator	Color separator	Indent cylinder	Gravity separator
Treater	Processing belts	Treater	Stoner
Bagging	Treater	Bagging	Roll mill
	Bagging		Cylinder separator
			Treater
			Bagging

TABLE 12. Taxonomic classification system in plants

Classification	Example
Kingdom (Regnum)	Plants
Subkingdom	Cryptogamia
Division	Fungi
Subdivision	Eumycotina
Class	Basidiomycetes
Subclass	Heterobasidiomycetes
Order	Uredinales
Suborder	Uredineae
Family	Pucciniaceae
Subfamily	Puccinioideae
ribe	Puccineae
Subtribe	Puccinia
Genus	*Puccinia*
Subgenus	*Hetero-Puccinia*
Section	–
Subsection	*Puccinia graminis*
Species	*Puccinia graminis* var. *phlei-pratensis*
Subspecies	–
Variety	–
Subvariety	–
Form	–
Special form *(forma specialis)*	–
Physiological race	*P. graminis* var. *phlei-pratensis* Race 3

TABLE 13. Decimal code for plant growth in cereals

Code	Stage	Code	Stage
0	Germination	5	Inflorescence emergence
00	Dry seed	50-51	1st spikelet visible
01-02	Start of imbibition	52-53	1/4 of inflorescence emerged
03-04	Imbibition completed	54-55	1/2 of inflorescence emerged
05-06	Radicle emerged	56-57	3/4 of inflorescence emerged
07-08	Coleoptile emerged	58-59	Inflorescence emerged
09	Leaf on coleoptile tip	6	Anthesis
1	Seedling growth	60-63	Beginning of anthesis
10	1st leaf through coleoptile	64-67	1/2 of anthesis
11	1st leaf unfolded	68-69	Anthesis completed
12	2nd leaf unfolded	7	Milk development
13-19	3rd leaf-9th leaves unfolded	70-71	Caryopsis waterripen
2	Tillering	73-74	Early milk ripe
20	Main shoot only	75-76	Medium milk ripe
21	Main shoot + 1 tiller	77-79	Late milk ripe
22	Main shoot + 2 tillers	8	Dough development
23-29	Main shoot + 3-9 tillers	80-84	Early dough
3	Stem elongation	85-86	Soft dough
31	Pseudoerection of stem	87-89	Hard dough
32	1st node detectable	9	Ripening
33	2nd node detectable	90	Caryopsis hard
34-36	3rd-6th node detectable	91	Caryopsis hard
37-38	Flag leaf detectable	92	Caryopsis hard
39	Flag leaf ligula + collar visible	93	Caryopsis loosening
4	Booting	94	Overripe, straw dead
40-41	Flag leaf sheath extending	95	Seed dormant
42-43	Boots just visibly swollen	96	Seeds giving 50% germination
44-45	Boots swollen	97	Seed not dormant
46-47	Flag leaf sheath opening	98	2nd dormancy
48-49	First awns visible	99	2nd dormancy lost

Source: After ZADOK et al., 1974.

TABLE 14. Genome relationships of oats (*Avena* spp.)

Ploidy level	Wild species	Cultivated species	Chromosome number (2*n* =)	Genome formula
Diploids	*pillosa*		14	$C_p C_p$
	clauda		14	$C_p C_p$
	ventricosa		14	$C_v C_v$
	prostrata		14	$A_p A_p$
	wiestii	*strigosa*	14	$A_s A_s$
	hirtula		14	$A_s A_s$
	longiglumis		14	$A_l A_l$
	damascena		14	$A_d A_d$
	canariensis		14	$A_c A_c$
		brevis	14	$A_s A_s$
		nudibrevis	14	$A_s A_s$
Tetraploid	*barbata*	*abyssinica*	28	AABB
	vaviloviana		28	AABB
	magna		28	AACC
	murphyi		28	????
Hexaploid	*sterilis*	*sativa*	42	AACCDD
	fatua	*byzantina*	42	AACCDD
		nuda	42	AACCDD

TABLE 15. The approximate protein composition (%) in some cereals

Cereal	Albumin	Globulin	Prolamin	Glutelin
Barley	13	12	(hordein) 52	(hordenin) 23
Maize	4	2	(zein) 55	39
Oats	11	56	(avenin) 9	23
Rice	5	10	(oryzin) 5	(oryzenin) 80
Sorghum	6	10	(kafirin) 46	38
Wheat	9	5	(gliadin) 40	(glutenin) 46

Source: After PAYNE and RHODES, 1982.

Note: In parentheses are the common names of some storage proteins.

TABLE 16. Food reserves of some crop plants

Crop	Average composition (%)			Storage organ
	Protein	Fat	Carbohydrate	
Barley	12	3	76	Endosperm
Maize	10	5	88	Endosperm
Oats	13	8	66	Endosperm
Rye	12	2	76	Endosperm
Wheat	12	2	75	Endosperm
Broad bean	23	1	56	Cotyledons
Pea	25	6	52	Cotyledons
Peanut	31	48	12	Cotyledons
Soybean	37	17	26	Cotyledons
Castor bean	18	64	–	Endosperm
Oil palm	9	49	28	Endosperm
Pine	35	48	6	Gametophyte
Rapeseed	21	48	19	Cotyledons

Source: After BEWLEY and BLACK, 1995.

TABLE 17. Taxonomic relationships of some tuberous *Solanum* spp.

Series	Species	(2*n* =)	Genome
Etuberosa	*brevidens, etuberosum, fernandezianum*	24	E^bE^b, E^eE^e, E^fE^f
Morelliformia	*morelliformae*	24	A^mA^m
Bulbocastana	*bulbocastanum, clarum*	24	A^bA^b
Pinnatisecta	*caridophyllum, jamesii,. pinnatisectum*	24	$A^{pi}A^{pi}$
Commersoniana	*chacoense, commersonii*	24	AA
Conicibacata	*santolallae, chromatophillum,*	24	$A^{c1}A^{c1}$, $A^{c2}A^{c2}$
	agrimonifolium, longiconicum, oxycarpum	48	$A^{c1}A^{c1}C^aC^a$, $A^{c1}A^{c1}C^lC^l$, $A^{c1}A^{c1}C^oC^o$
Piurna	*piurnae*	24	A^pA^p
	tuquerrense	48	A^pA^pPP
Acaulia	*acaule*	48	AAA^aA^a
	acaule ssp. *albicans*	72	$AAA^aA^aX^bX^b$

TABLE 17 *(continued)*

Series	Species	(2*n* =)	Genome
Demissa	*brachycarpum, demissum, guerreoense, spectabile*	72	$AADDD^bD^b$, $AADDD^dD^d$, $AADDD^gD^g$, $AADDD^sD^s$
Longipedicellata	*vallis-mexici*	36	AAB
	fendleri, polytrichon, stoliferum	48	AABB
Polyadenia	*polyadenium, infundibuliformae, megistacrolobum, raphanifolium, sanctae-rosae, toralapanum*	24	AA
Ingaefolia	*rachialatum*	24	A^iA^i
Olmosiana	*olmosense*	24	A^oA^o
Tuberosa (wild)	*abancayense, bukasovii, canasense, gourlayi, kurtzianum, leptophyes, maglia, microdontum, soukupii, sparsipilum, speganzinii, vernei, verrucosum*	24	AA
	sucrense	48	AAA^sA^s
Tuberosa (cultivated)	*ajanhuiri, goniocalyx, phureja, stenotomum*	24	AA
	chaucha, juzepszukii	36	AAA^t, AAA^a
	tuberosum ssp. *andigena*	48	AA A^tA^t
	tuberosum ssp. *tuberosum*	48	AA, A^tA^t
	cortilobum	60	$AAAA^aA^t$

TABLE 18. Types of flowers in higher plants

Flower type	Characteristics
Bisexual, hermaphroditic, monoclinous, perfect	Male and female in one flower
Protandry	Pollen shed before stigma is receptive
Protgyny	Stigma matures and ceases to be receptive before pollen is shed
Chaemogamy	Stigma receptive and pollen shed after flower opens
Cleistogamy	Stigma receptive and pollen shed in closed flower
Pin flower	Long styles and short stamens

Thrum flower	Short styles and long stamens
Diclinous, unisexual, imperfect	Males and females in separate flowers
Male, staminate	Male flower
Carpellate, female, pistillate	Female flower
Monoecious	Male and female flowers on one plant
Dioecious	Male and female flowers on separate plants
Mixed, polygamous	Presence of male, female, and perfect flowers
Polygamo-monoecious	Presence of male, female, and perfect flowers on the same plant
Polygamo-dioecious	Presence of male, female, and perfect flowers on separate plants

TABLE 19. Segregation of a single gene and/or alleles in subsequent generations

P		Parent 1 × Parent 2	
F1		Aa (1/1)	
F2	AA (1/4)	Aa (1/2)	aa (1/4)
F3	AA (3/8)	Aa (1/4)	aa (3/8)
F4	AA (7/16)	Aa (1/8)	aa (7/16)
F5	AA (15/32)	Aa (1/16)	aa (7/16)

Note: The fraction given for genotypes represents the expected frequencies in each generation after self-pollination

TABLE 20. Phenotypic relations of homozygotes and heterozygotes depending on different dominance levels in diploids

	Dominance level			
Genotype	**Additive interaction**	**Partial dominance**	**Complete dominance**	**Over-dominance**
AA	5	5	5	5
Aa	3	4	5	6
aa	1	1	1	1

TABLE 21. Phenotypic ratios in the F2 generation for two unlinked genes depending on the degree of dominance at each locus and epistasis between loci

<table>
<tr><th rowspan="2">Genetic explanation</th><th colspan="9">F2 genotypes</th></tr>
<tr><th>AA BB</th><th>AA Bb</th><th>Aa BB</th><th>Aa Bb</th><th>AA bb</th><th>Aa bb</th><th>aa BB</th><th>aa Bb</th><th>aa bb</th></tr>
<tr><td>Complete dominance lacking at either locus, no epistasis, phenotypic and genotypic ratios are equal</td><td>1.</td><td>2</td><td>2</td><td>4</td><td>1</td><td>2</td><td>1</td><td>2</td><td>1</td></tr>
<tr><td>Complete dominance lacking in A, complete dominance in B, no epistasis</td><td colspan="2">3</td><td colspan="2">6</td><td>1</td><td>2</td><td colspan="2">3</td><td>1</td></tr>
<tr><td>Complete dominance in A and B, no epistasis</td><td colspan="4">9</td><td colspan="2">3</td><td colspan="2">3</td><td>1</td></tr>
<tr><td>Recessive epistasis, aa epistatic to B and b</td><td colspan="4">9</td><td colspan="2">3</td><td colspan="3">4</td></tr>
<tr><td>Dominant epistasis, A epistatic to B and b</td><td colspan="6">12</td><td colspan="2">3</td><td>1</td></tr>
<tr><td>Dominant and recessive epistasis, A epistatic to B and b, bb epistatic A and a, A and bb produce identical phenotypes</td><td colspan="7">13</td><td colspan="2">3</td></tr>
<tr><td>Duplicate recessive epistasis, aa epistatic to Bb, bb epistatic to A and a</td><td colspan="4">9</td><td colspan="5">7</td></tr>
<tr><td>Duplicate dominant epistasis, A epistatic to B and b, B eptistatic to A and a</td><td colspan="8">15</td><td>1</td></tr>
</table>

TABLE 22. Calculation of recombination frequency between two loci from a cross between *AaBb* x *aaaa*

	Genotypes				
	AaBb	*Aabb*	*aaBb*	*aabb*	**Total**
Observed	190	38	35	201	464
Expected	116	116	116	116	464
Description	parental	non	non	parental	

Recombination (%) = (sum of nonparental class/total number of individuals) × 100

Recombination = (38 + 35 / 190 + 38 + 35 + 201) × 100 = 14.4%

Recombination = (*Aabb*+*aaBb*/*AaBb*+*Aabb*+*aaBb*+*aabb*) × 100

TABLE 23. Test crosses with monosomics to determine the location of a dominant allele

Possible crosses	F1
1 The critical gene is not located on the missing chromosome	
Cross: AA × aa	Aa (dominant)
2 The critical gene is located on the missing chromosome	
Cross: A0 × aa	Aa (dominant)
	A0 (recessive)

TABLE 24. Possible planting arrangements for a diallel crossing (six parents, no reciprocal cross, no self-pollination)

Unpaired parents

P1 P2 P3 P4 P5 P6

Paired parents

P1 × P2 P1 × P3 P1 × P4 P1 × P5 P1 × P6

P2 × P3 P2 × P4 P2 × P5 P2 × P6

P3 × P4 P3 × P5 P3 × P6

P4 × P5 P4 × P6

P5 × P6

Semi-Latin square

P1 × P2 × P6 × P3 × P5 × P4

P2 × P3 × P1 × P4 × P6 × P5

P6 × P1 × P5 × P2 × P4 × P3

Bulk design

P1 × P1 P2 × P2 P3 × P3 P4 × P4 P5 × P5 P6 × P6

TABLE 25. A randomized complete-block design for five entries and ten replications

	Replication									
Block	**1**	**2**	**3**	**4**	**5**	**6**	**7**	**8**	**9**	**10**
I	P5	P1	P5	P4	P2	P4	P1	P5	P2	P5
II	P1	P2	P1	P1	P1	P1	P4	P3	P1	P4
III	P3	P5	P4	P3	P5	P5	P5	P2	P3	P2
IV	P2	P3	P2	P2	P4	P2	P3	P4	P5	P3
V	P4	P4	P3	P5	P3	P3	P2	P1	P4	P1

TABLE 26. Lattice design (42 entries, 3 replications, no blocks within the replication, entries assigned at random to the 42 plots)

Replication 1

1	2	3	4	5	6
7	8	9	10	11	12
13	14	15	16	17	18
19	20	21	22	23	24
25	26	27	28	29	30
31	32	33	34	35	36
37	38	39	40	41	42

Replication 2

7	13	19	37	25	31
1	14	20	32	26	38
8	2	21	27	33	39
3	9	28	15	34	40
16	4	10	22	35	41
42	29	23	17	11	5
18	12	6	24	30	36

Replication 3

38	33	28	22	17	12
13	2	24	29	40	35
42	36	25	20	9	4
6	11	16	27	37	32
18	1	7	23	34	39
3	8	14	19	41	30
26	31	21	15	10	5

TABLE 27. Recovering of genes from recurrent parent during backcrossing

	% of parentage	
Generation	**Recurrent**	**Nonrecurrent**
F1	50	50
BC1	75	25
BC2	87.5	12.5
BC3	93.75	06.25
BC4	96.875	03.125
BC5	98.4375	01.5625

% homozygous individuals = $(2^m - 1/2^m)^n$
n = number of backcrosses

TABLE 28. Scheme of seed purification and increase

1. Harvest of individual plants	
2. Plant individual rows, discard off-types (or rows), bulk seeds of similar type	⋎
	Pedigree seed
3. Plant pedigree seeds, remove off-types, harvest breeder seeds	⋎
	Breeder seed
4. Plant breeder seeds, remove off-types, harvest foundation seeds	⋎
	Foundation seed
5. Plant foundation seeds, harvest registered seeds	⋎
	Registered seed
6. Plant registered seeds, harvest certified seeds	⋎
	Certified seed

TABLE 29. Types and characteristics of several markers in breeding and genetics

Characteristics	**Morphological traits**	**Isoenzymes**	**RFLPs**	**AFLPs RAPDs**	**Microsatellites**
Number of loci	Limited	Limited	Almost unlimited	Unlimited	High
Inheritance	Dominant	Codominant	Codominant	Codominant or dominant	Codominant
Positive features	Visible	Easy to detect	Utilized before the latest techniques were available	Quick assay with many markers	Well-distributed within the genome, many polymorphisms
Negative features	Possibly negative linkage to other characters	Possibly tissue-specific	Radio-activity required; rather expensive	High basic investment, patented	Long development of the marker; expensive

TABLE 30. Typical characteristics for identification of wheat, barley, or wild oat

Characteristic	Wheat	Barley	Wild oat
Ligule	Membranous	Membranous	Membranous
Auricle	Short + hairy or without hair	Long + clasping	Absent
Blades, collar	~ hairy	Without hair	Long hair on margin
Sheath	~ hairy	Without hair	~ without hair
Blade twist	Clockwise	Clockwise	Counterclockwise

TABLE 31. Taxonomy of the genus *Helianthus*

Section	Series	Species	Subspecies	(2n =)
I. ANNUUI		*niveus	*niveus, tephrodes, canescens*	34
		**debilis*	*debilis, vestitus, tardilorus, silvestris, cucumerifolius*	34
		**praecox*	*praecox, runyonii, hirtus*	34
		**petiolaris*	*petiolaris, fallax*	34
		**neglectus, *annuus, *argophyllus, *bolanderi, *anomalus, *paradoxus, *agrestis*		34
II. CILIARES	1. Pumili	*gracilenthus, pumilus, cusickii*		34
	2. Ciliares	*arizonensis, laciniatus*		34
		ciliaris		68
III. DIVARICATI	1. Divaricati	*mollis, divaricatus, decapetalus*		34
		occidentalis	*occidentalis, plantagineus*	34
		hirsutus, strumosus		68
		eggertii, tuberosus		102
		rigidus	*rigidus, subrhomboideus*	102
	2. Gigantei	*giganteus, grosseserratus*		34
		nuttallii	*nuttallii, parishii, rydbergii*	34
		maximiliani, salicifolius, californicus		34
		resinosus, schweinitzii		102
	3. Microcephali	*microcephalus, glaucophyllus, smithii, longifolius*		34
		laevigatus		68
	4. Angustifolius	*angustifolius, simulans, floridanus*		34
	5. Atrorubentes	*silphioides, atrorubens, hetero-phyllus, radula, carnosus*		34

*Annual growth habit; remaining species are perennial.

TABLE 32. About the evolution of Triticeae

Divergence	~ Million years ago
Earliest land plant fossils	420
Origin of angiosperms	200-340
Monocot-dicot divergence	160-240
Origin of grass family	65-100
Oldest known grass fossils	50-70
Divergence of the subfamilies	50-80
Pooideae (wheat, barley, oat)	
Bambusoideae (rice)	
Panicoideae (maize, sorghum)	
Earliest fossil of the rice lineage	40
Divergence of maize and sorghum lineages	15-20
Divergence of wheat and barley lineages	10-14
Divergence of wheat and rye lineages	7

Note: Panicoideae diverged from the Pooideae-Bambusoideae first, followed shortly by divergence of the latter two subfamilies, or authors suggest that either the Pooideae or the Bambusoideae branched off first. In general, it seems that all three subfamilies diverged about the same time. Some phylogenies divide Bambusoideae into two subfamilies, Oryzoideae and Bambusoideae.

TABLE 33. Inheritance and variation of several breeding characteristics

Breeding aims	Qualitative variation	Quantitative variation
Yield traits	Dwarf growth, monogermous fruits	Yield/ha; harvest index, nutritional efficiency, rate of photosynthesis
Quality traits	Linolenic acid, large seeds	Protein content, cooking ability, digestibility
Resistance traits	Monogenic resistance, tolerance to aluminum	Polygenic resistance, drought tolerance

TABLE 34. Some examples of heritability of breeding characteristics

Heritability	Characteristic	Begin selection from . . .
High	Certain resistance traits, ripening time, awns, grain color, glume shape	F2-F3
Medium	Seed or spike characters, straw length, lodging resistance	F4-F5
Low	Yield components, yield potential, physiological traits	F7-F8

TABLE 35. Proposed breeding schemes depending on the crop, reproduction system, and basic population features

	Crop										
	1	2	3	4	5	6	7	8	9	10	11
Reproduction											
Autogamous	+	−	−	+	−	−	+	−	+	−	−
Allogamous	−	+	+	−	+	−	+	+	−	+	+
Anemophilous	−	+	+	−	+	−	−	−	−	−	+
Entomophilous	−	−	−	−	−	−	+	+	−	+	−
Self-fertile	+	−	+	+	−	−	+	+	+	−	+
Self-sterile	−	+	−	−	+	−	−	−	−	+	−
Isolation required	−	+	+	−	+	+	+	+	−	+	+
Clonable	−	+	−	−	+	+	−	−	−	+	+
Basic population derived from											
Crossing	+	+	+	+	+	+	+	+	+	+	+
Wild origin	−	−	−	−	−	+	−	−	−	+	+
Polyploidization	−	+	−	−	+	−	+	−	−	+	+
Induced mutation	+	−	+	+	−	−	+	−	−	−	−
Breeding method											
Mass selection	−	+	−	−	−	+	+	−	−	+	+
Pedigree method	+	−	+	+	+	+	+	+	+	+	+
Combination breeding	+	+	+	+	+	+	−	+	+	+	+

Residue method	−	+	−	−	+	−	−	−	−	+	−
Couple method	−	+	−	−	+	−	−	−	−	−	+
Heterosis breeding	−	+	+	−	+	−	+	+	−	−	+
Polycross method	−	+	+	−	+	−	−	−	−	+	

Note: 1 = Wheat, oats, barley, rice; 2 = Rye; 3 = Maize; 4 = Millet; 5 = Sugar beet; 6 = Potato; 7 = Rapeseed; 8 = Sunflower; 9 = Broad bean; 10 = Clover; 11 = Forage grasses; + = Applicable.

FIGURES

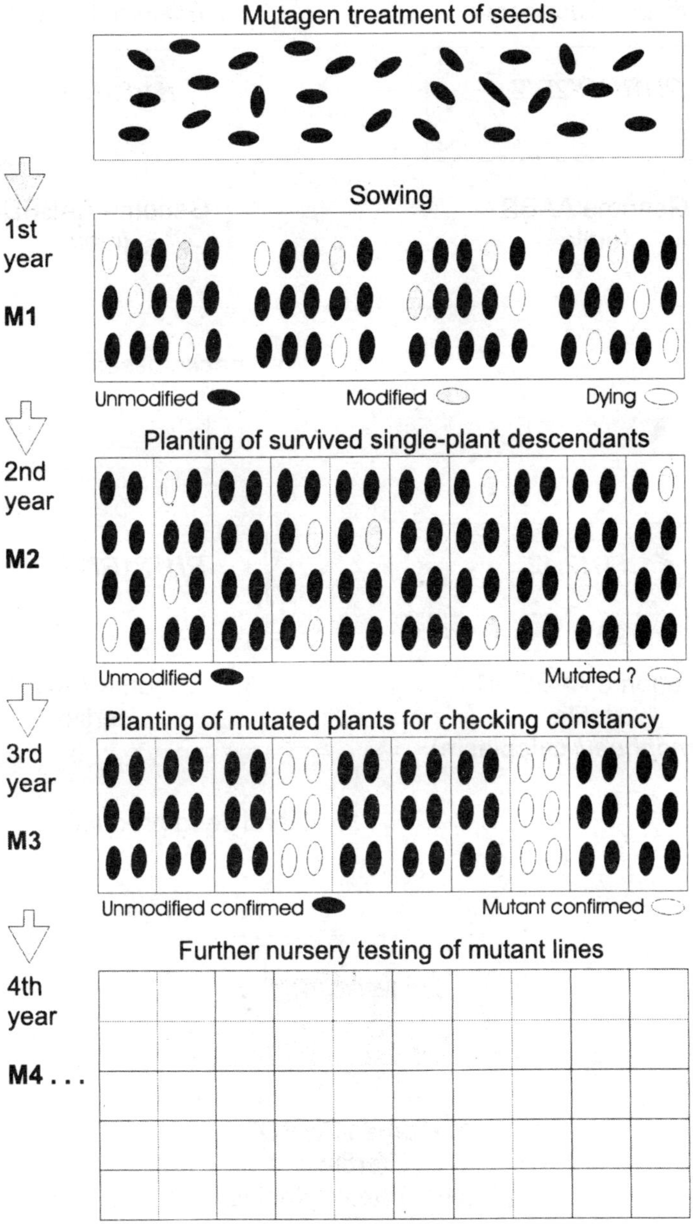

FIGURE 1. Basic scheme of selection in mutation breeding

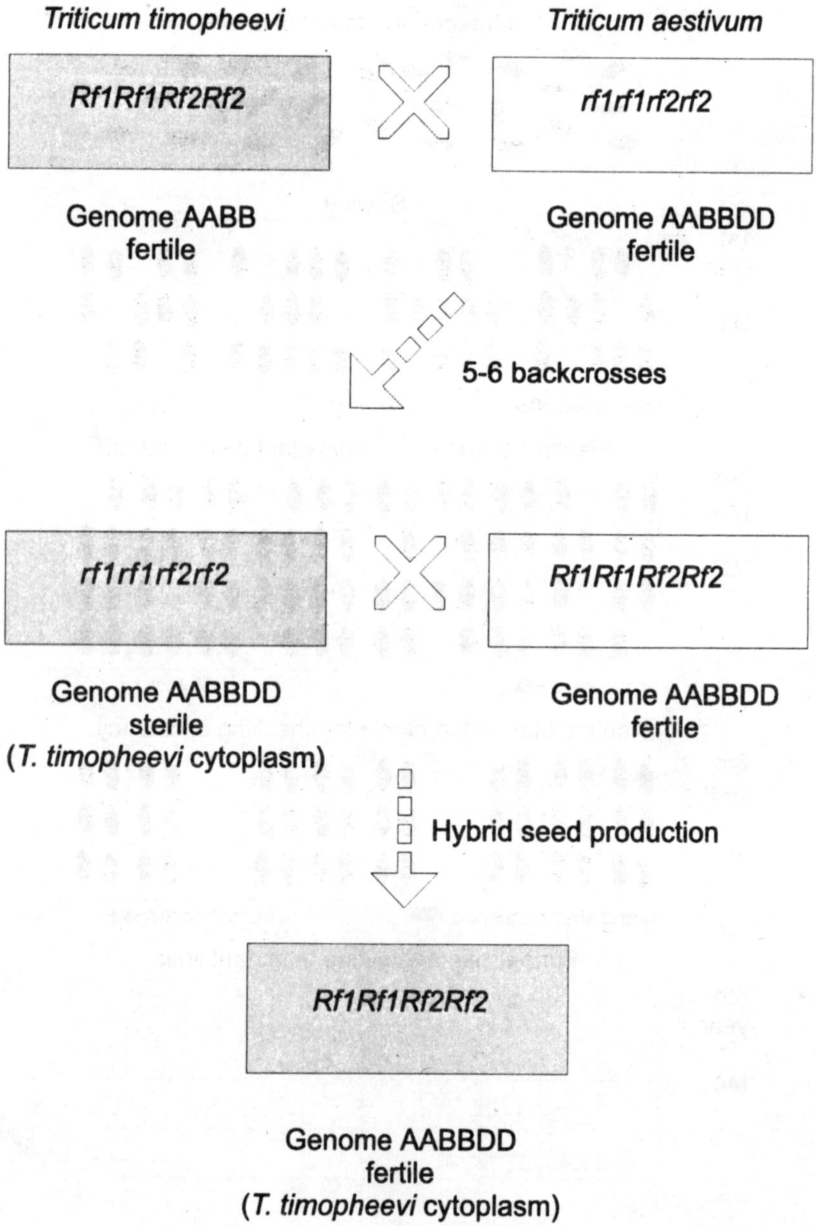

FIGURE 2. Main steps to establish a hybrid variety in wheat

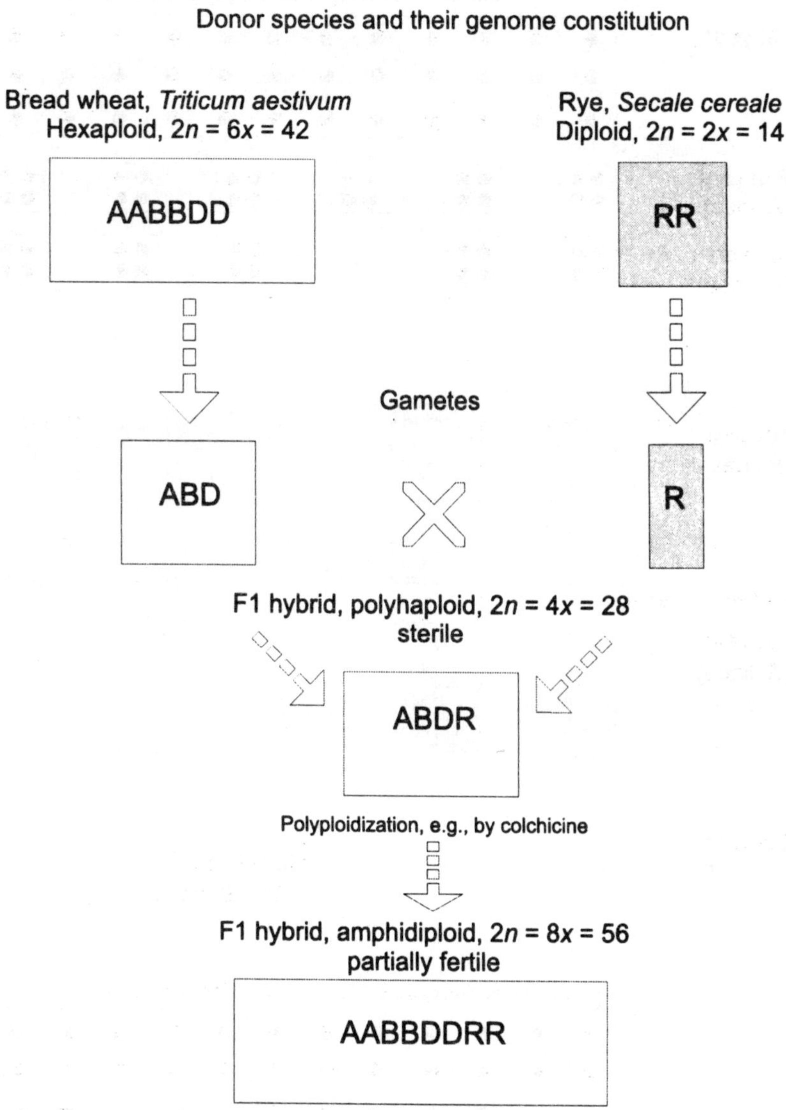

FIGURE 3. Development of allopolyploid (amphidiploid) hybrids (e.g., a wheat-rye hybrid [octoploid triticale])

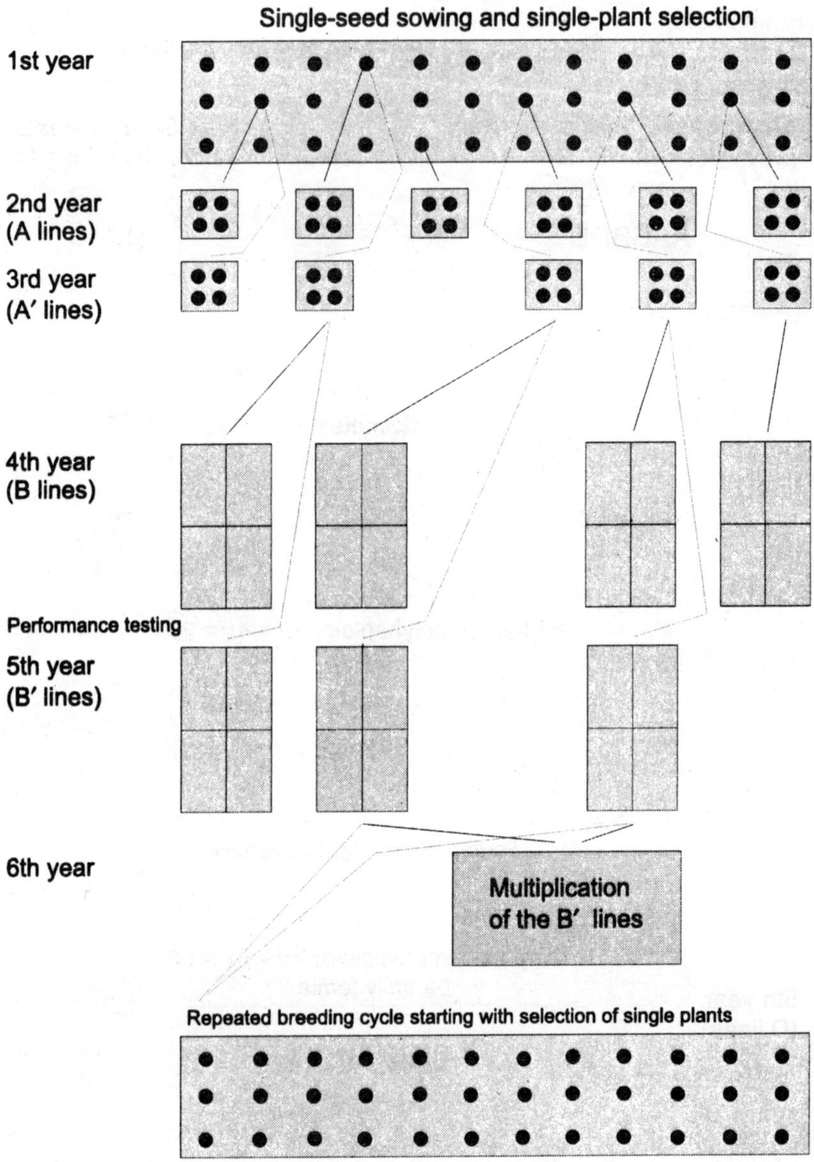

FIGURE 4. Residue seed method of breeding, half-sib progeny selection, or method of overstored seeds

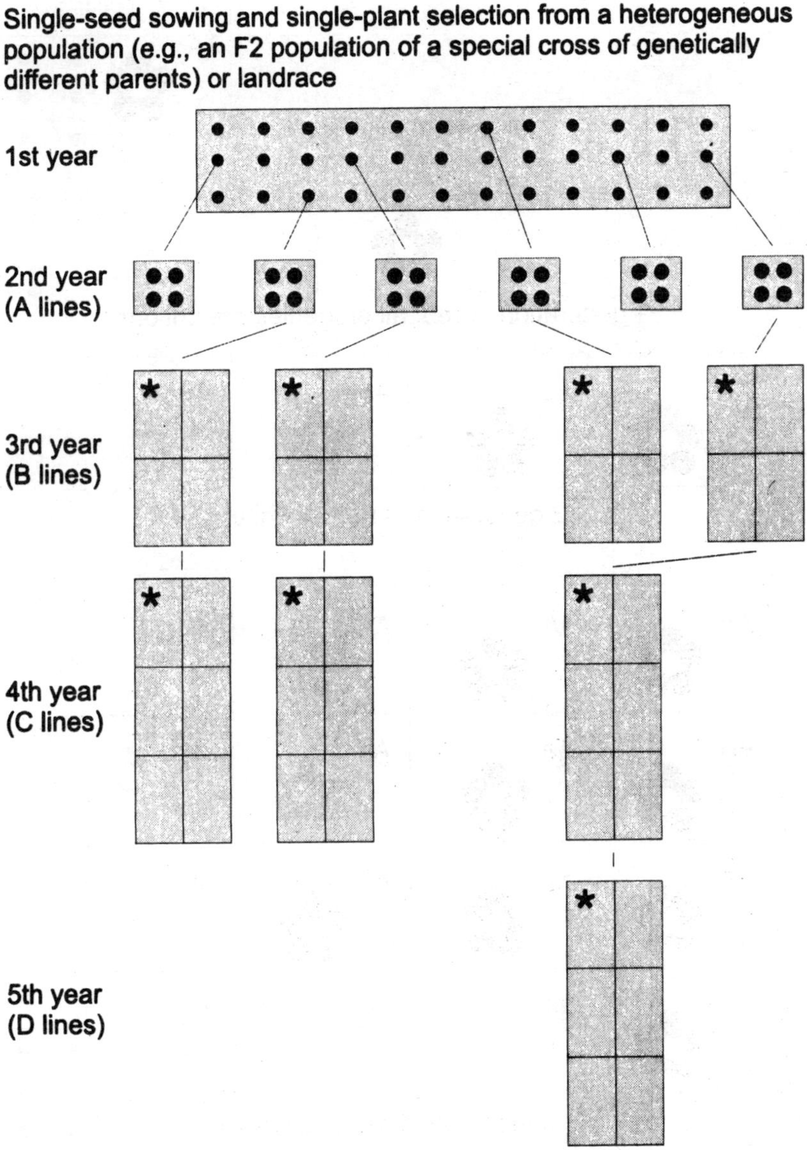

FIGURE 5. Single-plant selection including testing of progeny in autogamous plants (*Performance tesing with four or six replications.)

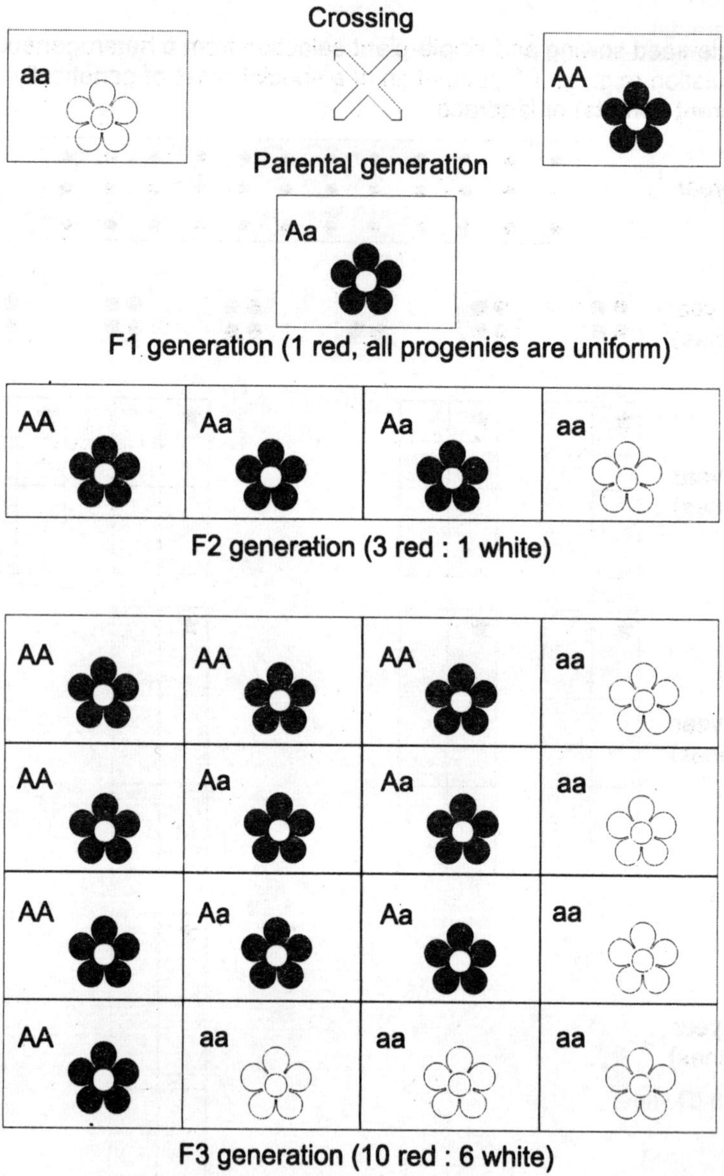

FIGURE 6. Inheritance and segregation patterns in subsequent generations of flower color from the cross of a red-flowered with a white-flowered plant, and with dominant inheritance of red flowers

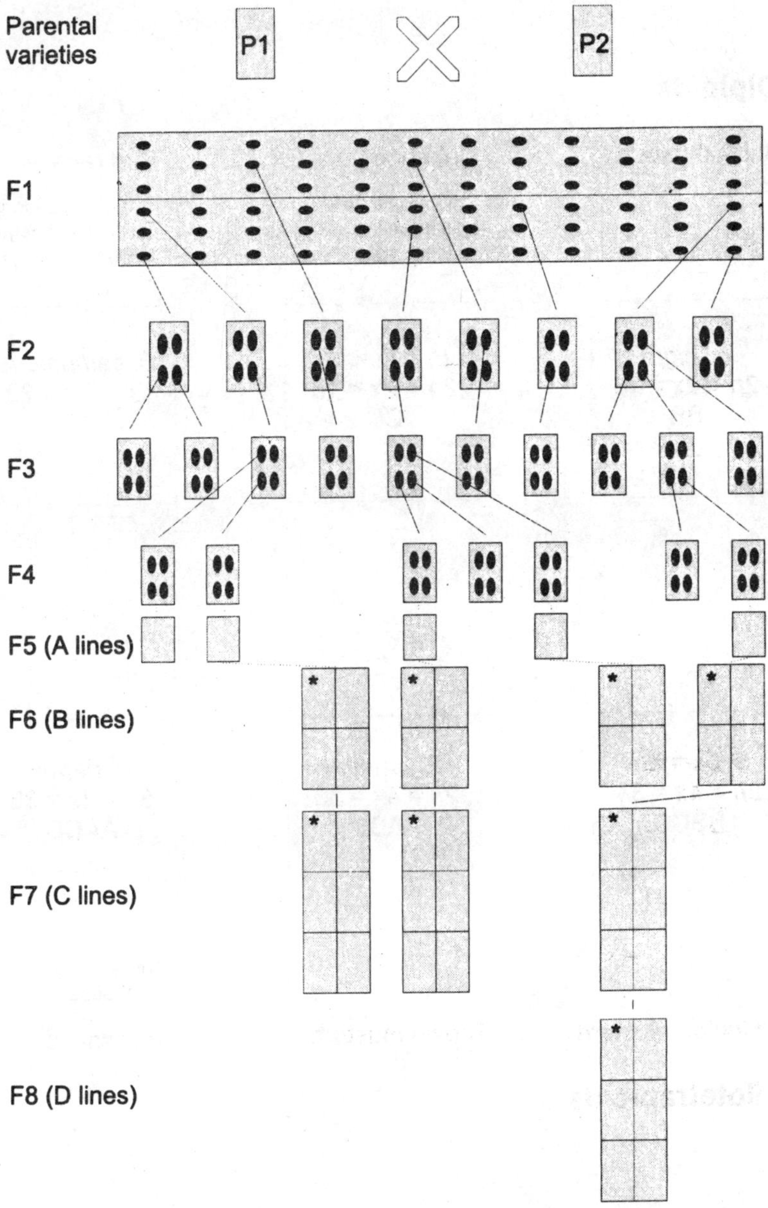

FIGURE 7. Pedigree breeding (cross-combination breeding) in autogamous plants (*Performance testing with four or six replications.)

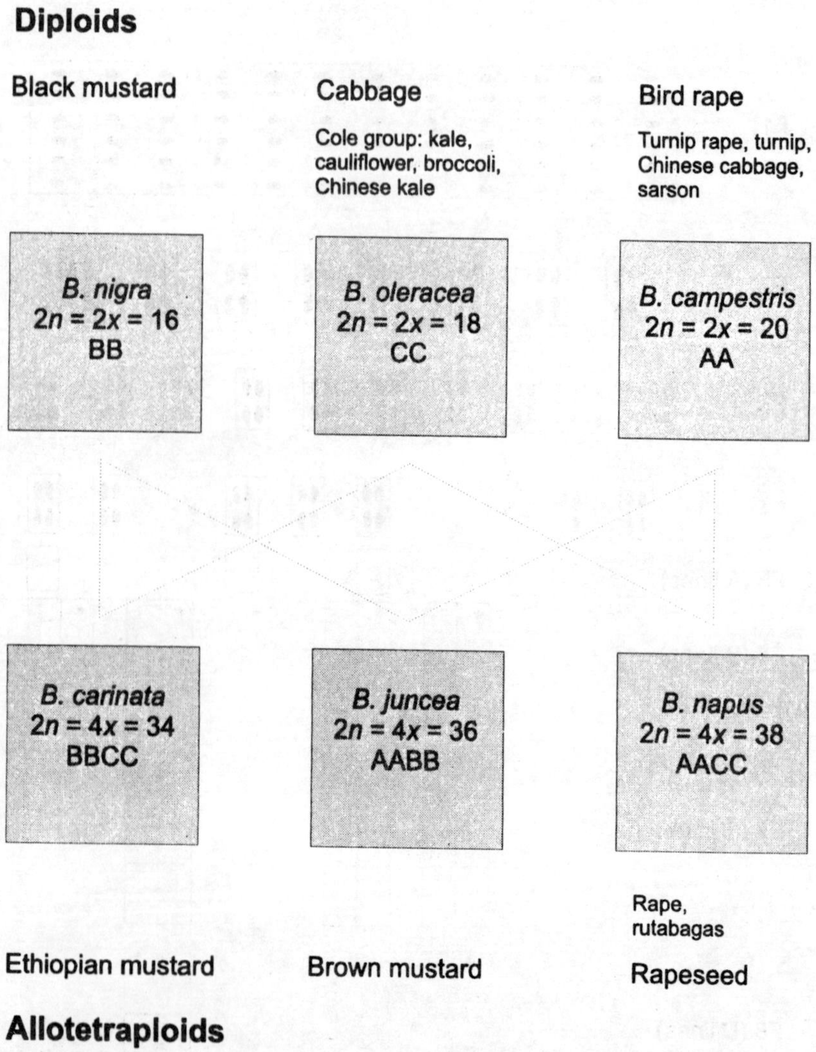

FIGURE 8. Some diploid and allopolyploid species of the genus *Brassica* with agricultural and breeding importance

Treatment/Variant columns

Replication blocks	I	II	III	IV
a	1	2	3	4
b	2	3	4	1
c	3	4	1	2
d	4	1	2	3

blocks	I α	I β	II α	II β	III α	III β	IV α	IV β	V α	V β
a	2	5	3	6	8	9	1	10	4	7
b	6	9	7	10	2	3	5	4	8	1
c	8	1	9	2	4	5	7	6	10	3
d	4	7	5	8	10	1	3	2	6	9
e	10	3	1	4	6	7	9	8	2	5

FIGURE 9. Experimental field design of a Latin square considering four variants and four replications (four blocks, four columns) and a Latin rectangle considering ten variants (five blocks, five columns)

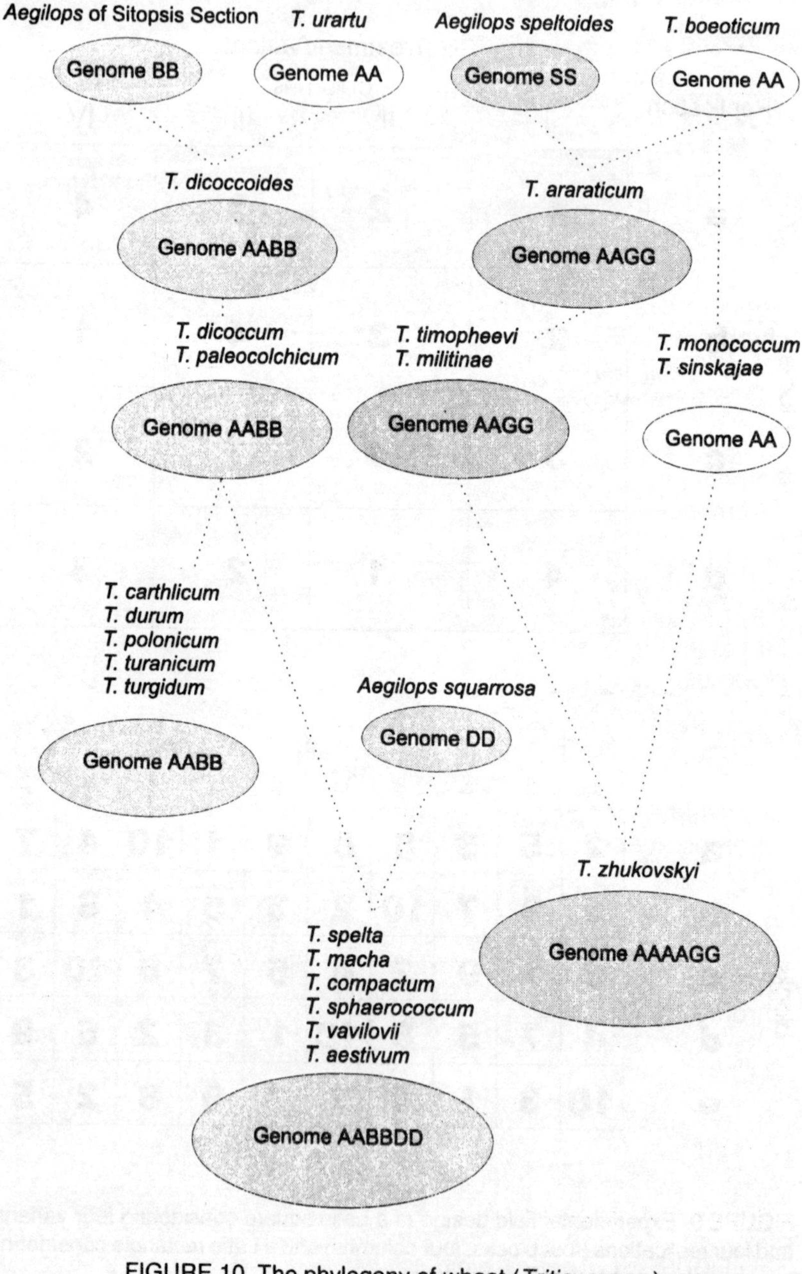

FIGURE 10. The phylogeny of wheat (*Triticum* spp.)

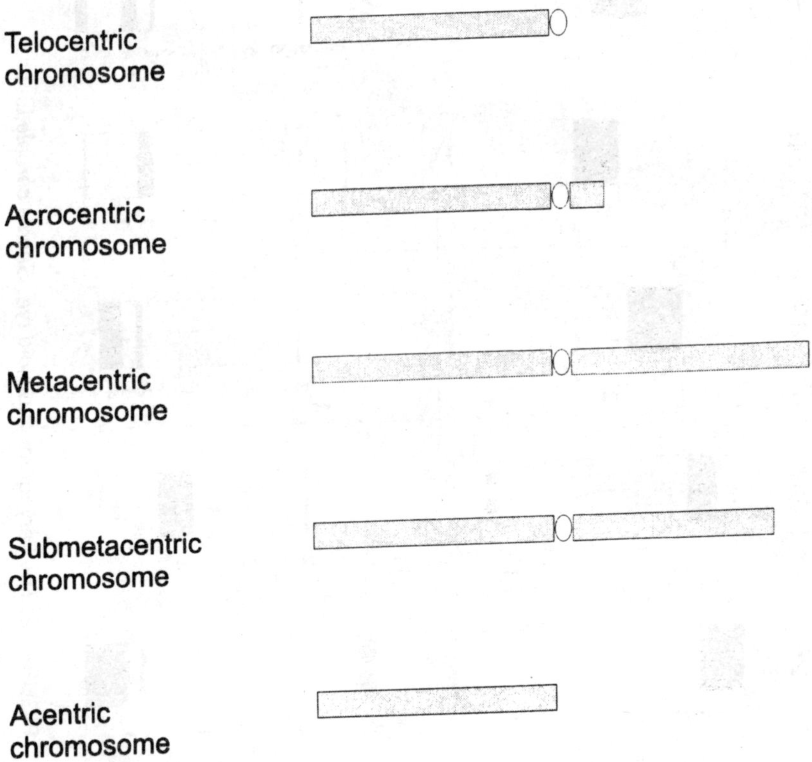

FIGURE 11. Types of chromosome and/or centromere constrictions

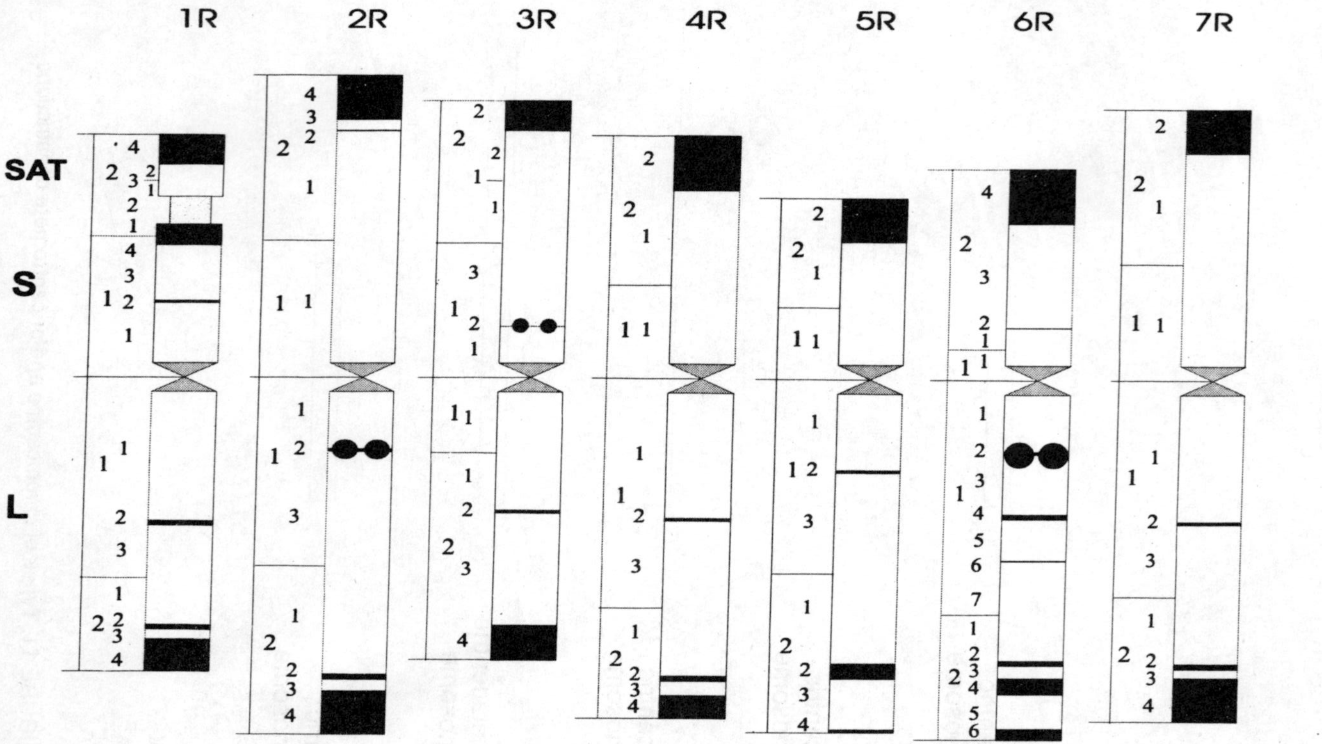

FIGURE 12. The karyogram of diploid rye, *Secale cereale* L.

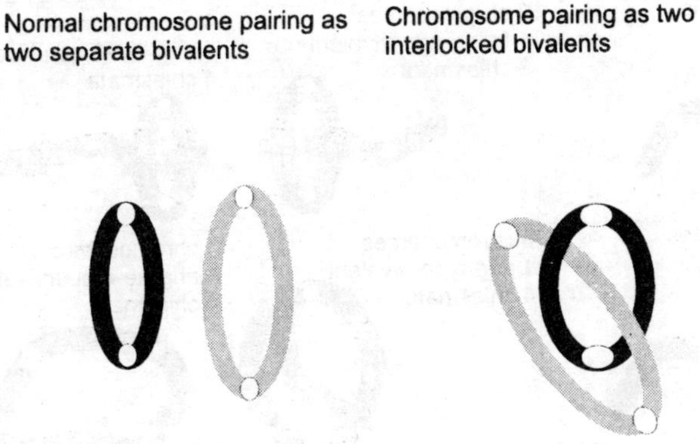

FIGURE 13. Pairing failure of meiotic chromosomes as an interlocked configuration

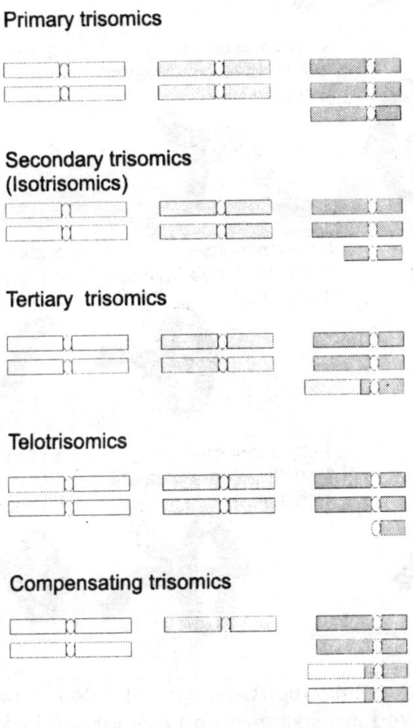

FIGURE 14. Different types of trisomics in plants

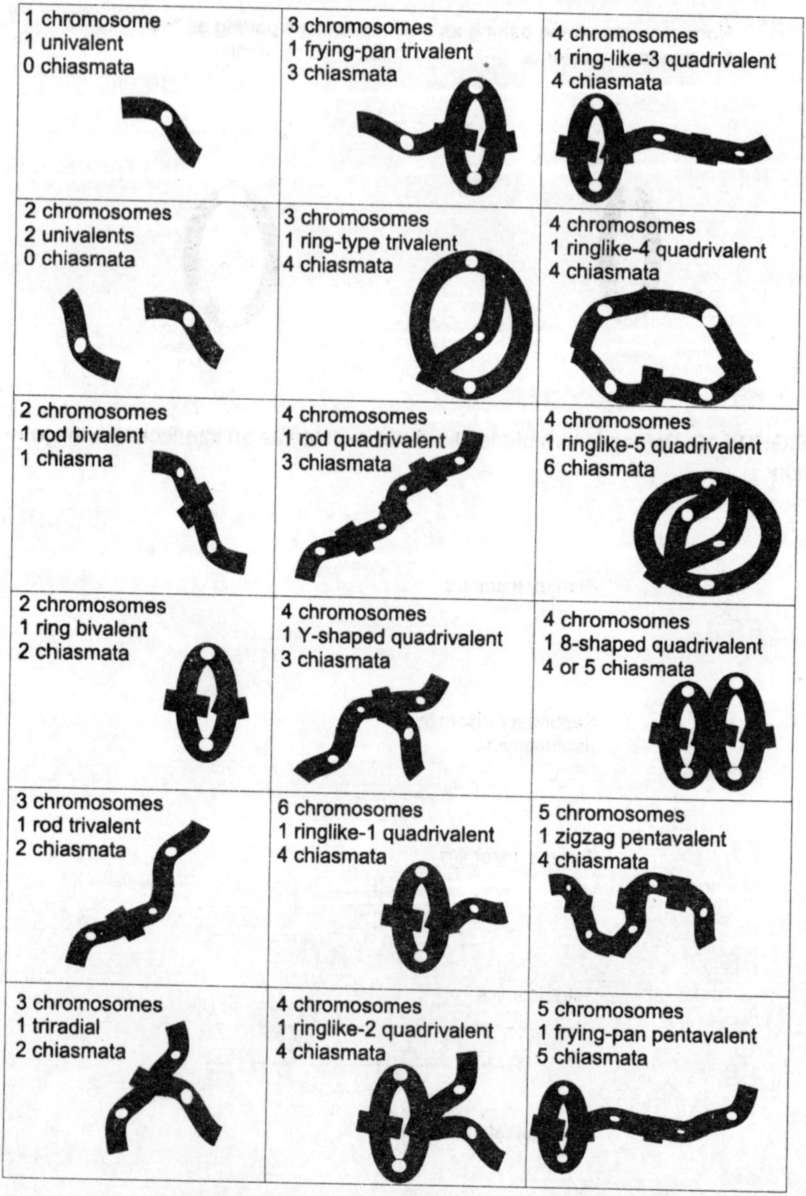

FIGURE 15. A diagrammatic representation of different meiotic chromosome configurations observed in diakinesis and metaphase including the minimum chaismata

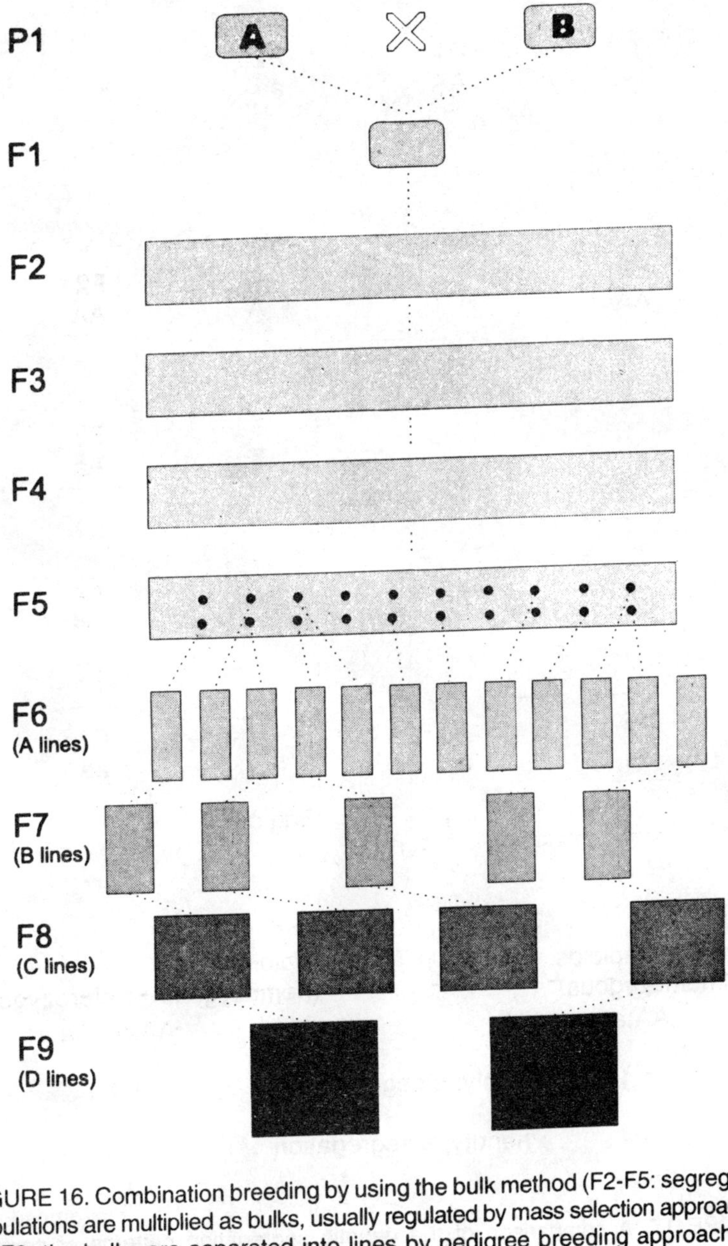

FIGURE 16. Combination breeding by using the bulk method (F2-F5: segregating populations are multiplied as bulks, usually regulated by mass selection approaches. F6-F9: the bulks are separated into lines by pedigree breeding approach and progeny testing.)

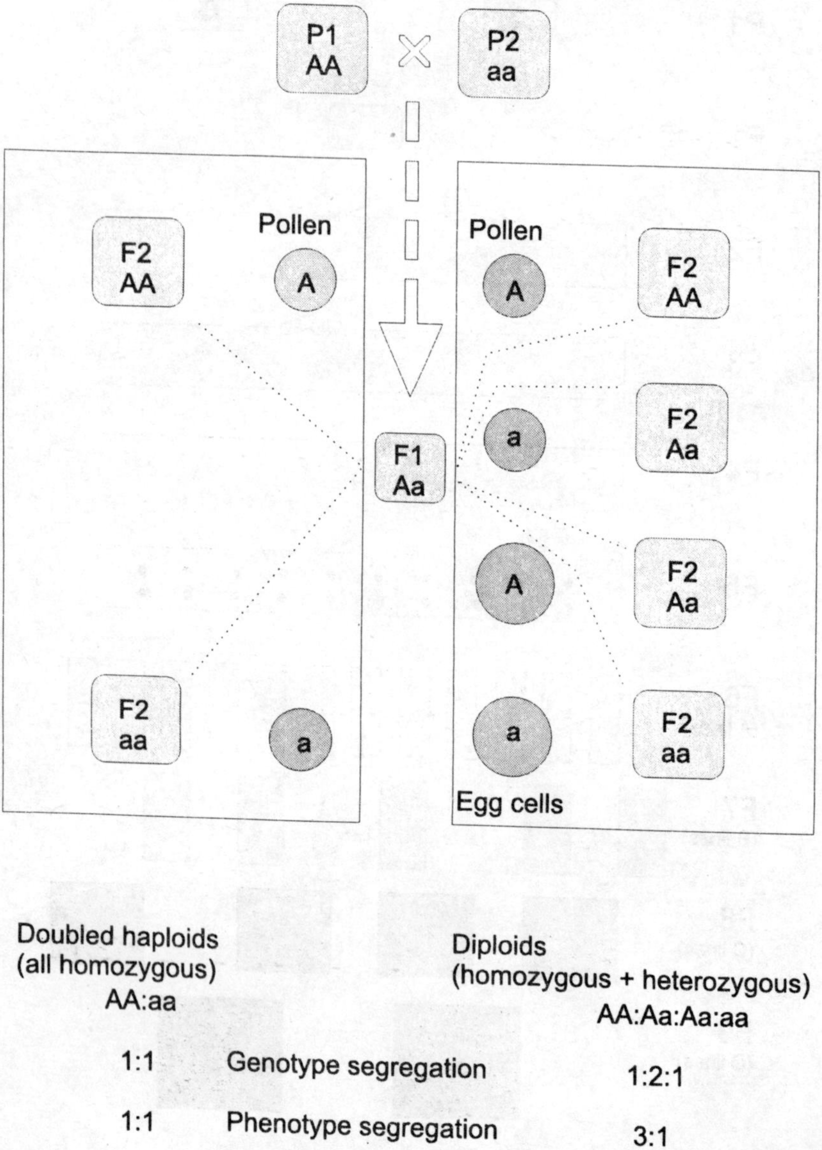

FIGURE 17. A comparison of the genetic segregation patterns of sexually derived and double-haploid-derived F2 progenies from F1 heterozygotes

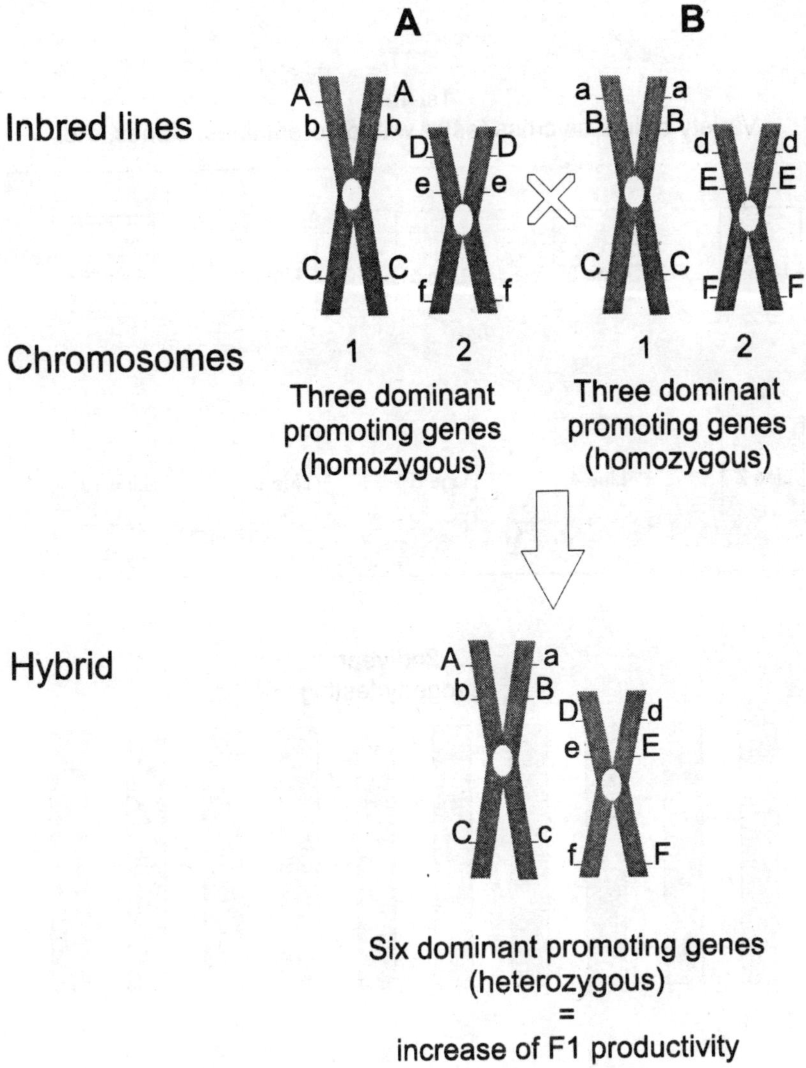

FIGURE 18. Schematic interpretation of "dominance hypothesis" considering inbred lines and an F1 hybrid

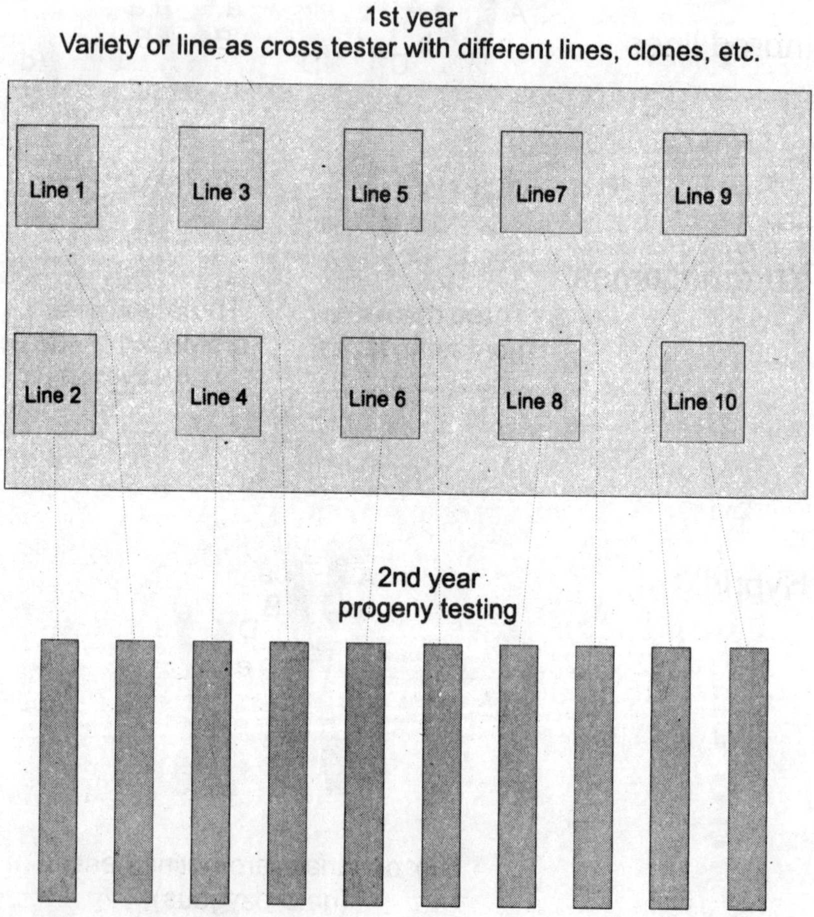

FIGURE 19. Schematic drawing of topcross design including progeny testing for general combining ability

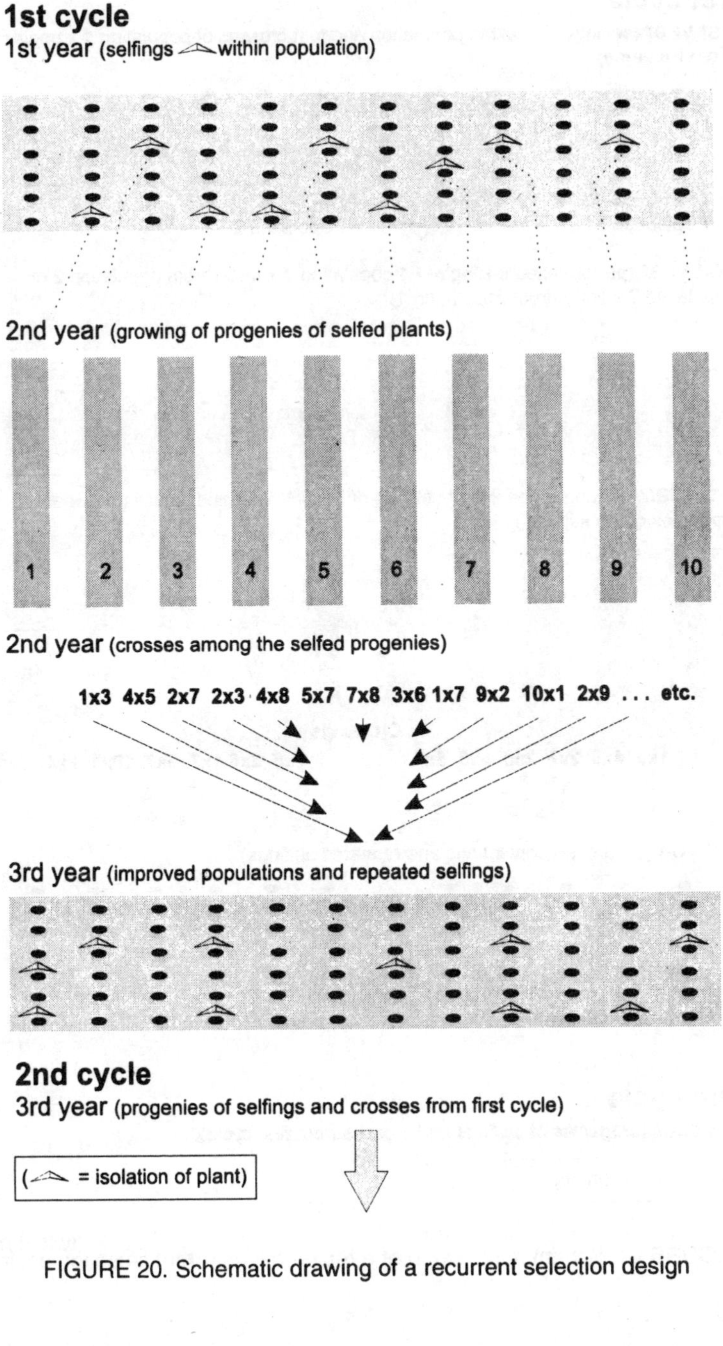

FIGURE 20. Schematic drawing of a recurrent selection design

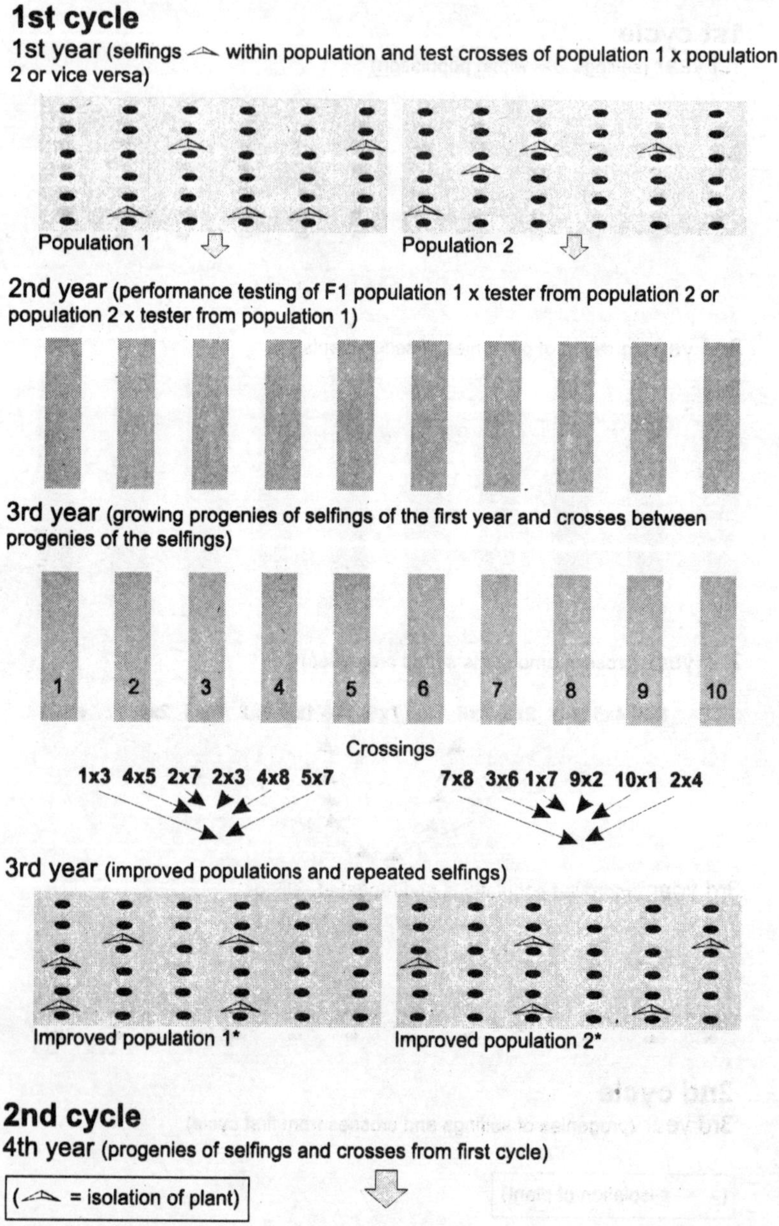

FIGURE 21. Schematic drawing of a reciprocal recurrent selection design

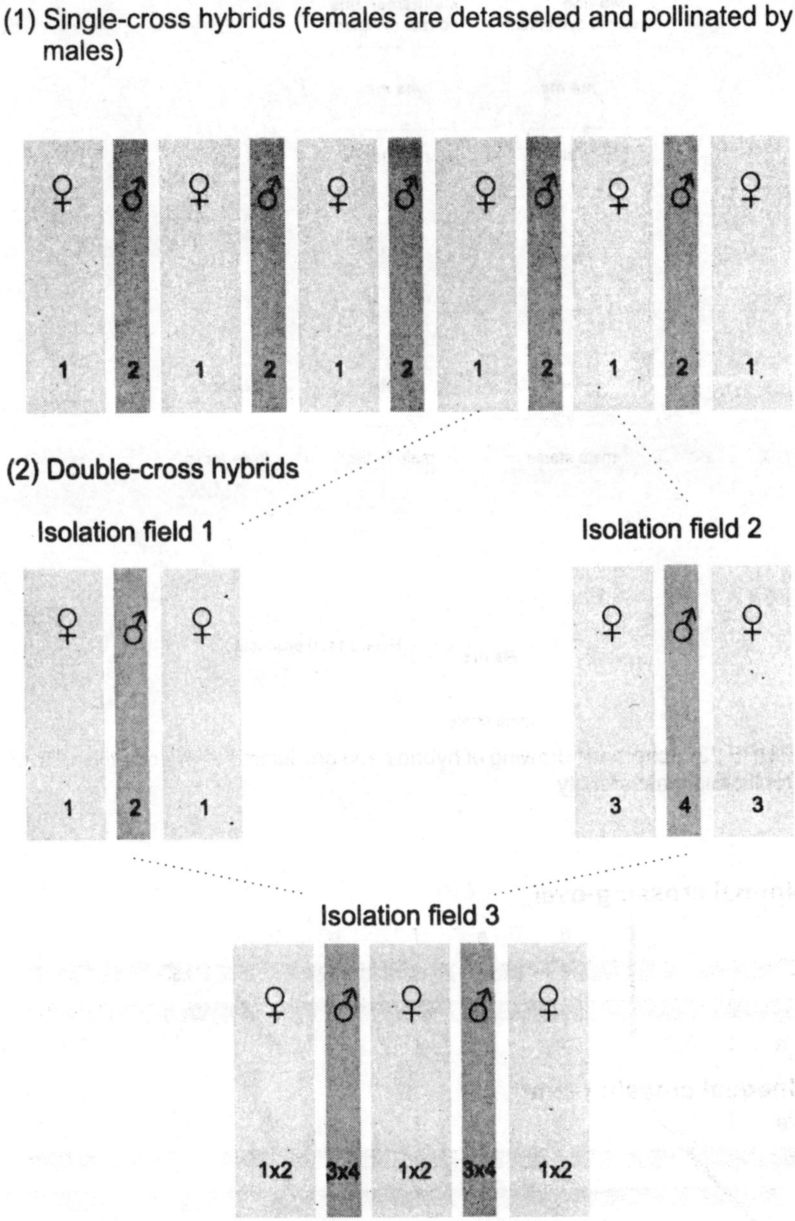

FIGURE 22. Schematic drawing of hybrid seed production in maize

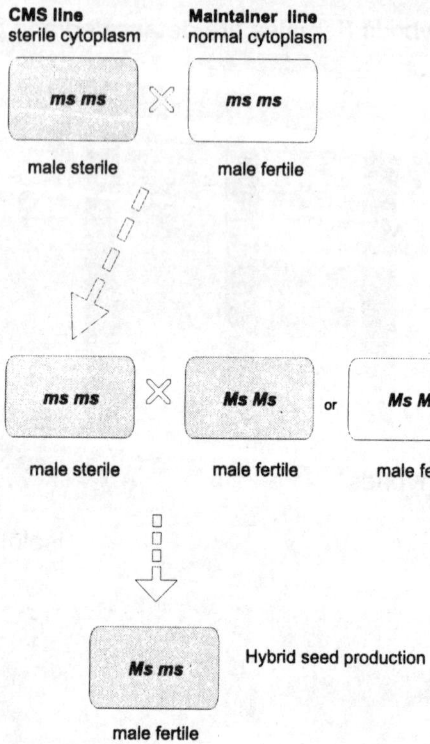

FIGURE 23. Schematic drawing of hybrid seed production after and utilization of cytoplasmic male sterility

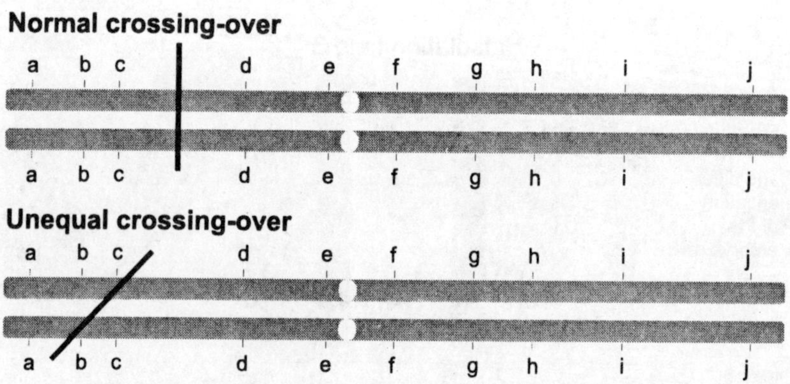

FIGURE 24. Schematic drawing of unequal crossing-over

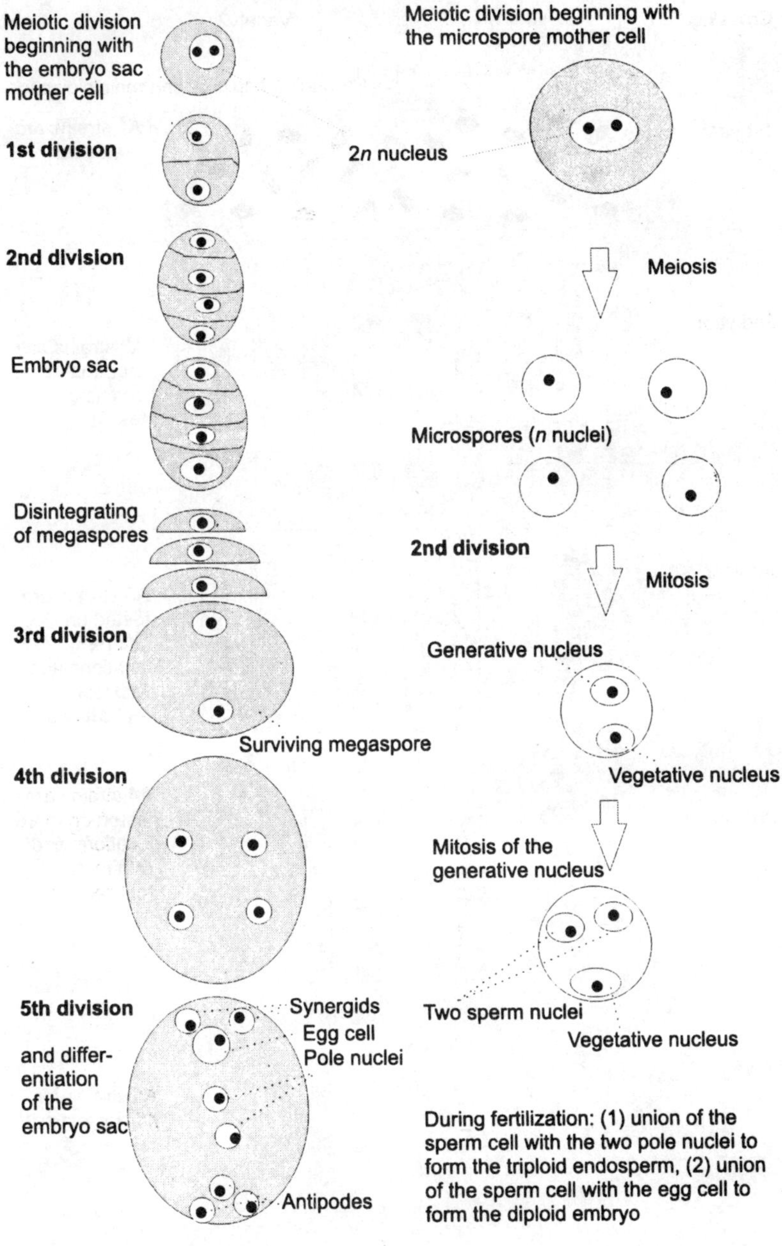

FIGURE 25. Schematic drawing of embryo sac and pollen formation

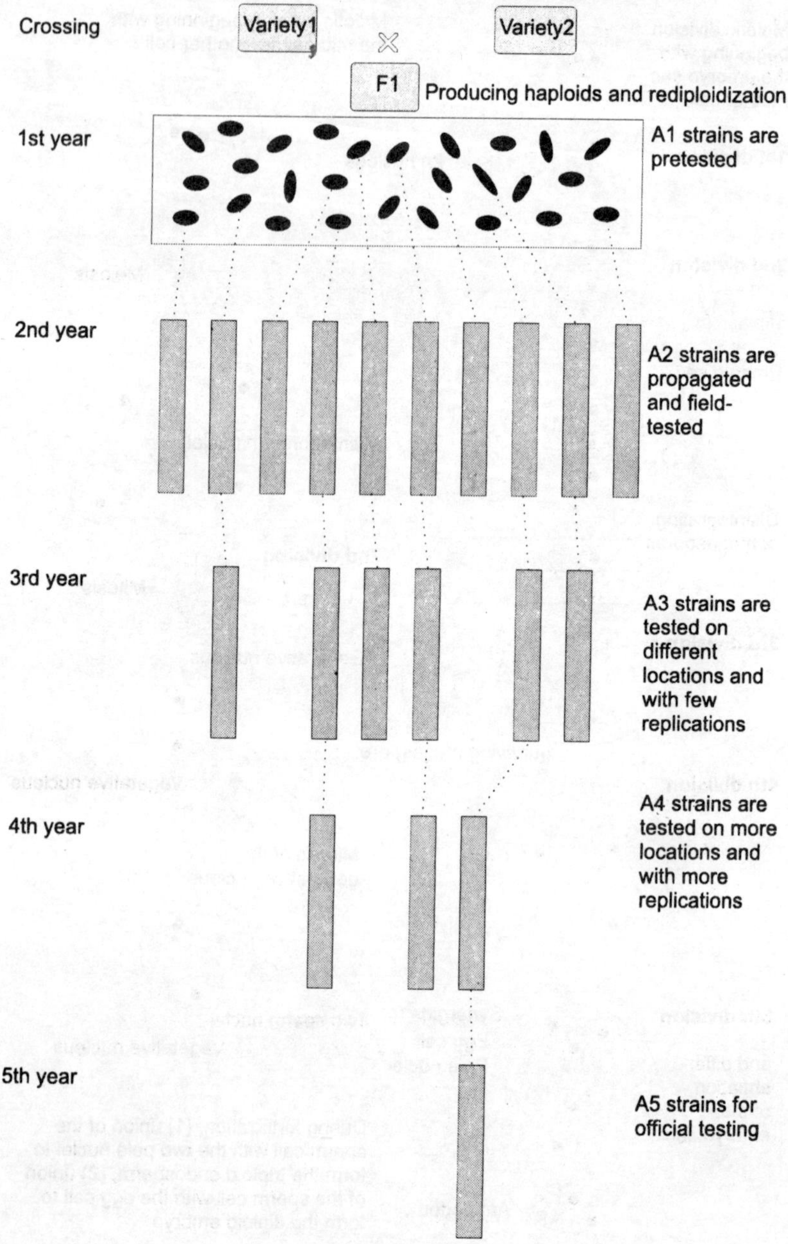

FIGURE 26. Breeding scheme using doubled haploids

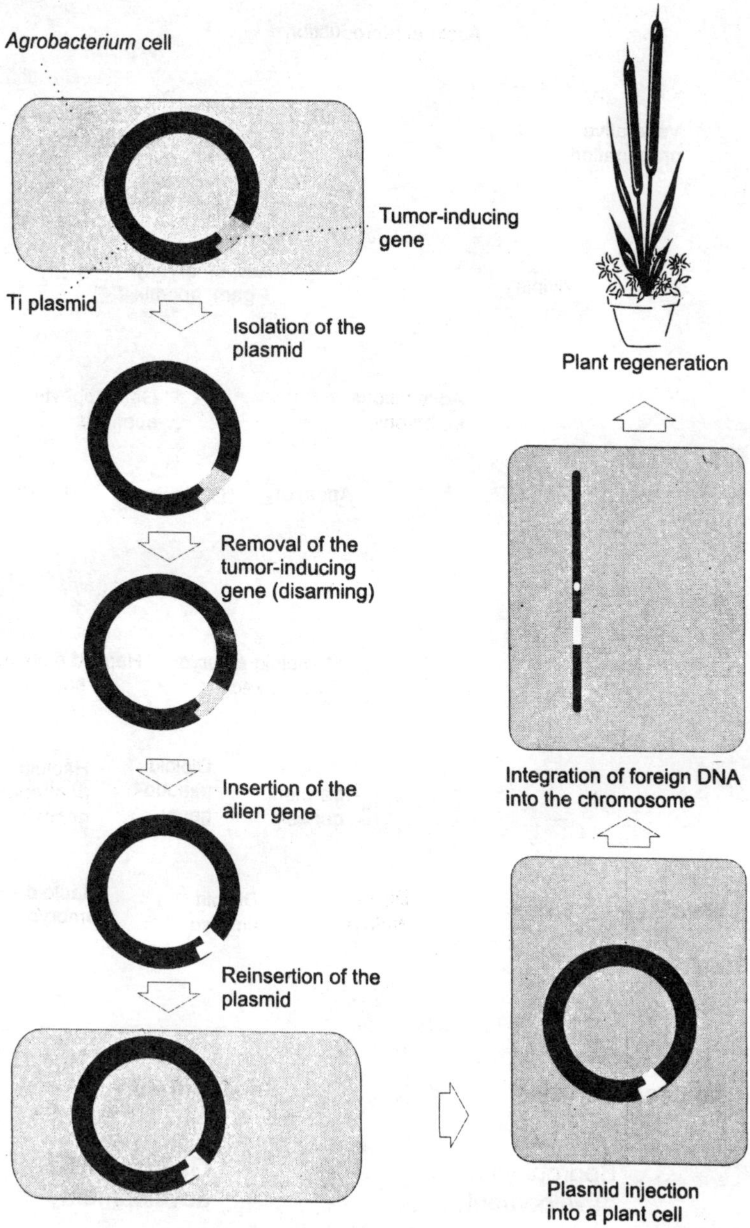

FIGURE 27. Schematic drawing of *Agrobacterium*-mediated gene transfer

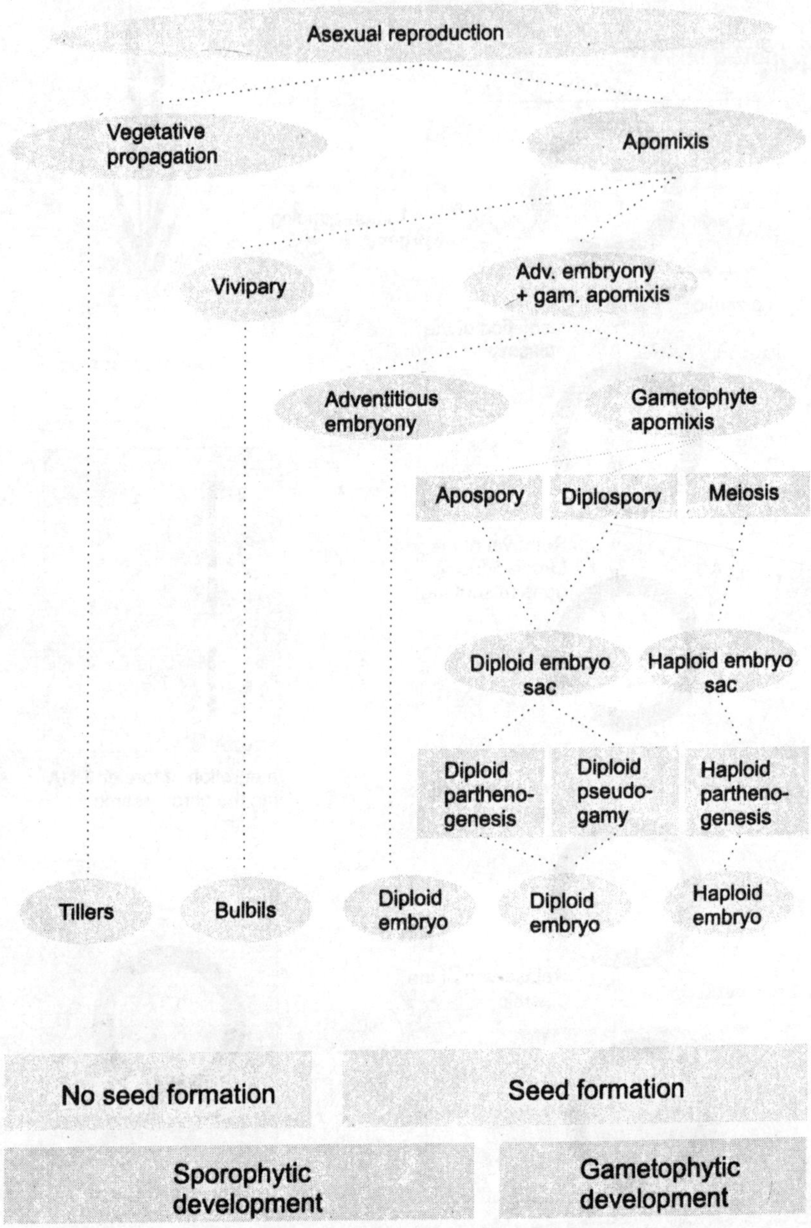

FIGURE 28. Different ways of asexual reproduction in plants

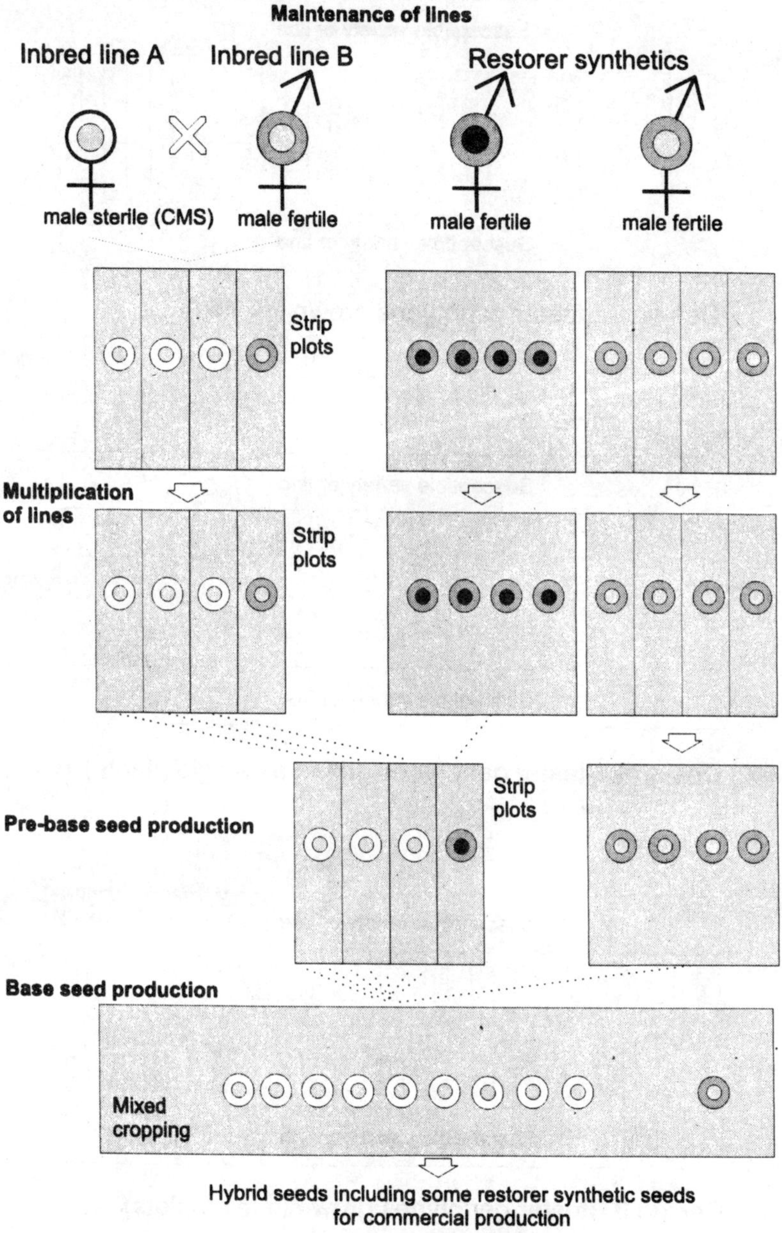

FIGURE 29. Hybrid seed production in allogamous rye

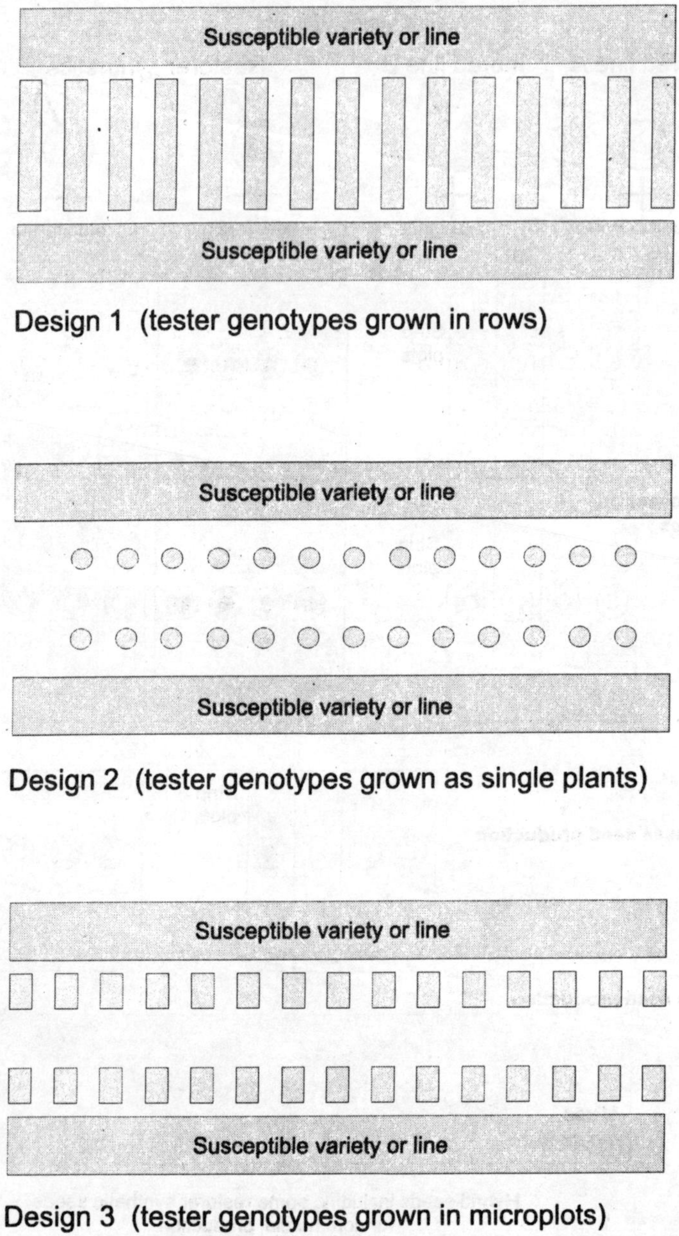

FIGURE 30. Designs of spreader nurseries

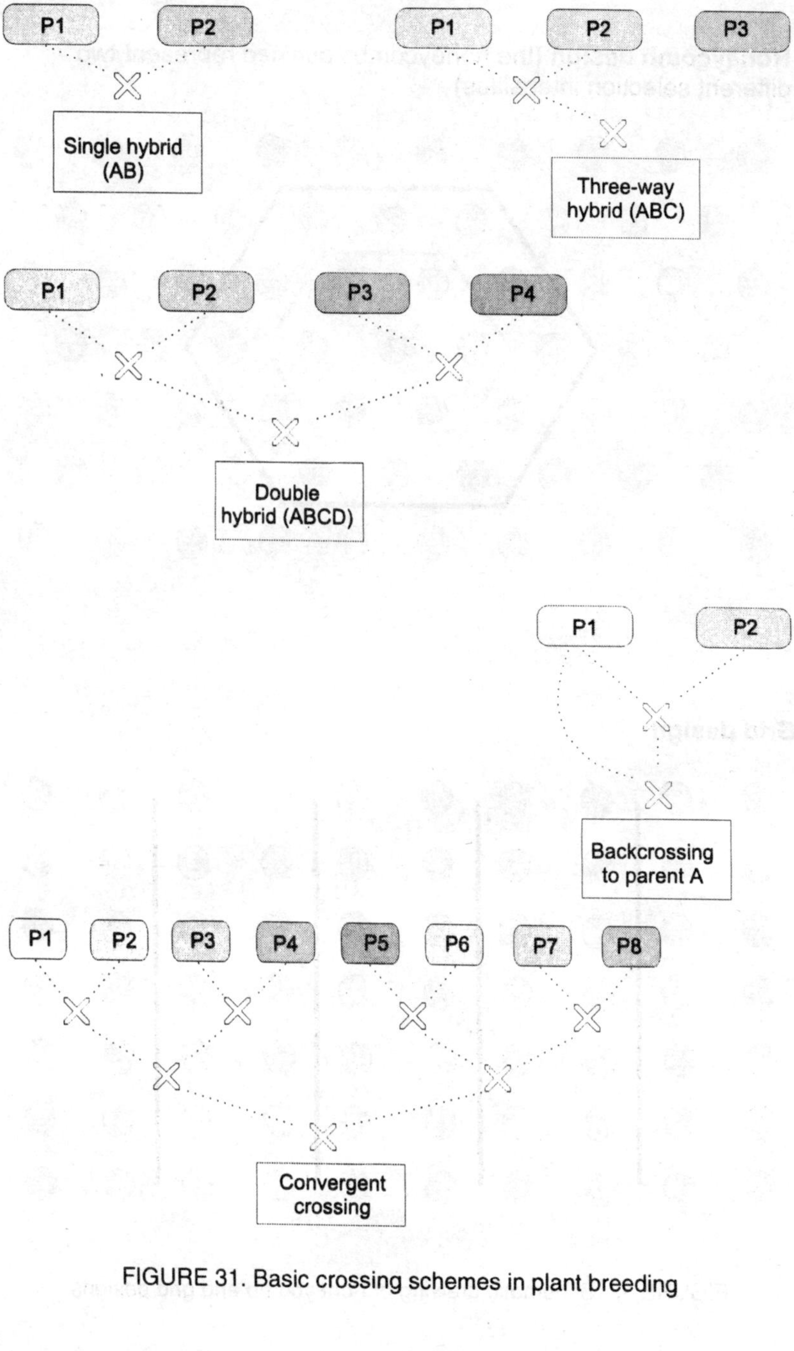

FIGURE 31. Basic crossing schemes in plant breeding

Honeycomb design (the honeycombs outlined represent two different selection intensities)

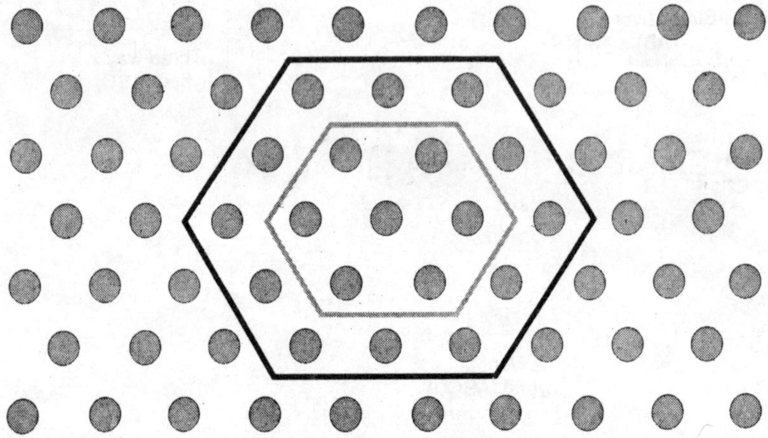

Grid design

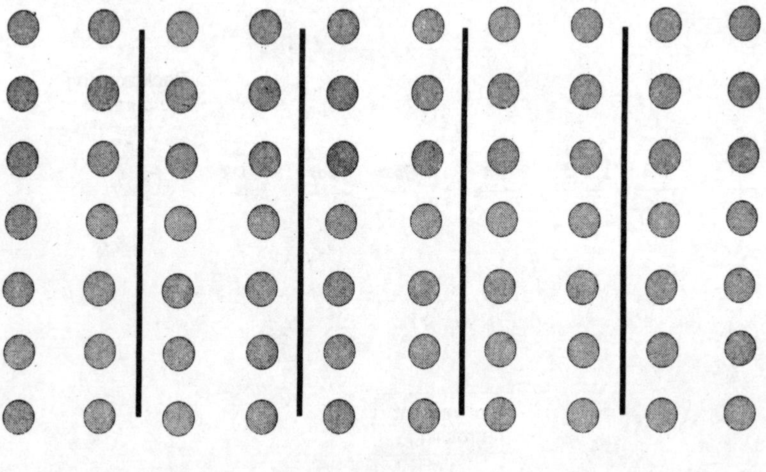

FIGURE 32. Schematic drawing of honeycomb and grid designs

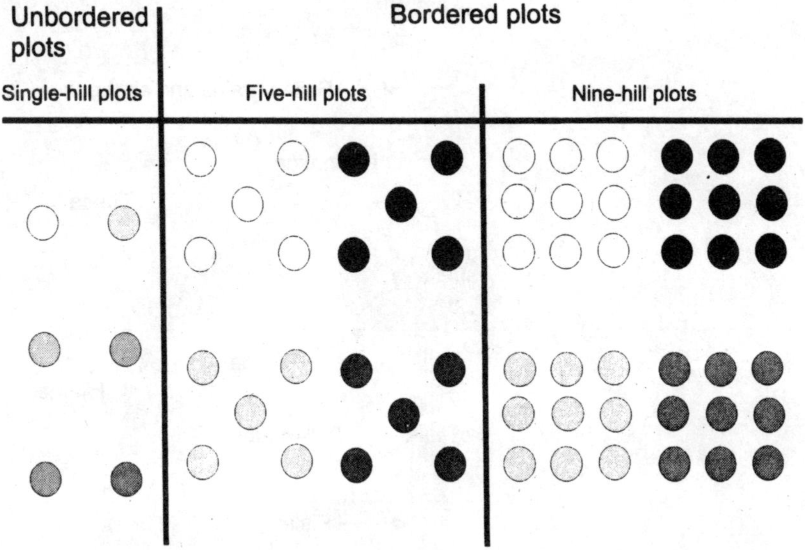

FIGURE 33. Several types of hill plots

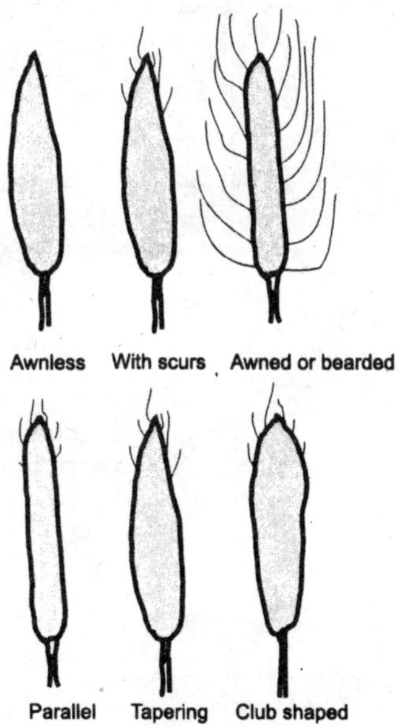

FIGURE 34. Basic shapes and types of spikes in wheat

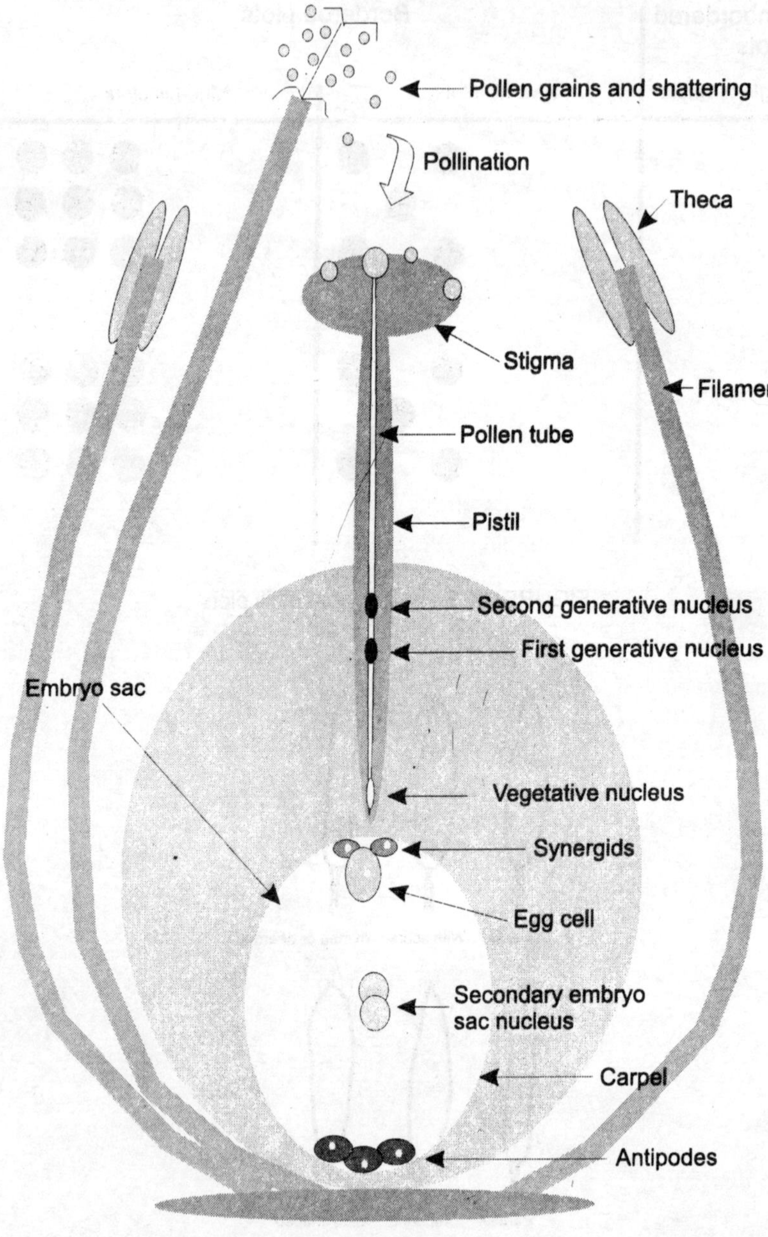

FIGURE 35. Schematic drawing of sexual organs of a plant flower

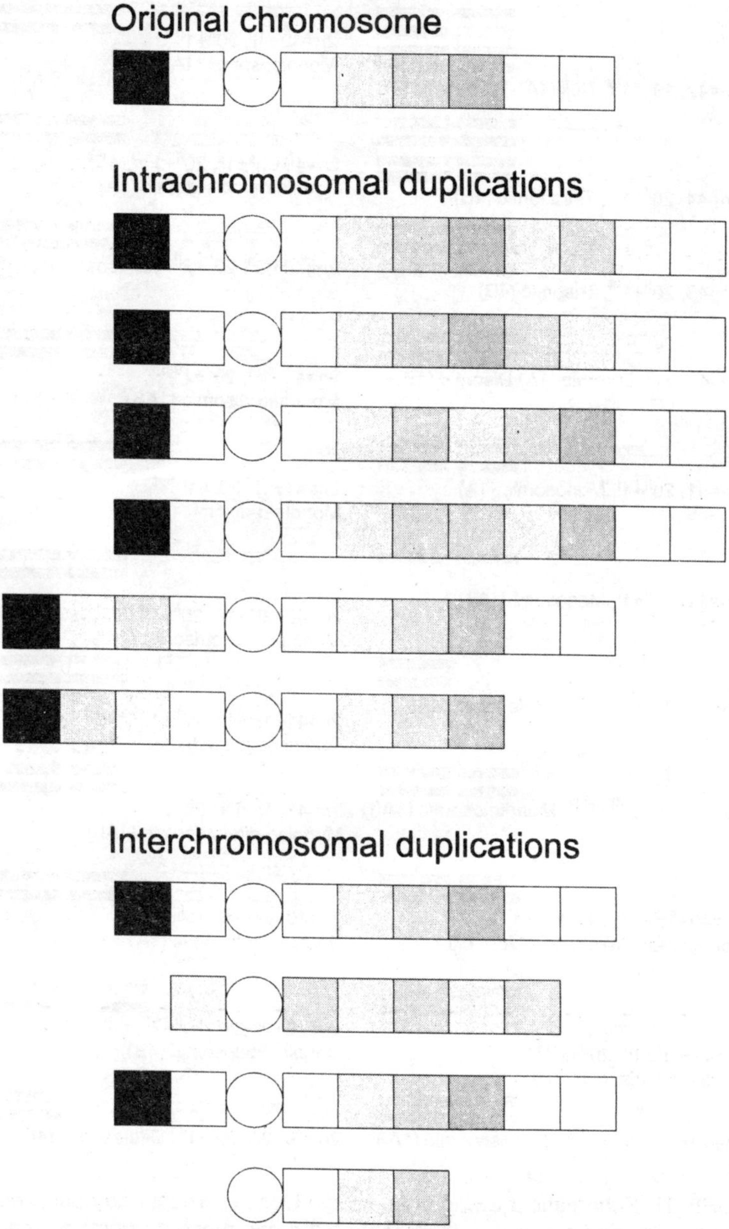

FIGURE 36. Types of duplicated chromosome segments

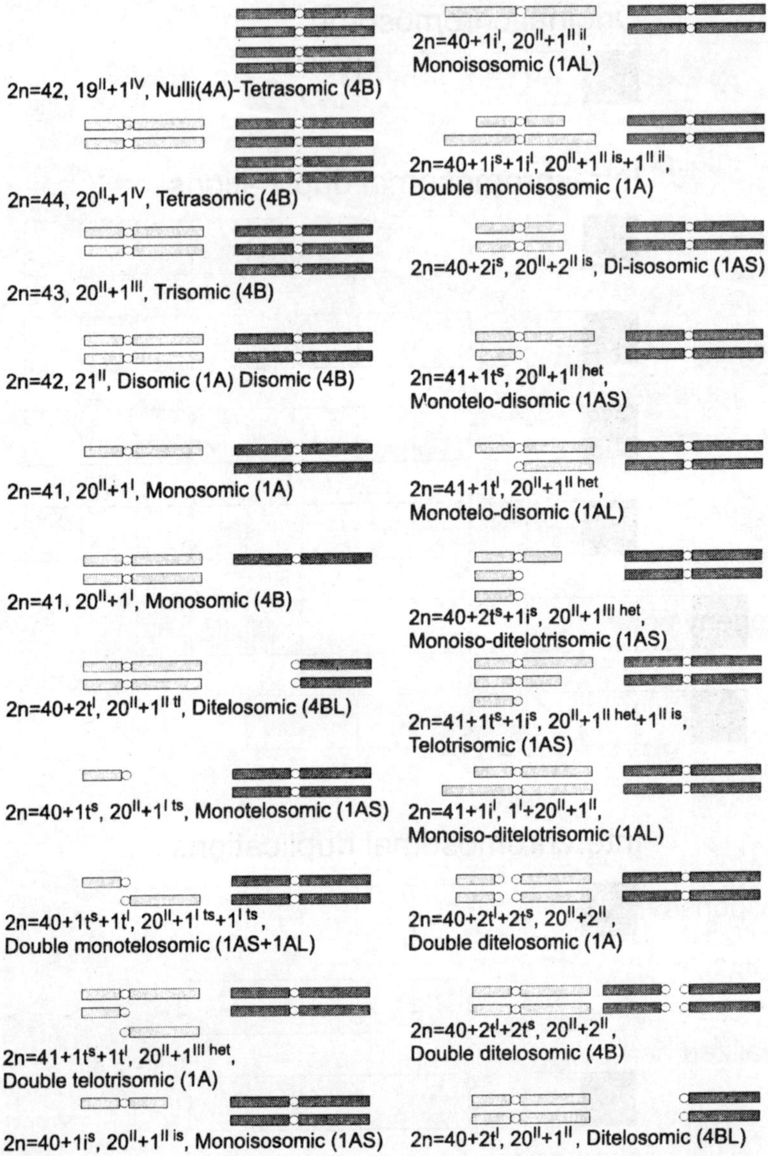

FIGURE 37. Schematic drawing of aneuploid types in wheat and applicable to other diploids and polyploids (It describes the chromosome number, meiotic configuration, and name of the aberration; in isosomics 1" means a ringlike structure.)

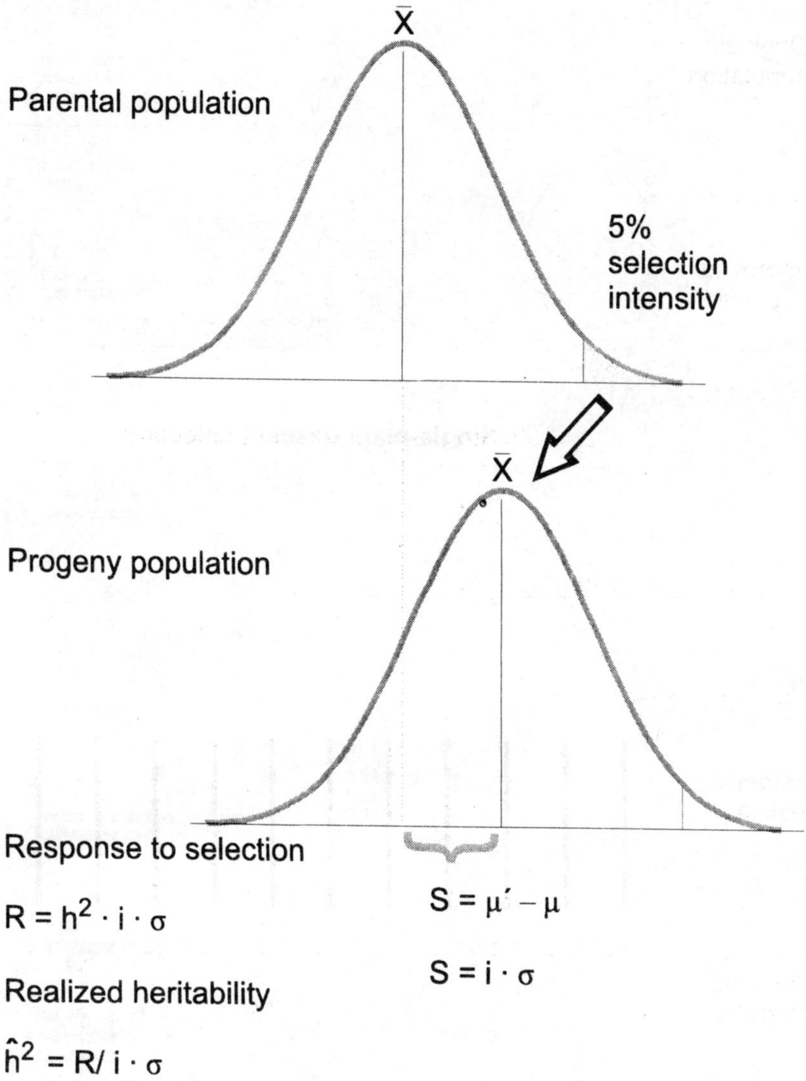

FIGURE 38. Estimation of heritability h^2 based on selection advantage "R," selection coefficient "S" and selection intensity "i"

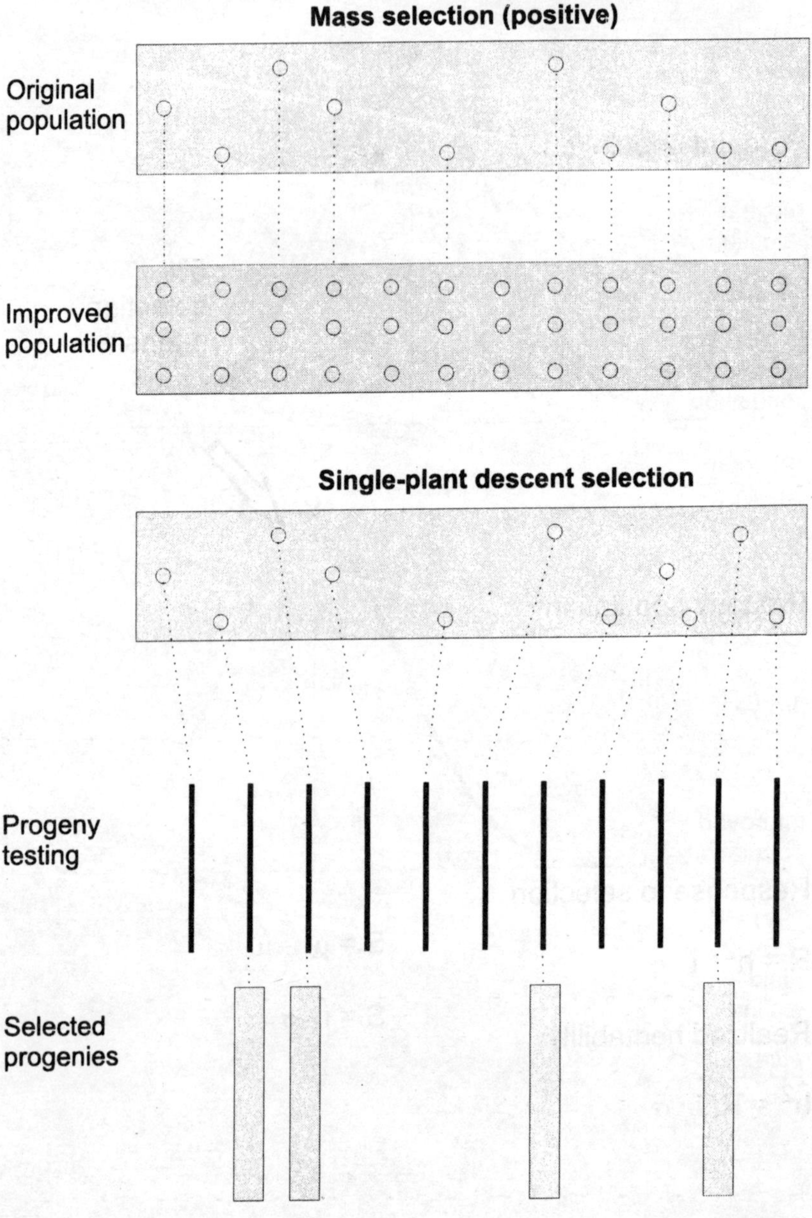

FIGURE 39. Schematic drawing of mass selection and single-plant selection

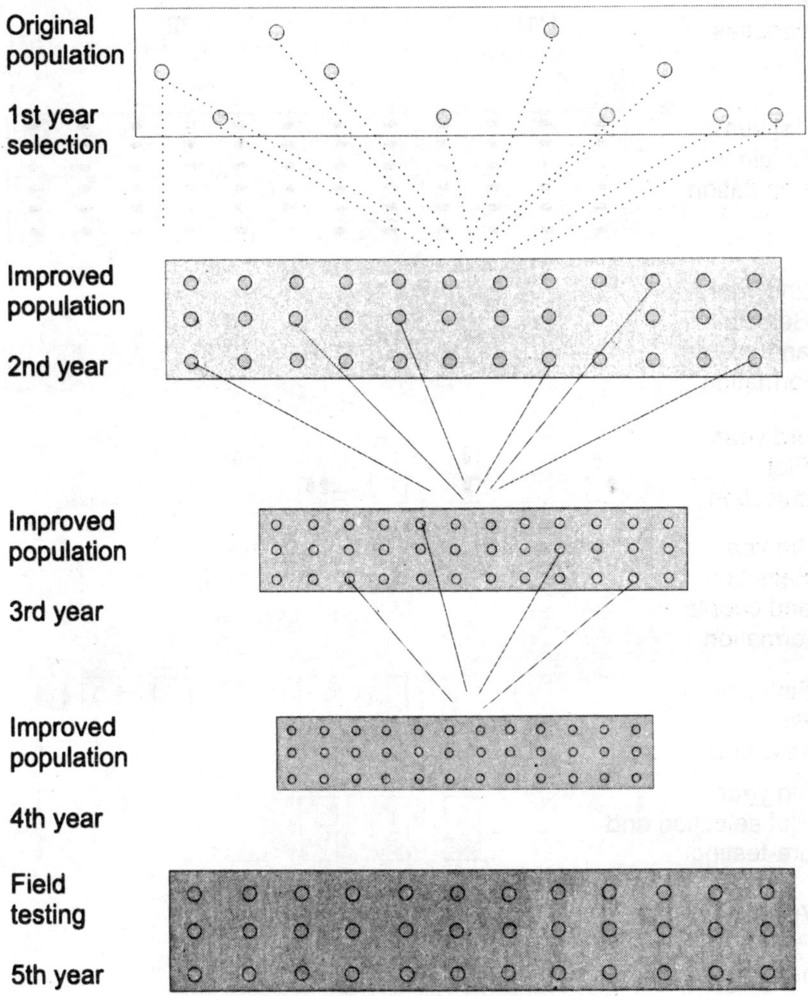

FIGURE 40. schematic drawing of positive mass selection

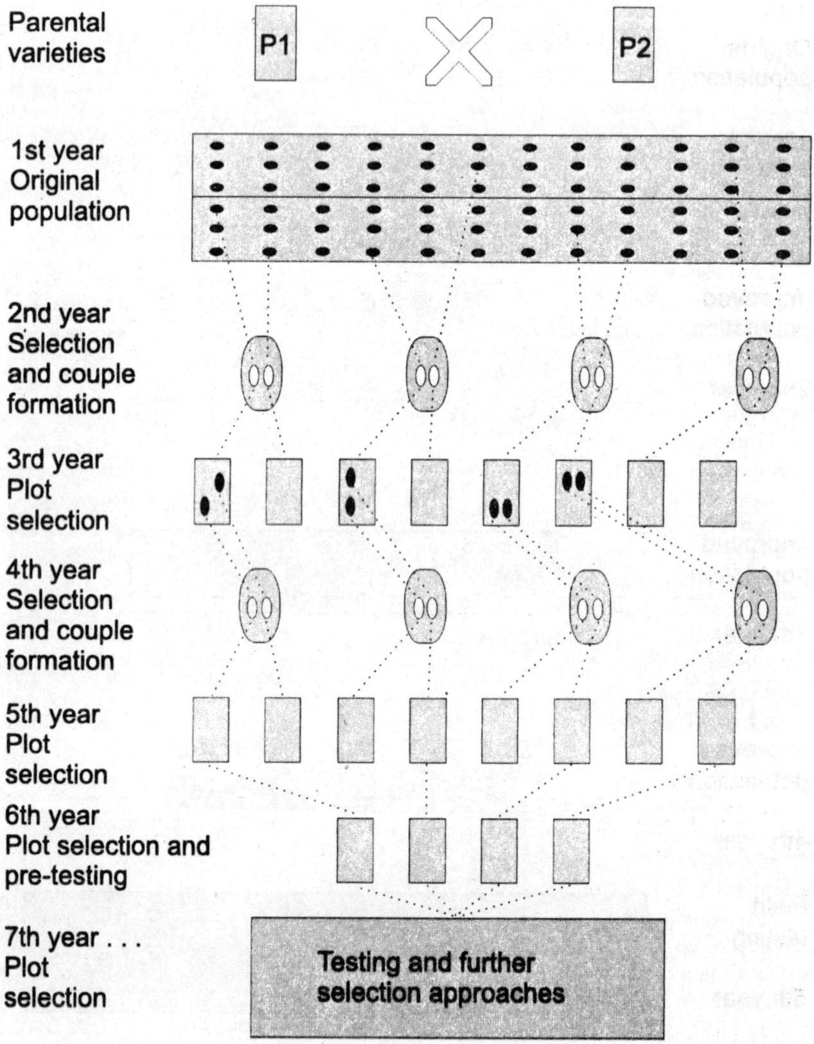

FIGURE 41. Couple method of breeding in allogamous plant

Bibliography

Agrios, G. N. 1997. *Plant pathology*. San Diego, CA: Academic Press.

Allard, R. W. 1960. *Principles of plant breeding*. New York: John Wiley and Sons, Inc.

Anonymous. 1952. *Manual for testing agricultural and vegetable seeds*. Agricultural Handbook 30. Washington, DC: USDA.

Anonymous. 1989. *Intellectual property rights associated with plants*. Madison, WI: Crop Science Society.

Anonymous. 1990a. *Detailed description of varieties of wheat, barley, oats, rye and triticale*. Cambridge, United Kingdom: National Institute of Botany.

Anonymous. 1990b. *Guidelines for soil description*. Rome: Food and Agriculture Organization of United Nations.

Barigozzi, C. 1981. *The origin and domestication of cultivated plants*. Amsterdam: Elsevier Science Publisher.

Becker, H. 1993. *Pflanzenzüchtung* (Plant breeding). Stuttgart: Verlag Eugen Ulmer.

Bennett, W. F. 1994. *Nutrient deficiencies and toxicities in crop plants*. St. Paul, MN: APS Press.

Bewley, J. D. and Black, M. 1995. *Seeds—Physiology of development and germination*. New York: Plenum Press.

Borojevic, S. 1990. *Principles and methods of plant breeding. Developments in crop science 17*. New York: Elsevier Science Publishing Company, Inc.

Bos, I. and Caligari, P. 1995. *Selection methods in plant breeding*. London: Chapman and Hall.

Briggs, F. N. and Knowles, P. F. 1967. *Introduction to plant breeding*. Reinhold Publishing Corporation.

Chapman, G. P., Mantell, S. H., and Daniels, R. W. 1985. *Experimental manipulation of ovule tissues*. New York: Longman.

Clement, S. L. and Quisenberry, S. S. 1999. *Global plant genetic resources for insect-resistant crops*. Danvers: CRC Press LLC.

Cobley, L. S. 1976. *An introduction of the botany of tropical plants*. London: Longman.

Cooke, D. A. and Scott, R. K. 1993. *The sugarbeet crop*. London: Chapman and Hall Ltd.

Copeland, L. O. and McDonald, M. B. 1985. *Principles of seed science and technology*. Minneapolis, MN: Burgess Publishing Company.

Dahlgreen R. M. T., Clifford, H. T., and Yeo, P. F. 1985. *The families of the monocotyledons: Structure, evolution and taxonomy*. Stuttgart: Springer Verlag.

Falconer, D. S. 1989. *Introduction to quantitative genetics*. New York: Longman Scientific Technology.

Fehr, W. R. 1987. *Principles of cultivar development.* Volume 1: *Theory and technique.* New York: Macmillan Publishing Company.

Fehr, W. R. and Hadley, H. H. 1980. *Hybridization of crop plants.* Madison, WI: American Society of Agronomy, Inc.

Frey, K. J. 1981. *Plant breeding 11.* Ames, IA: Iowa State University Press.

Gassen, H., Martin, G., and Bertram, S. 1985. *Gentechnik.* Jena, Germany: G. Fischer Verlag.

Graham, R. D. 1984. *Breeding for nutritional characteristics in cereals. Advances in plant research.* New York: Praeger.

Gu, W. K., Weeden, N. F., Yu, J., and Wallace, D. H. 1995. Large-scale, cost-effective screening of PCR products in marker-assisted selection applications. *Theoretical and Applied Genetics* 91: 465-470.

Heinz, D. J. 1987. *Sugarcane improvement through breeding.* Amsterdam: Elsevier.

Heyne, E. G. 1987. *Wheat and wheat improvement.* Madison, WI: American Society of Agronomy Publisher.

Karp, A., Isaac, P. G., and Ingram, D. S. 1998. *Molecular tools for screening biodiversity.* London: Chapman and Hall.

Kuckuck, H., Kobabe, G., and Wenzel, G. 1991. *Fundamentals of plant breeding.* Berlin: Springer Verlag.

Lampeter, W. 1982. *Saat- und Pflanzgutproduktion. Dt.* (Seed and plant production). Berlin, Germany: Deutscher Landwirt Schaftsverlag.

Leitch, A. R., Schwarzacher, T., Jackson, D., and Leitch, I. J. 1994. *In situ hybridization.* Oxford, United Kingdom: Bios Scientific Publishers.

Lupton, F. G. H. 1987. *Wheat breeding.* London: Chapman and Hall Ltd.

Martin, J. M., Leonard, W.H., and Stamp, D.L. 1976. *Principles of field crop production.* New York: Macmillan.

Mayo, O. 1987. *The theory of plant breeding.* Oxford: Clarendon Press.

Metcalfe, D. S. and Elkins, D. M. 1980. *Crop production.* New York: Macmillan.

Morris, C. F. and Geroux, M. J. 2000. Modification of cereal grain hardness via expression of puroindoline proteins. Beltsville, MD: USDA, Office of Technology Transfer. 1998, Patent No. 61580.

Oka, H. I. 1988. *The origin of cultivated rice.* Amsterdam: Elsevier Science Publisher.

Oliver, S. G. and Ward, J. M. 1988. *Wörterbuch der Gentechnik.* Jena, Germany: G. Fischer Verlag.

Payne, P. I. and Rhodes, A. P. 1982. *Encyclopaedia of plant physiology 14.* Berlin and Heidelberg: Springer Verlag.

Peters, J.H. and Baumgarten, H. 1992. *Monoclonal antibodies.* Berlin: Springer Verlag.

Poehlman, J. M. 1966. *Breeding field crops.* New York: Holt, H. and Company Inc.

Poehlman, J. M. and Sleper, D. A. 1995. *Breeding field crops.* Ames, IA: Iowa State University Press.

Reed, C. D. 1977. *The origin of agriculture.* The Hague, The Netherlands: Mouton Publisher.

Reidie, G.P. 1982. *Genetics.* New York: Macmillan.

Schlegel, R. 1990. Effektivität sowie Stabilität des interspezifischen Chromosomen- und Gentransfers beim hexaploiden Weizen (Efficiency and stability of interspecific chromosome and gene transfer in hexaploid wheat), *Triticum aestivum* L. *Kulturpflanze* 38, 67-78.

Schlegel, R. 1996. Triticale—Today and tomorrow. In H. Guedes-Pinto, N. Darvey, and V. Carnide (Eds.), *Triticale today and tomorrow, developments in plant breeding.* Volume 5 (pp. 21-32). Dordrecht, The Netherlands: Kluwar Academic Publishers.

Schlegel, R. and Cakmak, I. 1997. Micronutritional efficiency in crop plants—A new challenge for cytogenetic research. In T. Lelley (Ed.), *Current topics in plant cytogenetics related to plant improvement* (pp. 91-102).Wien, Austria: Wiener Universitatsverlag Facultas.

Schlegel, R., Cakmak, I. Ekiz, H., Kalayci, M., Kalayci, and Braun, H. J. 1998. Screening for zinc efficiency among wheat relatives and their utilisation for an alien gene transfer. *Euphytica* 100: 281-286.

Schlegel, R., Melz, G., and Korzun, V. 1997. Genes, marker and linkage data of rye (*Secale cereale* L.). Fifth updated inventory. *Euphytica* 101: 23-67.

Schoonhoven, A. V. and Voysest, O. 1993. *Common beans—Research for crop improvement.* Wallingford, United Kingdom: CAB International.

Siegel, A. F. 1988. *Statistics and data analysis.* Toronto: J. Wiley and Sons.

Simmonds, N. W. 1976. *Evolution of crop plants.* London: Longman, Green and Co. Ltd.

Simmonds, N. W. 1981. *Principles of crop improvement.* New York: Longman Publisher.

Snedecor, W. G. and Cochran, W. G. 1980. *Statistical methods.* Ames, IA: Iowa State University Press.

Steel, R. G. D. and Torrie, J. H. 1980. *Principles and procedures of statistics: A biometrical approach.* Tokyo, Japan: McGraw-Hill Inc.

Strickberger, M. W. 1988. *Genetik.* München and Wien: C. Hanser Verlag.

van Harten, A. M. 1998. *Mutation breeding, theory and practical application.* Cambridge, United Kingdom: University Press.

Vavilov, N. I. 1928. Geographische Zentren unserer Kulturpflanzen (Geographic centers of our crop plants). *Zeitschrift für induklive Abstammungs und Vererbungslehre,* 342-369.

Webster, C. C. and Wilson, P. N. 1980. *Agriculture in the tropics.* London: Longman.

Zadok, J. C., Chang, T. T., and Konzak, C. F. 1974. A decimal code for the growth stages of cereals. *Weed Research* 4: 415-421.

Zeven, A. C. and deWet, J. M. J. 1982. *Dictionary of cultivated plants and their regions of diversity.* Wageningen: Centre for Agricultural Publications.

Zhukovski, P. M. 1964. *Cultivated plants and their wild relatives.* Leningrad: Kolos.